Ecological Variability: Separating Natural from Anthropogenic Causes of Ecosystem Impairment

Other titles from the Society of Environmental Toxicology and Chemistry (SETAC):

Avian Effects Assessment: A Framework for Contaminants Studies
Hart, Balluff, Barfknecht, Chapman, Hawkes, Joermann, Leopold, Luttik, editors
2001

Ecotoxicology of Amphibians and Reptiles
Sparling, Linder, Bishop, editors
2000

Multiple Stressors in Ecological Risk and Impact Assessment: Approaches to Risk Estimation
Ferenc and Foran, editors
2000

Linkage of Effects to Tissue Residues: Development of a Comprehensive Database for Aquatic Organisms Exposed to Inorganic and Organic Chemicals
Jarvinen and Ankley, editors
1999

Multiple Stressors in Ecological Risk and Impact Assessment
Foran and Ferenc, editors
1999

Advances in Earthworm Ecotoxicology
Sheppard, Bembridge, Holmstrup, Posthuma, editors
1998

Ecological Risk Assessment Decision-Support System: A Conceptual Design
Reinert, Bartell, Biddinger, editors
1998

Ecological Risk Assessment for Contaminated Sediments
Ingersoll, Dillon, Biddinger, editors
1997

Whole Effluent Toxicity Testing: An Evaluation of Methods and Prediction of Receiving System Impacts
Grothe, Dickson, Reed-Judkins, editors
1996

For information about any SETAC publication, including SETAC's international journal, *Environmental Toxicology and Chemistry*, contact the SETAC Office nearest you.

1010 North 12th Avenue
Pensacola, Florida, USA 32501-3367
T 850 469 1500
F 850 469 9778
E setac@setac.org

Avenue de la Toison d'Or
B-1060 Brussels, Belgium
T 32 2 772 72 81
F 32 2 770 53 83
E setac@setaceu.org

http://www.setac.org

Environmental Quality Through Science®

Ecological Variability: Separating Natural from Anthropogenic Causes of Ecosystem Impairment

Edited by

Donald J. Baird
University of Stirling
Institute of Aquaculture
Stirling, Scotland

G. Allen Burton, Jr.
Wright State University
Institute for Environmental Quality
Dayton, Ohio, USA

Current Coordinating Editor of SETAC Books
Andrew Green
International Lead Zinc Research Organization

Publication sponsored by the Society of Environmental Toxicology and Chemistry (SETAC) and the SETAC Foundation for Environmental Education

Cover by Michael Kenney Graphic Design and Advertising
Copyediting and typesetting by Wordsmiths Unlimited
Indexing by IRIS

Library of Congress Cataloging-in-Publication Data

Ecological variability: separating natural from anthropogenic causes of ecosystem impairment / edited by Donald J. Baird, G. Allen Burton.

p. cm.

Includes bibliographical references (p.).

ISBN 1-880611-43-0 (alk. paper)

1. Ecology. 2. Nature--Effect of human beings on. 3. Ecological disturbances. I. Baird, Donald J. II. Burton, G. Allen.

QH541 .E3174 2001
577--dc21

2001049906

International Standard Book Number 1-880611-43-0
Printed in the United States of America
08 07 06 05 04 03 02 01 10 9 8 7 6 5 4 3 2 1

∞ The paper used in this publication meets the minimum requirements of the American National Standard for Information Sciences—Permanence of Paper for Printed Library Materials, ANSI Z39.48-1984.

Reference Listing: Baird DJ, Burton Jr GA. 2001. Ecological variability: Separating natural from anthropogenic causes of ecosystem impairment. Published by the Society of Environmental Toxicology and Chemistry (SETAC), Pensacola FL, USA. 336 p.

SETAC Publications

The publication of books by the Society of Environmental Toxicology and Chemistry (SETAC) provides in-depth reviews and critical appraisals on scientific subjects relevant to understanding the impacts of chemicals and technology on the environment. The books explore topics reviewed and recommended by the Publications Advisory Council and approved by the SETAC Board of Directors for their importance, timeliness, and contribution to multidisciplinary approaches to solving environmental problems. The diversity and breadth of subjects covered in the series reflect the wide range of disciplines encompassed by environmental toxicology, environmental chemistry, and hazard and risk assessment. These volumes attempt to present the reader with authoritative coverage of the literature, as well as paradigms, methodologies, and controversies; research needs; and new developments specific to the featured topics. The books are generally peer reviewed for SETAC by acknowledged experts.

SETAC Publications, which include Technical Issue Papers (TIPS), workshop summaries, and newsletter (*SETAC Globe*) and journal (*Environmental Toxicology and Chemistry*) articles, are useful to environmental scientists in research, research management, chemical manufacturing and regulation, risk assessment, and education, as well as to students considering or preparing for careers in these areas. The publications provide information for keeping abreast of recent developments in familiar subject areas and for rapid introduction to principles and approaches in new subject areas.

SETAC would like to recognize the past SETAC Special Publication Series editors:

C.G. Ingersoll, Midwest Science Center
U.S. Geological Survey, Columbia, Missouri, USA

T.W. LaPoint, Institute of Applied Sciences
University of North Texas, Denton, Texas, USA

B.T. Walton, U.S. Environmental Protection Agency
Research Triangle Park, North Carolina, USA

C.H. Ward, Department of Environmental Sciences and Engineering
Rice University, Houston, Texas, USA

Contents

List of Figures

List of Tables

Acknowledgments

We wish to thank the following organizations for their financial support of the Pellston Workshop from which this book developed:

- Exxon Corporation
- International Copper Association
- Monsanto Company
- Rio Tinto PLC
- Rio Algom Limited
- U.S. Geological Survey (USGS)
- U.S. Environmental Protection Agency (USEPA)
- Zeneca Corporation.

The workshop would, of course, not have been possible without the logistical and technical support of the Society of Environmental Toxicology and Chemistry (SETAC), specifically Rod Parrish, Greg Schiefer, and Linda Longsworth. We also appreciate the editorial efforts of Mimi Meredith of Wordsmiths Unlimited.

Finally, we express our gratitude for the commitment and valuable input of our colleagues, Wayne Landis and Michael Swindoll, who participated in the peer review of *Ecological Variability: Separating Natural from Anthropogenic Causes of Ecosystem Impairment.*

About the Editors

Donald J. Baird is a Senior Lecturer at the University of Stirling, Scotland. He obtained his Ph.D. at the University of Glasgow, Scotland, and subsequently carried out post-doctoral research at the Universities of Calgary, Canada and Sheffield, England. In his current position, Dr. Baird is Director of the Freshwater Science degree program at Stirling and manages a large research group studying the interface between aquatic ecology and applied water management issues. His group has a strong international focus in its research, with current projects active in Scotland, Portugal, and Canada, in addition to a particular interest in the tropics (Mexico, Brazil, Colombia, Sri Lanka, and Thailand). He regularly teaches overseas and has been instrumental in developing courses for the Masters Programme in Ecology at the University of Coimbra, Portugal. His major research interests include the development of modeling approaches to link individual toxic responses to higher-order ecological effects and the role of food webs in understanding risks posed by contaminants.

G. Allen Burton, Jr. is the Brage Golding Distinguished Professor of Research and Director of the Institute for Environmental Quality at Wright State University, Dayton, Ohio, USA. Since coming to WSU, he has held positions as a NATO Senior Research Fellow in Portugal and Visiting Senior Scientist in Italy and New Zealand, and he has been involved in projects on 7 continents. Dr. Burton's research during the past 20 years has focused on ecological risk assessments and development of effective methods for identifying water quality stressors, particularly where sediment or stormwater contamination is a concern. He has been active in the development and standardization of toxicity methods for the U.S. Environmental Protection Agency (USEPA), American Society for Testing and Materials (ASTM), Environment Canada, and the Organization of Economic Cooperation and Development (OECD). Dr. Burton has served on numerous national and international scientific committees and review panels and has contributed to more than 100 publications dealing with aquatic systems.

Workshop Participants*

Donald J. Baird
University of Stirling
Institute of Aquaculture
Stirling, Scotland, UK

Alistair B.A. Boxall
Soil Survey and Land Research Centre
Cranfield University
Shardlow, Derby, UK

Theo C.M. Brock
DLO-Winand Staring Centre
Wageningen, The Netherlands

G. Allen Burton, Jr.
Wright State University
Institute for Environmental Quality
Dayton, Ohio, USA

William H. Clements
Colorado State University
Department of Fishery & Wildlife Biology
Fort Collins, Colorado, USA

Susan M. Cormier
U.S. Environmental Protection Agency
Cincinnati, Ohio, USA

Joseph M. Culp
National Water Research Institute
Environment Canada
Saskatoon, Saskatchewan, Canada

Peter C. de Ruiter
Department of Environmental Studies
University of Utrecht
Utrecht, The Netherlands

Elaine J. Dorward-King
Rio Tinto plc
Castlemead, Bristol, UK

Scott D. Dyer
The Proctor & Gamble Company
Miami Valley Labs
Cincinnati, Ohio, USA

Peter Eldridge
U.S. Environmental Protection Agency
Newport, Oregon, USA

Valery E. Forbes
Department of Life Sciences and Chemistry
Roskilde University
Roskilde, Denmark

Jeroen Gerritsen
Tetra Tech, Inc.
Owings Mills, Maryland, USA

Keith Grasman
Wright State University
Department of Biological Sciences
Dayton, Ohio, USA

Audrey Hatch
Department of Zoology
Oregon State University
Corvallis, Oregon, USA

Udo Hommen
Ecological Modeling and Statistics
Alsdorf, Germany

Robbert G. Jak
TNO Institute of Environmental Sciences, Energy Research and Process Innovation
Den Helder, The Netherlands

Paul Jepson
Department of Entomology
Oregon State University
Corvallis, Oregon, USA

Jan E. Kammenga
Laboratory for Nematology
Wageningen University
Wageningen, The Netherlands

Lawrence A. Kapustka
Ecological Planning & Toxicology, Inc.
Corvallis, Oregon, USA

Tim J. Kedwards
Zeneca Agrochemicals
Ecological Risk Assessment Section
Brackennell, Berkshire, UK

Karen A. Kidd
Department of Fisheries and Oceans
Freshwater Institute
Winnipeg, Manitoba, Canada

Matthew G. Luxon
Institute of Environmental
Toxicology & Chemistry
Western Washington University
Bellingham, Washington, USA

Samuel N. Luoma
U.S. Geological Survey
Menlo Park, California, USA

Lorraine Maltby
University of Sheffield
Animal and Plant Sciences
Sheffield, UK

David R. Mount
U.S. Environmental Protection
Agency / ORD
Mid-Continent Ecology Division
Duluth, Minnesota, USA

Wayne R. Munns, Jr.
U.S. Environmental Protection Agency
Atlantic Ecology Division
Narragansett, Rhode Island, USA

Rodney Parrish, *ex officio*
SETAC/SETAC Foundation
Pensacola, Florida, USA

Donna K. Reed-Judkins
U.S. Environmental Protection Agency
Office of Science & Technology
Washington, DC, USA

Trefor Reynoldson
Environment Canada
National Water Research Institute
Burlington, Ontario, Canada

Amy H. Ringwood
Marine Resources Research Institute
Charleston, South Carolina, USA

Gregory E. Schiefer, *ex officio*
SETAC/SETAC Foundation
Pensacola, Florida, USA

Glenn W. Suter II
U.S. Environmental Protection Agency
National Center for Environmental
Assessment
Cincinnati, Ohio, USA

Ronald M. Thom
Battelle Marine Sciences Laboratory
Pacific Northwest National Laboratory
Sequim, Washington, USA

Judith S. Weis
Rutgers University
Biological Sciences
Newark, New Jersey, USA

Simon N. Wood
The Mathematical Institute
St. Andrews, Fife, UK

* Affiliations were current at the time of the Pellston Workshop.

Contributing Authors*

Ted DeWitt
U.S. Environmental Protection Agency
ORD/NHERL
Newport, Oregon, USA

Bryan Griffiths
Scottish Crop Research Institute
Dundee, UK

John C. Moore
Department of Biology
University of Northern Colorado
Greeley, Colorado, USA

* Affiliations were current at the time of the Pellston Workshop.

Introduction

At the beginning of the 1990s, the Ecological Society of America (ESA) produced an agenda for future research, the Sustainable Biosphere Initiative (Lubchenco et al. 1991), which attempted to provide research priorities for ecologists and their funding agencies against a background of increasing evidence for global change in ecosystems. The report identified an urgent need to carry out research in the following areas, among others:

- Understanding ecological processes in natural and human-dominated ecosystems in order to prescribe restoration and management strategies
- Describing patterns of diversity and their critical ecological and evolutionary determinants
- Gauging the plasticity of organism response to environmental stress
- Evaluating the influence of internal structure of population in response to stresses
- Elucidating the factors that govern the assembly of communities and ecosystems and their response to stress
- Measuring the feedbacks between the biotic and abiotic portions of ecosystems and landscapes, including regulation of biogeochemical processes, at different spatial and temporal scales
- Determining the consequences of environmental variability, including natural and anthropogenic disturbances, for individuals, populations, or communities.

Since that time, considerable ecological research effort has been allocated to answer these questions, with major research programs devoted to understanding the potential consequences of global warming, multi-stressor impacts on watersheds, spatial and temporal dynamics of ecosystems, use of indicator approaches in ecosystems, transport and uptake of contaminants in food chains, food web perturbations, scaling effects, and field and laboratory uncertainty.

In the mid-1990s, the Society of Environmental Toxicology and Chemistry's (SETAC) Scientific Initiatives Committee was developing a similar research agenda. As a group, the scientists who form SETAC encompass a broad range of disciplines and geographies, and as a group devoted to the investigation of complex, multidisciplinary environmental science issues, they have much to offer ecologists (and indeed many SETAC members are practicing ecologists) — not least in the sense that their studies are a rich proving ground for ecological theory. With this in mind, a workshop was convened at Pellston, Michigan, USA from 11–17 September 1999.

A previous SETAC Pellston workshop (Foran and Ferenc 1999) identified and addressed the problem of managing multiple stress agents and indicated the need for a more realistic approach to stressor management in the natural environment.

This need arises at least in part from the fact that regulatory risk assessment has emerged from disciplines (i.e., toxicology, chemistry) that focus on the effects of predefined chemicals on a few laboratory organisms in simple, uniform environments. In contrast, ecologists routinely deal with variability, change, and complexity, though often, it is perceived, with limited success.

The objective of our Pellston workshop was, therefore, to bridge the chasm between the disciplines of ecology and toxicology, and in particular, to consider how new techniques and approaches developed in ecological research can be applied in ecological risk assessments to better identify natural and anthropogenic stressors and their effects in complex ecosystems. For this reason, and in contrast with previous Pellston workshop on the multiple stress issue (i.e., Foran and Ferenc 1999), the steering committee's selection of workshop participants was deliberately biased toward active ecological and ecotoxicological researchers from academia, industry, and regulatory agencies. Also in contrast with previous Pellston workshops in this area, we deliberately drew from the pool of researchers on both sides of the Atlantic, with a significant number of European scientists participating. Again, the steering committee felt that this was essential, in order to ensure that the broadest possible range of scientific opinions was brought to bear on the subject.

These workshop proceedings are based on intensive discussions, and while they have been edited to improve clarity, there is some overlap of content between chapters. In order to preserve the spontaneity of the discussions, the flow of argument, and the weight of complementary viewpoints, however, some overlap in content is inevitable, and indeed desirable.

The workshop itself was convened in plenary session, with 4 breakout groups charged to deal with multiple stressors in an ecological context, as follows:

1) Identifying and Separating Multiple Stressors within Ecosystems
2) Linking Individual Responses to Stress with Population Consequences
3) Role of Environmental Variability in Evaluating Stressor Effects
4) Assessing Multiple Stress Impacts in Food Webs

In addition to the work of the breakout groups, 4 invited presentations were solicited to stimulate discussion in each of the 4 topic areas. With one exception (Chapter 4), these invited papers were subsumed within the multi-author chapters presented here. The resulting discussions generated a further 7 chapters on the workshop theme of complex stressor interactions within ecological systems.

With increasing population growth and economic development on a global scale, the human race faces difficult choices regarding the prioritizing of environmental problems and necessary remediation. As a science, ecological risk assessment aims to develop tools and approaches to effectively manage the environmental impact of human activities on our environment. However, much of the effort in this area has focused on the development of methods that address the potential impacts of single stressing agents, such as individual chemicals. While this has arisen partly from the dynamic relationship between governmental agencies seeking to regulate individually marketed substances produced by industry, there is a clear and urgent need to shift emphasis away from single-chemical effects measures toward more process-oriented techniques that realistically assess complex systems. This forms the central issue of Chapter 1, in which the philosophy, primary issues, and overall approaches of identifying and separating different stress agents in a complex stress regime are discussed.

Chapter 2 considers one of the most active areas of current ecotoxicological research: the linkage of effects measured at the level of the individual (the focus of traditional toxicological study) with population-level phenomena. In particular, the question of how to link biomarker responses (see also Chapter 1) to higher-order ecological responses at population, community, and ecosystem levels and the need to develop approaches that can be generalized across a range of species and ecosystems are considered. Also contained in this chapter is a technically detailed review of approaches used by population ecologists to study change in natural populations, including the pros and cons of alternative modeling techniques, the use of population growth as an ecotoxicological endpoint, the need to conserve genetic diversity within populations, and a consideration of the phenomenon of quasi-extinction — the concept that stressed populations may reach a point where, although viable individuals exist, the population has declined to the extent that recovery may not be achievable.

Chapter 3 develops the theme of how contaminant effects become pervasive among interacting populations within ecological communities and makes a plea for the need to consider food web interactions within regulatory frameworks dealing with risk assessment, not merely as an afterthought, but as the underlying framework behind perceived changes in species composition (biodiversity) within ecosystems. Particular issues addressed in Chapter 3 are the potential rewards to be gained from improved classification of species by function as an alternative to traditional taxonomic categorization and the need for improved models of trophic transfer of contaminants, which explicitly consider feeding interactions among species and mechanistically link fate of toxic substances with ecologically important effects. Using detailed case studies from marine, freshwater, and terrestrial systems, the authors underscore the importance of dealing with indirect effects of stressors, in addition to the traditional focus on direct effects, and give examples of alternative modeling approaches that allow quantitative assessment of complex interactions in the area of food web ecology. Chapter 4, which was submitted in advance to the workshop participants as a discussion document, complements the themes developed in Chapter 3 and uses the example of species interactions in soil ecosystems to illustrate how apparently counterintuitive phenomena can be explained through a detailed understanding of trophic interactions and, in particular, how the loss of species of apparently minor importance can have important consequences for ecosystem function.

The next 3 chapters address the key question: How do we detect adverse ecological change in naturally variable ecosystems? Chapter 5 attempts to provide guidance on this question by considering how we might attempt to incorporate our knowledge of system variability (e.g., from existing environmental monitoring databases) into the management of increasingly scarce resources. The need to design impact studies and their components at an appropriate geographic scale is discussed, as is the question of how to include environmental variability and uncertainty in models of contaminant bioavailability. The pros and cons of the use of indicator species are considered, illustrated by the example of the decline of amphibian populations and its linkage to multiple stress factors.

Chapter 6 continues the case study approach of the previous chapters by detailing several case studies that encompass marine, freshwater, and terrestrial systems, and it clearly identifies the need to avoid basing environmental management decisions on short-term assessments (i.e., single field season), which often leads to an incorrect diagnosis of causality. The chapter further illustrates the need to provide proof of underlying cause by adopting multiple lines of inquiry (i.e., weight of evidence) to identify mechanistic relationships between putative stressing agents and observed ecosystem damage.

Chapter 7 concludes the theme of naturally variable systems by addressing the issue of multiple stressors acting against a fluctuating ecoystem baseline response. Specifically, the issues of toxic substance mixtures, global climate change, and increasing urbanization and industrial development are considered, with the conclusion that in some cases (e.g., agricultural runoff) it may not always be necessary to understand complex stressor interactions in detail in order to address an obvious problem.

Finally, Chapter 8 concludes with a detailed case study on a North American river system, where reports of impacted fish populations have indicated a need to understand the relative contribution of specific causal agents against a background of complex stressor interactions. Using a tiered approach and employing environmental monitoring databases coupled with targeted field and laboratory investigations, the authors illustrate the strengths and pitfalls of existing approaches. They demonstrate the benefits of a tiered approach for identifying probable stressors with increasing certainty in a multi–land-use watershed. This approach and the associated methods can be customized for use in any watershed, identifying the dominant natural and/or anthropogenic stressors that impact aquatic communities.

The objectives of this Pellston Workshop were met, thanks to the enthusiastic participation and contributions of the many scientists who attended. We believe that the contents of this book will provide readers with a timely introduction to the methods and approaches for the identification and separation of complex stressors in ecosystems, and thereby provide a useful bridge between the linked disciplines of ecology and ecotoxicology.

— Donald J. Baird

— G. Allen Burton, Jr.

Chapter 1

Distinguishing Among Factors that Influence Ecosystems

Elaine J. Dorward-King, Glenn W. Suter II, Lawrence A. Kapustka, David R. Mount, Donna K. Reed-Judkins, Susan M. Cormier, Scott D. Dyer, Matthew G. Luxon, Rodney Parrish, G. Allen Burton Jr.

Increasingly, it is recognized that environmental management decisions, whether natural resource management or contaminated site remediation, must be long range in their view and based on a comprehensive understanding of the system to be managed. Many of the problems to be solved involve factors that are difficult to control, either because of the magnitude or scale of the exposure, the expense of control, or the social impediments to controlling them. As a result, managers and society are placing higher demands of proof that these concerns are worthy of being addressed, that is, that the concerns present an unacceptable risk to a valued resource. Society is also less inclined to invest in environmental improvements "just because we can." Environmental managers must predict with reasonable certainty that the outcome of an action will be an improvement in the environmental properties of concern. For example, it may be difficult to justify installation of a higher-level water treatment facility to improve water quality if no concomitant efforts are made to address other point and nonpoint sources that affect the system.

Weight-of-evidence approaches provide the most powerful means for estimating adverse effects on ecosystems. These approaches use laboratory and field techniques, including chemical, physical, and biological characterizations; toxicity and other experimental tests; diagnostics such as biomarkers; and retrospective evaluations of existing databases. Yet it is not clear that the existing tools, as they are widely applied now, effectively distinguish individual environmental factors from each other or that such tools are used in ways that appropriately integrate results.

Environmental managers who seek advice from scientists may receive different guidance depending on the discipline, training, and expertise of the scientists. Environmental toxicologists and chemists familiar with the tools of risk assessment and accustomed to focusing on chemical contaminants may provide very different

Ecological Variability: Separating Natural from Anthropogenic Causes of Ecosystem Impairment.
D.J. Baird and G.A. Burton, Jr., editors.
ISBN 1-880611-43-0

advice from that of ecologists concerned with system or regional landscape-level changes, for example, loss of habitats. Managers are likely to view ecological risk assessors and ecologists as having equivalent expertise. To the extent their advice is conflicting, however, both are likely to be ignored in the final decision.

This chapter suggests ways in which knowledge and tools from the ecological and toxicological–chemical disciplines can be combined more effectively to describe the status of ecosystems, to define the factors that contribute to that status, and to allow predictions of response to changes in magnitude or intensity of environmental factors. We must integrate all available tools if we are to achieve the greater levels of ecosystem management required by increasing human populations with greater resource demands. Along with a better understanding of the influences of chemical factors, we need a better understanding of the role of nonchemical factors (e.g., habitat loss, sediment deposition) and natural disturbances (e.g., drought, flooding, earthquakes, erosion from landslides) in perturbing ecosystems. In addition, separating the adverse effects of these natural and anthropogenic factors is difficult at present and requires both improved measurements and the ability to quantify the variation of each.

What is a Complex Stressor?

The term "complex stressor" is used throughout this book. A complex stressor is defined as one or more environmental factors that potentially can cause a significant change in an organism, population, community, or ecological system. Stressors may act simultaneously or sequentially.

Use of the term "complex stressor" has the advantage of highlighting the complexities of factor interactions in the environment, such as these:

- Multiple stressors—Such stressors interact in ways that make it particularly difficult to distinguish the relative contribution of each individual factor to a measured response.
- Classes of single stressors—Such classes have markedly different consequences in different situations. A factor may alter a system from 1 stable domain to any of 2 or more stable domains. For example, a pesticide may remove a top predator in one situation but grazers in another, or a chemical's toxicity may increase or decrease at higher and lower temperatures, depending on its concentration.

Most complex stressors can be categorized as

- mixtures of chemicals,
- combinations of potential chemical and habitat factors,
- complex arrays of habitat alterations independent of chemical stressors, or
- sequentially acting stressors.

Considerations in Understanding Complex Stressors

Assessing Chemical Stressors

Much progress has been made in identifying, measuring, and assessing individual environmental factors that may cause adverse effects. A majority of the resources expended in the past 30 years, beginning with the National Environmental Policy Act (NEPA) signed in the U.S. in 1970 and similar laws enacted in many nations and other jurisdictions worldwide, have been directed toward resolving environmental pollution from human activities. This focus on chemical effects was appropriate and necessary to address the rapid increase in chemical use and the unregulated disposal of wastes that accelerated after World War II. Regulations developed in response to NEPA proscribed consideration of environmental impacts through environmental impact assessments.

Virtually all of the follow-on legislation was aimed at regulating the use, production, and disposal of chemicals.[1] Regulatory authorities in various jurisdictions have developed different approaches to assessing environmental conditions and imposing restrictions on the use, release, and disposal of chemicals. Procedures to set criteria, standards, thresholds, and benchmarks; to develop hazard quotients (probable effect concentration [PEC] and probable no-effect concentration [PENC]); and to conduct exposure assessments, effects assessments, impairment assessments, injury assessments, damage assessments, ecological risk assessments, and others evolved during the past 3 decades. Our collective knowledge of the fate, movement, and effects of chemicals has advanced steadily. Forensic and predictive methods have enabled us to assign causality to effects from contaminants, to evaluate alternative remedial actions, and to predict likely consequences of new releases. These efforts have paid off, resulting in measurably cleaner surface waters, air, and soils in much of North America and Europe. Throughout much of the world, concerted efforts to stem the degradation of air and water have achieved great improvements in the quality of these resources.

What we have learned through these efforts to address chemical pollutants, especially chemical mixtures, can be applied to other complex stresses. A great deal of experimentation has been completed in this area, and some general principles have emerged. Overall, it appears that joint toxicity occurs among chemicals with similar modes of action. Within similar modes of action, the concentration–addition model (often called the "toxic unit [TU] concept") seems to describe the interaction; in

[1] Examples in the U.S. were the Clean Water Act (1972); the Clean Air Act (1972); the Federal Insecticide, Fungicide and Rodenticide Act (FIFRA 1947); the Toxic Substances Control Act (TSCA 1976); the Comprehensive Environmental Response Compensation Liability Act (CERCLA; Superfund Act) of 1981; the Superfund Reauthorization Act (1986); the National Contingency Plan (1988); and the Resource Conservation Recovery Act (RCRA 1976).

essence, the toxicity of a mixture is the sum of the fractional potencies of its components so that for a mixture with *n* components,

$$\text{TU mixture} = \sum_{i=1}^{n} \text{TU}_i \qquad \text{(Equation 1-1)},$$

where TU is defined as the concentration in the mixture divided by the index concentration for effect.

For example, an acute TU is typically calculated as concentration / LC50. Additivity or near additivity has been demonstrated for many groups of chemicals, such as narcotics, organophosphate pesticides, polynuclear aromatic hydrocarbons (PAHs), major ions, and some metals (e.g., Bailey et al. 1997). In contrast to mixtures of chemicals with similar modes of action, chemicals with dissimilar modes of action (e.g., Zn and diazinon) show little or no interaction, such that the toxicity of a binary mixture shows toxicity equal to that of the most toxic component (Shirazi and Lowrie 1990).

Interactions of chemical mixtures, such as synergy (TU mixture >> $\sum \text{TU}_i$), have been demonstrated (e.g., Pape-Linstrom and Lydy 1997; Forget et al. 1999). Synergy usually occurs when 1 chemical has a potentiating effect on the physiological pathway of a second toxicant. The classic example is piperonyl butoxide and pyrethroid pesticides; piperonyl butoxide blocks the detoxification pathway for pyrethroids, thereby greatly exacerbating their toxicity. In fact, this interaction is used intentionally in pyrethroid pesticide formulations.

While laboratory experiments have demonstrated approaches for mixture assessment, the test of an approach lies in its effectiveness when applied to mixtures in the field. Experience suggests that assuming addition within a mode of action and independence between different modes of action is adequate in most cases. For example, in studies of more than 80 municipal and industrial effluents, toxicity identification studies showed no instances where observed toxicity was greater than would be predicted by this approach (Mount et al. 1992). Important cases of additive toxicity have been demonstrated in field situations for PAHs and dioxin-like compounds.

The finding that mixture models are necessary to account for the potency of PAHs and dioxin-like compounds in the field provides excellent insights into the circumstances necessary for the expression of interactive toxicity in the environment. In addition to sharing a common mode of action (narcosis for PAHs; Ah-receptor agonism for dioxins, furans, and polychlorinated biphenyls [PCBs]), the sources for these contaminants and their environmental fate are such that they occur in mixture compositions where multiple components contribute meaningfully to the toxicity. The absence of the latter attribute greatly simplifies the assessment of many mixtures. From a practical perspective, accounting for additive toxicity is important only when the magnitudes of toxicity contributed by components of the mixture are of comparable magnitudes (Figure 1-1). In cases where 1 component of the mixture dominates, ignoring toxic interactions within the mixture adds little uncertainty to the overall assessment. Metals provide an excellent example. Experimental work has clearly indicated that the toxicity of many metals (e.g., Cu, Cd, and Zn) approximates additivity, and it is also true that most sources of metal contamination contain multiple metals. However, in practice, many metal mixtures are dominated by a particular metal (Figure 1-2). Hence, assessing the potency of a mixture on the basis of its most potent component is often adequate, even though there may be small contributions from other components. In the case of PAHs, however, multiple individual PAHs contribute substantially to toxicity (Figure 1-3), and the additive toxicity must be taken into account to adequately assess the mixture.

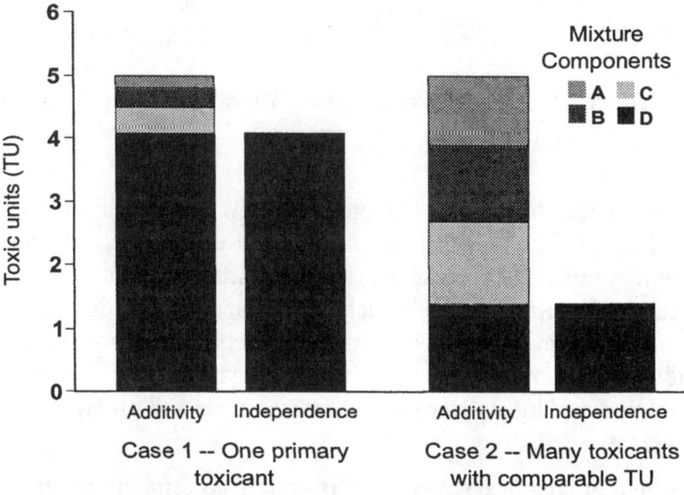

Figure 1-1 Importance of considering additive toxicity when mixture components exhibit toxicities of comparable magnitude

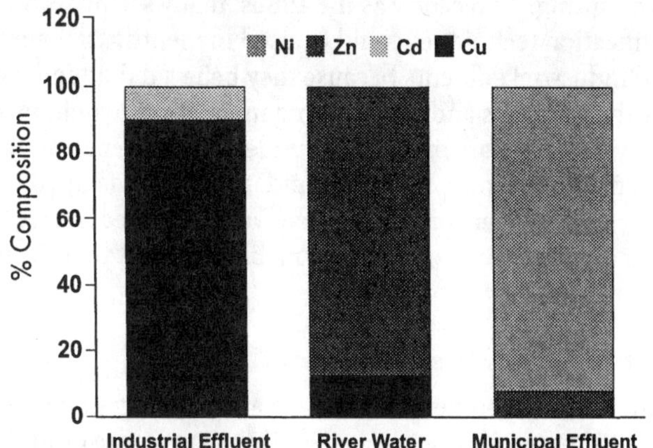

Figure 1-2 Examples of individual mixture component dominance in environmental media

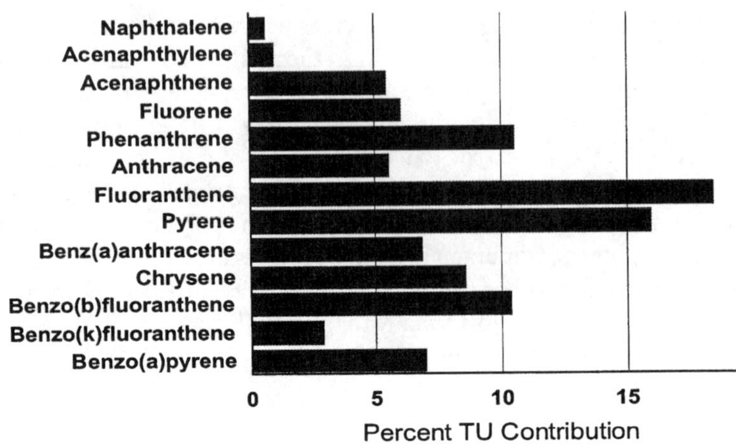

Figure 1-3 Contribution of individual PAHs to total toxicity of mixture

There remains a need to focus on chemicals in a few heavily contaminated areas and in regions of the world where little attention has been given to contaminants thus far. These environments still require assessment, intervention, and remediation of chemical contamination, many of which can be successfully addressed using well-tried approaches for evaluating exposure and effects of environmental factors individually. In addition, there is still much to be gained by further elucidating the chemical-to-chemical interactions in the environment and the significance of those interactions.

The assessment of chemical stressors seems relatively straightforward after 30 years of experience. However, before the research was done and the knowledge was applied to solve many problems, the prospect of predicting the effects of chemical mixtures was daunting. As recently as the 1980s, many scientists doubted that toxicant identification techniques could be used to identify the causes of toxicity in municipal and industrial effluents because they believed that understanding the complexity of the solutions and the resulting interactions would be beyond our abilities. However, these fears proved unfounded (Grothe et al. 1996), in part because the fears were born more from trepidation over imagined complexity than from the reality of the lesser complexity that was uncovered. As we approach the assessment of complex stressors, it is reasonable to expect that the path forward may be similar to our previous experiences.

Assessing Nonchemical Stressors

Many scientific and environmental management challenges now lie in determining the significance of potential chemical effects in the presence of other environmental stressors and interacting factors. Overemphasizing potential chemical effects without paying proper attention to other environmental factors and realistic exposures can result in wasteful pollution control investments that will not tangibly improve environmental quality. This is particularly the case as environmental management concerns expand to regional or global scales. There is growing consensus that physical alteration of landscapes has not been given adequate attention. Channelization of streams, fragmentation of plant communities, harvest of timber, control of fire, increase of impervious surfaces by paving or compaction, and other uses of land have much greater influence on ecological resources than do chemicals whose concentrations are near or slightly above toxicity thresholds (Burton and Pitt 2001). Moreover, compliance with regulations that focus on biological resources (e.g., Marine Mammal Protection Act of 1972; Migratory Bird Treaty 1972; Endangered Species Act 1973; Fish and Wildlife Conservation Act of 1980; and the proliferation of multilateral accords on biodiversity and sustainable development) requires more than assessments of chemicals.

Many of the tools exist in the ecological community to describe natural and managed ecosystems. Ecologists, regardless of their specialty, have struggled with the multitude of interrelated and interconnected parameters that affect organisms, populations, communities, and ecosystems. Measures of ecosystem condition, structure, function, and population and community responses to perturbations are well established (Krebs 1978; Grieg-Smith 1983; McIntosh 1985; Sheail 1987; Levin et al. 1989; Bailey 1998). The ability to use the information and interpretations developed from these tools has been demonstrated (Mitsch and Jørgensen 1989; Munshower 1993; Suter 1993b; Saunier and Meganck 1995). Ecologists are accustomed to dealing with complex suites of environmental parameters that increase variability and uncertainty. However, despite the many techniques available, relatively few have been used in ecological risk assessments, site characterization, or risk management. This probably is due to the narrow focus of most risk assessments on toxicology and engineering technologies, with relatively little attention given to ecology.

Scientists and engineers engaged in toxicological- or chemical-oriented environmental research and management may know little of exposure–response relationships for many nonchemical environmental factors. In some cases, the information may not exist. In other cases, it may not be documented or may be published in formats with which these scientists are not familiar. Efforts are underway at the U.S. Environmental Protection Agency (USEPA) to compile this information for dominant aquatic stressors. Some of the most important factors to be considered, such as habitat loss, nutrient flux, and pathogen flux in ecosystems, have not been integrated into the environmental regimes that have been used traditionally to define exposure–response relationships. Yet, the information is needed for thorough environmental assessments.

Ecological risk assessment methods were written largely to address chemical effects (USEPA 1992, 1993, 1994a, 1994b; United Nations Environmental Program–

International Environmental Technology Center [UNEP–IETC] 1996). Though the fundamental procedures are sufficiently flexible to accommodate biological and physical agents, there are few widely known examples in which biological or physical agents have been addressed with the same rigor that is applied to chemical agents. Traditional ecological methods (observational ecology and statistical methods) are well developed for identification of dominant or controlling environmental factors (biological, chemical, and physical agents). Acquiring the information that is necessary to characterize biological or physical factors involves greater emphasis on field-based investigation than is common for chemical agents. Typically, a combination of field observation, field experimentation, and lab experimentation is needed to strengthen conclusions regarding linkage of different factors and effects. Incorporating these factors into the ecological risk assessment process effectively ties them to social values. Perhaps one of the most important contributions of ecological risk assessment is the recognition of society's role in articulating which resources are valued.

Though risk assessments have not yet been adopted widely in many emerging countries, the framework for adapting risk assessments to enhance environmental management procedures exists. By endorsing risk assessment as an important part of modern environmental management and multilateral agreements, to conform with the agreed principles from the Earth Summit (held in Rio de Janeiro in 1992 and attended by political and environmental leaders from more than 100 nations), the United Nations (UNEP-IETC 1996) provides an opportunity for these emerging nations to achieve rapid improvement in environmental quality.

Natural Variability

Many nonchemical factors (e.g., suspended solids, flow, temperature) are a natural part of the environment, and considerable variability in their intensities and distribution occurs. However, it is possible that human modification of the intensity or pattern of a factor in the environment can cause subtle but important changes. Because these environmental factors show a range of intensities naturally, it can be difficult to predict when intensities or patterns of stress are significantly outside the boundaries of the baseline range for a particular system. In many cases, the techniques that are available for quantitating these factors are not adequate.

Before we can plan environmental management actions, we must separate the natural environmental effects within a system from the anthropogenic factors and their associated variation imposed from outside the system. Before we can establish the significance of chemical effects, we must understand natural variability and baseline conditions and their relationship to biological responses.

Baseline versus reference conditions

Detecting significant differences in ecosystem quality depends on establishing an appropriate background or reference condition for comparison. There are a variety of ways in which to evaluate data for change across space and time. Toxicologists are familiar with hypothesis testing using control groups, and comparisons among multiple tests over time may be evaluated using control-charting methods. Ecological evaluations may use analogous methods to establish baseline or reference conditions.

"Baseline" refers to a set of conditions that exist at a particular site at a specified space and time. A baseline is measured or otherwise described at the place of interest and used subsequently as the condition of comparison. Alternatively, baseline conditions may be inferred from measurements or descriptions of neighboring places that are thought to be representative or similar to the place of interest.

Reference conditions may be inferred from a specified range of conditions derived from similar areas in the region (e.g., the 20th percentile of species richness measures for first-order, low-gradient streams in an ecoregion). Another valid reference condition could be a target defined by stakeholders or resource managers (e.g., 20 deer per km^2).

Careful consideration must be given to specifying the conditions that will be used for comparison. This requires careful selection of the measurement endpoints and determination of the spatial density and temporal frequency of measurement. The endpoints should be indicative of potentially important controlling factors and relevant to the valued resources of interest. Endpoint selection and interpretation should take into consideration life-history characteristics, successional stage, indirect effects (predation, competition), and sensitivity or acclimation to other environmental factors. It also is crucial to recognize that ecological systems are characterized by inherent spatial and temporal variation. It is equally important to recognize that ecological systems do not achieve equilibrium. Because some system components fluctuate about a flat trend line (i.e., slope of 0) that appears essentially the same for decades, many scientists have inferred equilibrium. Critical historical examination of system components and experimentation with system models demonstrate that equilibrium does not occur. Different components (both abiotic and biotic) have unique patterns of change (e.g., diurnal to years) that may shift at differing rates and along different trend lines.

Near-threshold Effects, Nutrients, and Essentiality

Many of the anthropogenic environmental factors we are interested in may be present at near-threshold levels (i.e., in the gray zone between adverse effect and no effect). Even if the factor is well studied, uncertainty in predicting response can be high. When 1 stressor is present at near-threshold levels, other factors may be important in determining whether adverse effects will or will not occur.

Nutrients (N, P, S, and other essential elements) are present in ecosystems at a great variety of natural background levels over different geographic regions. These levels may be relatively high for macronutrients and quite low for certain minerals. This variation is well documented and appears to be so large for essential elements that the "window of essentiality" does not encompass it (i.e., regions that are toxic to some organisms are below nutrient requirement levels for others). This complicates establishing a "normal" range of variability that has broad applicability and emphasizes the importance of establishing system-specific determinations of variability.

Determining Effects

Because "zero effect" is not achievable for most factors across regional scales (e.g., temperature, essential nutrients), we should not ask where to draw a single line to delineate no effect but should focus on distinguishing acceptable from unacceptable. Approaches historically used to categorically define adverse effects (e.g., establish a toxicity threshold for the most sensitive species) do not provide the kind of discrimination (or predictive ability) we are seeking to accurately assess biological impairment or unacceptability. This is due to a variety of factors, such as an inability to discern natural stressor effects, causality, variability, or reference conditions.

Simplifying Assumptions

As we consider systems in the context of a broad array of factors that may be influencing them, we can envision almost endless scenarios under which 2 or more factors may interact to produce responses not predictable from the individual factors. While such interactions have been shown to occur under certain circumstances, deriving a diagnostic or risk characterization approach that accounts for all of these potential interactions a priori is beyond current science or state of practice. One way forward, then, is to derive approaches based on simplifying assumptions that describe most (but not all) situations sufficiently to guide management decisions in the near term.

An example of this simplification approach lies in the science of predicting the fate and bioavailability of nonpolar organic chemicals. Over the last 20 years, there has been marked progress in our ability to predict the activity of these chemicals in water, sediment, and biota (Mackay, DiGuardo, Paterson, Cowan 1996; Mackay, DiGuardo, Paterson, Kicsi, Cowan 1996). These approaches recognize organic C as a primary partitioning phase for nonpolar organic chemicals. In most models, organic C is quantified generically as total C (particulate or dissolved), even though it is widely recognized that organic carbons from different sources actually have differing composition and chemical properties. Despite the fact that these bioavailability and fate models are based on a premise that is, strictly speaking, false (i.e., that all organic C is the same), their simplified approach to partitioning was highly effective in moving this area of science forward. Though generic treatment of organic C ignored some suspected causes of variability, it was successful in reducing a large part of variability associated with environmental fate. As our understanding of partitioning behavior has improved, we are now better positioned to recognize where the simplified approach is not sufficiently representative of actual chemical behavior, and we can consider more complicated (and more realistic) models that recognize different types of organic C with substantively different partitioning behavior (e.g., soot C; McGroddy and Farrington 1995; Gustafsson et al. 1997).

Drawing from this example, we can consider a range of approaches to the multiple stressor questions that vary in the degree to which they consider interactions among environmental factors. The simplest of these models is one that assumes no interaction among factors, that each exerts its influence on the system without regard to the intensity of stress from other factors (see Figure 1-4). While this approach may seem hopelessly naive at first glance, there is no evidence to suggest that it would be grossly misrepresentative in many cases. As was the case for organic C, adopting this approach as a starting point is not to suggest in any way that all systems and environmental factors behave independently and that interactions do not occur; it is simply an entry point from which to begin building an approach that invokes complexity where it is needed but avoids it where added complexity adds relatively little to the power of overall assessment.

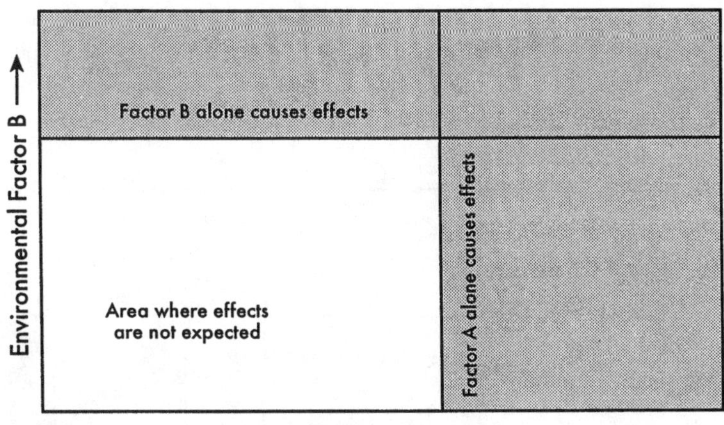

Figure 1-4 No Interaction Model: Individual stressors act independently in causing environmental effects

In considering the utility of this approach, it is important to examine which types of responses to environmental factors are the result of interactions and which are the result of multiple independent effects. Take as an example a stream that is influenced by both elevated temperature and sedimentation and where both trout and certain mayflies are reduced in abundance. Assume further that the temperature stress is severe enough to account for the reduction in trout abundance but that other species of invertebrates less tolerant of high temperature than the affected mayflies are still abundant, suggesting that high temperature alone cannot explain the decreased abundance of mayflies. Increased sedimentation might then be considered as a possible explanation for the decreased abundance of mayflies. If the degree of sedimentation is sufficient to explain the effects on mayflies, then it is not necessary to consider the interaction of the 2 stressors as an explanation for the observed effects. This is not to say that interaction has been disproven, only that an effective diagnosis could be made based on a model that assumes no interaction. Even though both factors were involved in altering the state of the system, temperature and sedimentation stresses did not necessarily interact; they can be viewed as having acted simultaneously but independently. Understanding the individual effects of both factors was necessary to understanding the cause of the change, but knowledge of potential interactions was not. On the other hand, knowing the interaction may be important for some other stressors, such as temperature and toxicants.

The key, then, lies in recognizing when interactions among stressors will occur, particularly to the degree that ignoring this interaction (i.e., assuming independence) leads to predictions that are markedly different from the actual outcome (Figure 1-5). Many factors may interact to a degree, but if the interaction is relatively weak, it may be practical to place lower priority on its quantification and subsequent incorporation into the general assessment approach. Alternatively, if these interactions are large, then there is higher priority for expanding the assessment approach to include consideration of that interaction.

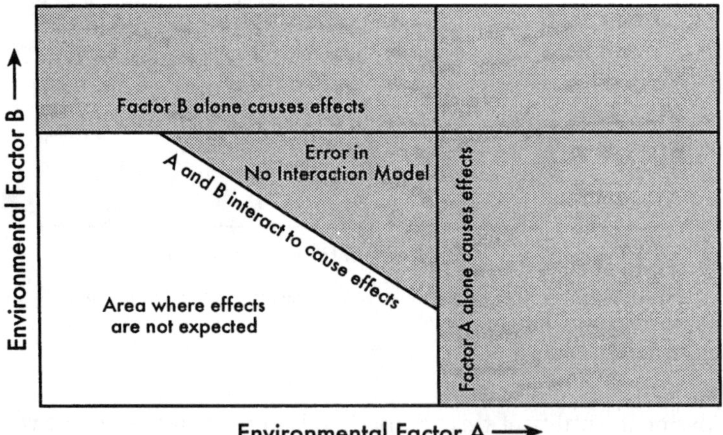

Figure 1-5 Error in No Interaction Model when stressors interact to cause effects

A Way Forward

This section illustrates some approaches for determining the causes of impairments to ecosystems, especially complex or multiple stressors. Although not exhaustive, it is meant to create an awareness of information and inferential methods that can be used to successfully identify key stressors and related factors. Other methods are discussed in the following chapters.

Understanding complex stressors presents a challenge that is best accepted when we are equipped with a flexible strategy and a diverse set of analytical tools. It is important to have training and experience as well as a team that can contribute knowledge of ecology, chemistry, hydrology, toxicology, geography, and geology. The larger the geographic scale or the greater the complexity of issues, the more important it is to have a multidisciplinary team that can communicate and collaborate. The scope of the assessment project should be commensurate with the importance of the decision to be made.

The temporal sequence for determining causes of impairments is as follows:

1) The impairment must be clearly defined.
2) Causal scenarios must be developed.
3) Field and laboratory methods must be applied to generate needed information.
4) Inferential approaches must be applied to determine the cause.

Defining Impairment

Defining impairment requires 1) a property of an entity (i.e., an endpoint) such as the number of species in a fish assemblage or the frequency of gross anomalies and 2) a degree of modification of that endpoint. The endpoints are often defined primarily by legal or policy considerations based on societal values. The degree of modification of an endpoint that is necessary to declare impairment may be defined in various ways. In the simplest case, a manifest impairment such as a fish kill, extirpation of a local population of an endangered species, or massive chlorosis or necrosis of leaves of trees is reported. Such events may be natural and uncommon, but typically they require investigation. In other cases, the impairment may be a reduction in a continuous or count variable such as abundance of fish or a multimetric index. In those cases, impairment must be defined relative to natural variation in the endpoint (i.e., based on a time series), the range of the variable at reference sites (Yoder and Rankin 1995), or regulatory or policy goals (e.g., fishable waters).

To determine the cause of impairments to valued resources, it is helpful to describe desired environmental conditions in terms of the range of states (e.g., classifications of environmental quality) that will be tolerated by society. The range that can be realistically achieved is limited by natural variability. However, because of the need to use the environment as a source of resources, for recreation, and as a location for housing, industry, and transportation, environmental conditions beyond the range of natural variance may be acceptable and even desirable. Input from many societal opinions and perspectives, including those of ecologists, can be combined to define acceptable boundaries (i.e., the range) of environmental structure and function. Acceptable limits include many attributes of ecosystem structure, function, and societal values. The combination and variety of possible acceptable environmental conditions can be depicted in a multidimensional volume, or "hypervolume." Impairment occurs when the set of valued ecological attributes is outside the acceptable environmental boundaries, that is, outside the hypervolume. The likelihood of future impairments may also be predicted if the trajectory of the set of valued ecological attributes projects outside the acceptable limits. However, in some situations, the multidimensional state space reduces to a simple numerical goal. Examples include the action level for PCBs in fish muscle tissues or the abundance of an endangered species that constitutes the restoration goal.

It is important for ecologists and environmental managers to acknowledge that societal values are prone to shift with time and with economic and political will. Consequently, we should expect that the status of a resource and the dimensions of acceptable environmental limits will shift with time. Nevertheless, it is the recognition of the impairment that triggers an investigation into the potential causes.

In order for causal analysis to be successful, we must clearly define impairment in terms of real properties of real entities. Many bioassessment programs have been soundly criticized (Suter 1993a; Kapustka and Landis 1998) because they use ecosystem health or integrity as endpoints and define them in terms of an index that is an arithmetic combination of multiple metrics. While this approach may be acceptable in terms of defining impairment, for causal analysis the indexes must be disaggregated, and the actual responses that constitute the impairment must be identified. For example, frequencies of fish deformity, erosion, lesion, and tumor anomalies should be converted to frequencies of each anomaly such as scoliosis or dermal lesions.

Developing Potential Causal Scenarios

Once impairment has been defined, it is necessary to determine potential causes. A cause may be a single chemical or other simple factor such as channelization or a set of factors such as high temperature and low dissolved oxygen. It is important to develop causal scenarios that are sufficiently complex to explain why the impairment occurred but not more complex than necessary. This rule may be explained by the concept of necessary and sufficient causes. If a release of 95 °C cooling water occurred prior to a fish kill in which the proteins of the gills were denatured, that hot water alone is both necessary and sufficient to cause the observed impairment. However, if the fish kill is caused by a combination of a metal in an effluent and low pH from mine drainage into a circumneutral stream, neither alone is sufficient to cause the impairment, but they are both necessary and together they are sufficient. Causal scenarios must constitute a sufficient cause, so they must contain all necessary causes but none that are unnecessary.

Causal scenarios may come from various sources. Ecologists who perform a biological survey may observe and report potential causes. Local residents, potentially responsible parties, or resource management agencies may propose causes. However, the assessors who perform the causal analysis must conduct their own investigation. This difficult task is rendered more so by diffusely specified impairments that make more potential causes seem plausible. In addition, if impairments occur over large spatial scales, then the number of candidate causes increases. The first step in the analysis, then, will be to identify all factors that may potentially constitute necessary causes. These may come from examination of remotely sensed data, review of permitted effluents, or examination of the site. The second step is to analyze the potential factors to generate a set of causal scenarios. This requires that the assessor either know how potentially causal factors influence the endpoint and how they interact or be willing to frame hypotheses based on assumptions. At this point in the process, it is more important to include all reasonably plausible scenarios than to demand evidence. In the absence of evidence concerning the nature of interactions among factors, it is generally appropriate to begin by assuming additive action for factors with the same or similar modes of action and independent action for others.

Obtaining Site Information

Because of the diversity of impairments, stressors, and ecosystems, a large suite of methods in ecology, toxicology, chemistry, and pathology may be useful in determining whether stress is occurring and which stressors are important. A combination of methods (to provide a weight of evidence) that describe the physical and chemical exposures and the responses of resident or surrogate biota is often necessary to identify significant stressors. Many of these methods are well known and have been widely used to assess ecosystem impairment (Foran and Ferenc 1999). However, a few techniques that are particularly useful are discussed below and in Chapters 2, 3, and 8.

In situ toxicity and bioaccumulation indicators

In situ toxicity and bioaccumulation indicators refer to the exposure of surrogate or resident species in relatively small, confined chambers at the study site. There are many other useful in situ methods that involve the collection of resident organisms for laboratory assessments, enumeration of populations and communities, tagging of organisms, or placement of organisms in large mesocosms (e.g., limnocorrals), but these methods are not discussed here.

In situ exposures of confined organisms provide useful information that laboratory bioassays and characterizations of indigenous communities cannot. This is not to say these 2 assessment approaches are unimportant; quite the contrary, each provides unique information on the exposure and effects of aquatic organisms in the target ecosystem. The laboratory provides more control over experimental variables such as temperature, light, and manipulations of contaminant concentrations or exposures. However, these controlled conditions inhibit our ability to provide realistic exposures to both fluctuating stressors and interactions of stressors with key variables such as suspended or dissolved solids, light, temperature, and hardness. In addition, sampling and processing artifacts can alter both bioavailability and exposure (e.g., Burton 1991; Sasson-Brickson and Burton 1991; Burton et al. 1996). While indigenous biota reflect the integrated effects of all environmental factors, the indigenous community has a relatively high degree of spatial and temporal variability; is strongly affected by natural factors such as habitat, flow, food, and predation; and may consist of monitored species that are relatively insensitive to stressor pulses. However, tolerance to environmental conditions is a fundamental aspect of ecology and should not be viewed negatively. The weaknesses of in situ exposures are the strengths of laboratory bioassays and community characterizations. The strengths of in situ approaches lie in their ability to improve the certainty of stressor–effect relationships. For variable control and ecological realism, in situ approaches are the middle ground between the laboratory and field surveys.

No one design or standard method can be prescribed for in situ testing. The peer-reviewed literature offers numerous examples of successful applications involving marine and freshwater systems at all life stages with a myriad of chamber designs (see reviews by Burton et al. 1996; Chappie and Burton 2000). The selection of appropriate species, life stages, exposure periods, and chamber design or placement is best made to answer site-specific questions. For example, if a site has PAH contamination, then chambers should be placed both in the water column and at the sediment interface to allow both ultraviolet (UV) and no-UV exposures. The key questions may require that exposures be conducted at low flow and also during high flow, providing information on the influence of sediment versus water column contamination and on the role of photoinduced PAH toxicity (Ireland et al. 1996). If stormwater runoff is a concern, then chamber design, placement, and exposure periods can target low and high flows with varying mesh sizes to separate suspended solids and flow stressors (Burton and Moore 1999; Tucker and Burton 1999). In situ exposures have often been used to show bioaccumulation, either through tissue residue analyses or other biomarkers, particularly with bivalves along coastal areas and with fish below pulp and paper mills. Recent studies by Burton et al. (2002) showed rapid accumulation of PCBs in the freshwater oligochaete *Lumbriculus variegatus* exposed to sediments but not overlying waters and differential toxicity to *Hyalella azteca, Chironomus tentans*, and *Daphnia magna.* Another design that allows more control over chamber deployment and manipulation is to establish a bank-side field station where stream waters are pumped continuously through exposure chambers.

In situ diffusion samplers

A wide variety of abiotic samplers have been used for assessing bioaccumulation or exposure to chemicals. These samplers have varied widely in their design and placement into water or sediments. Some of the first diffusion samplers were "peepers," cellulose dialysis membranes covering small, thin frames and buried in surficial marine sediments (Carignan et al. 1985). These peepers allowed sediment pore waters to be collected and analyzed for chemical constituents after an equilibration period of approximately 2 weeks. In the past few years, the peepers have been modified to be larger and possess a variety of membrane coverings ranging from 1- to 250-micron pore size (Burton 1992; Fisher 1992; Sarda and Burton 1995). Davison and Zhang (1994) have used a double gel-lined membrane chamber for assessing the presence of dissolved metals in sediment pore waters. If the surficial sediments contain oxygen, then organisms can actually be placed within the peepers after they equilibrate, to provide realistic in situ exposures (Greenberg et al. 1998; Burton et al. 2000). Comparing the chemistry of pore water collected with peepers to that collected by more traditional methods (grab or core sampling followed by centrifugation) showed ammonia concentrations became significantly higher when the sediments were collected and manipulated (Sarda and Burton 1995).

The term "semipermeable membrane device" refers to polymer bags containing various solvents that are placed in the water column. These bags are left in the water column for a period of weeks and then retrieved and analyzed for nonpolar organics. They have been shown to accumulate most of the same compounds that resident fish species do and at similar concentrations. This approach is therefore a useful surrogate for aquatic species and allows for long-term exposure evaluations of nonpolar organic chemical presence.

Biologically directed fractionation

Toxicity identification evaluation (TIE), or more broadly biologically directed fractionation, is a relatively well developed set of techniques used to identify the causes of toxicity in environmental samples or other matrices of interest. In these approaches, samples are manipulated using a variety of physical or chemical treatments intended to isolate, neutralize, or otherwise influence toxicants with particular properties (e.g., volatile, chelatable). Rather than measure the effects of these changes using chemical analysis, changes are assessed using biological responses such as toxicity. Examining the aggregate responses observed over a suite of manipulations provides evidence as to the property of the causative toxicants. Knowledge of physical and chemical properties can in turn be used to design fractionation procedures, which simplify the sample chemically by isolating chemicals that cause toxicity. By using biological responses to track the causative toxicants through the fractionation, subsequent chemical analysis can focus on only a subset of chemicals known to be associated with toxicity.

While much of the research on and application of biologically directed fractionation has been in the evaluation of municipal and industrial effluent discharges, the technology is by no means restricted to that application. Biologically directed fractionation has also been applied to ambient water samples, sediment samples, tissue extracts, and commercial chemical products (USEPA 1991). The biological responses are not limited to traditional water-column toxicity tests; fractionations have also been based on bacterial assays, biochemical assays, sediment toxicity, and toxicity induced by UV light. The only real requirement to use the approach is our ability to produce the index effect (e.g., toxicity, biochemical assay response) experimentally using an environmental sample taken from the site of interest.

Limitations of the approach stem from this same requirement. We must have some assurance that the response measure used as the basis for fractionation (e.g., toxicity test) has some relationship to the response of interest in the system being studied. For example, if the impairment being assessed is reduced fish populations, fractionation studies that use bacterial luminescence as an endpoint are relevant only if we are confident that the bacterial assay responds in an interpretable way relative to fish. When TIEs are conducted on effluents, this connection is often assured because the response of interest is toxicity itself. When causes of impairment in field populations are diagnosed, this connection may be less clear.

Another limitation to biologically directed fractionation is the scope of stressor that can be addressed. In general, fractionation schemes are designed around chemical compounds, although pathogens have also been characterized by TIE when the effect of the pathogen was expressed in the course of a toxicity test. In addition, the method is not conducive to longer, chronic toxicity assessments because of the volumes of pore water required and the possible artifacts related to sample handling and storage.

Although traditional fractionation may be limited primarily to chemical stressors, the logic used to make inferences in many ways parallels the diagnosis of impairments in natural systems caused by other stressors. For example, steps to confirm a suspected cause of toxicity as the actual toxicant include not only correlation of exposure concentration and effect but also coherence of symptoms and relative species sensitivity between the candidate toxicant and the sample being studied. These approaches have direct parallels in the evaluation of natural systems.

Biomarkers

Advances in biochemistry, cell biology, and physiology during the last several decades, when applied to toxicity studies, have greatly increased our understanding of toxicological mechanisms and have provided scientists with additional tools for environmental assessments. Many of these applications are referred to as "biomarkers" or "biological markers." Biomarkers are molecular and cellular (and possibly physiological) responses that have been proposed as potentially valuable indicators of exposure and effects of chemical contaminants, as well as potential susceptibility. Within the past decade, increasing emphasis has been placed on the potential use of biomarkers as indicators of anthropogenic impacts (Huggett et al. 1992; Decaprio 1997). Molecular and biochemical responses (i.e., biomarkers) are believed to be among the most sensitive and earliest detectable responses.

Because a wide variety of techniques have been used for many diverse purposes under the classification of biomarkers, there is confusion about what biomarkers can and cannot tell us in environmental assessments. It is important to note that while biomarkers have a number of advantages and uses, not all advantages or uses apply to all biomarkers. The key to interpreting biomarker information lies in understanding the underlying mechanisms involved in altered responses, the degree of specificity, how the responses relate to exposure parameters (habitat or tissue contaminant data), and potential higher-order effects. Some molecular and biochemical responses to environmental contaminants are part of an organism's normal metabolic processes. The potential effects on alterations in responses that are essential for normal metabolic and physiological processes may be readily appreciated. However, many biomarker responses represent detoxification mechanisms. In these cases, simple changes may not represent significant adverse effects, particularly if they effectively ameliorate potential adverse effects of toxic chemicals.

Some biomarkers may be useful for assessing exposure. Such biomarkers may substitute for chemical analysis, especially if the measurement of the biomarker is less expensive than analytical chemistry. Biomarkers of exposure have the advantage of demonstrating whether chemical exposure is sufficient to elicit a biological (i.e., biochemical or physiological) response. A change in a biomarker of exposure may suggest the presence of a certain type of chemical. Analytical chemistry can then be used to identify particular chemicals that are present. Good biomarkers of effect should be biologically relevant, that is, they will lead to effects at a higher level of biological organization). Some biomarkers may be very specific to particular chemicals (e.g., dicholordiphenyldichloroethylene [DDE]–induced eggshell thinning, organophosphate- or carbamate-induced inhibition of acetylcholinesterase activity of organophosphates and carbamates). Some biomarkers can be used to predict effects at the individual or population level; others may be early warnings of effects at higher levels. Biomarkers of exposure and effect can be used to monitor changes over time (e.g., in Great Lakes birds) or as indicators of health; others may be used in a diagnostic manner to determine causation.

Inferential Assessment Approaches

Field and laboratory methods may be used for either falsification or verification of stressor identification (i.e., causality), significant effects, and model predictions. Falsification provides a strong sense of certainty and is usually used to cull the number of potential causal pathways to a manageable number. Verification techniques can be used to confirm the likelihood that a factor is shaping conditions in the ecosystem. In a weight-of-evidence analysis, verification techniques help to eliminate some pathways and emphasize others.

Falsification approaches

Any one of the following lines of evidence can independently eliminate a proposed cause for impairment. If stronger arguments are needed, multiple lines of evidence may be combined.

- Toxicology—Media or effluents that are not toxic in sufficiently sensitive toxicity tests are not causes of impairments. Note that this does not eliminate the possibility that the effluent or medium is episodically toxic or toxic under particular circumstances (i.e., when another necessary cause is present) or that it contains chemicals that do not act through toxicity (e.g., nutrients).

- Temporality—If the putative cause began after the effect, it can be rejected. If the effect is episodic and the putative cause is continuous, that cause can be rejected. (However, the continuous cause may be acting with another factor that is episodic, possibly including the occurrence of susceptible life stages.)
- Experiments—If the factor that constitutes a necessary cause is removed or significantly reduced and the receiving systems do not begin to recover within an appropriate time, the cause can be rejected.

Verification approaches

Unlike falsification approaches, verification approaches increase confidence that a particular causal pathway represents a true cause. However, verification may not be as conclusive as falsification. For example, verifying that 50 mg/L of a chemical is toxic to all species is nearly impossible to prove, as opposed to showing that 1 species is sensitive. This is why classical scientific methods attempt to disprove, not prove, hypotheses. Multiple lines of evidence, therefore, can increase confidence in the assessment and provide a convincing argument about the cause of the impairment.

- Chemistry—The occurrence of chemicals at toxic concentrations supports their causal status.
- Toxicology—The toxicity of media or effluents in relevant tests is supportive of toxicity as a cause.
- Diagnostics—If histopathologies, physiological abnormalities, or other characteristic symptoms of a putative cause are present in organisms that exhibit the impairment, that cause is acting and may be contributing to the impairment.
- Temporality—If the putative cause began before the effect, it may be a cause. If the effect is episodic and the episodic appearance of the putative cause corresponds in time to the effects (taking lag times into consideration), that cause is supported.
- Internal exposure—If tissue burdens, metabolites, or adducts of a chemical are present at potentially effective levels, that cause is supported (e.g., PCBs found in fatty tissues at critical threshold levels).
- Experiments—If the factor that constitutes a necessary cause is removed or significantly reduced (e.g., a power plant is shut down for repairs) and the receiving system begins to recover within an appropriate time, the cause is supported. Alternatively, if the exposure of organisms in the field to one of the causal agents is experimentally reduced (see Chapter 8) and effects are reduced, the causal status of that agent is supported.

- Modeling—Models may confirm that a hypothesized indirect causal pathway is plausible or that known effects at 1 level or organization may cause the observed impairment at a higher level of organization.

Approaches for reaching a conclusion

Approaches for inferring causation in epidemiology and ecoepidemiology vary depending on the problem addressed and the evidence available. In the simplest case, the problem is to determine whether a particular factor is the cause. For example, the regulations for Natural Resource Damage Assessment (NRDA) specify rules for determining whether a particular spill or waste is the cause of an observed injury (U.S. Department of Interior [USDOI] 1986). Similarly, effluent toxicity testing can serve to determine whether an effluent is the cause of impairment of a receiving community (Grothe et al. 1996). Alternatively, the problem may be to determine the cause of impairment, whatever it may be. This problem is more difficult because a simple process of elimination is unlikely to provide a solution.

The most generally applicable method for inferring causation is the analysis of the weight of evidence for all available evidence. This approach has been used in epidemiology and is the basis for judgment in civil law. A standard way to perform this analysis is to consider the evidence in terms of a set of causal considerations (Hill 1965; Susser 1986a, 1986b; Suter 1998). The following set was developed for analyzing the cause of impairment of ecological systems; a subset of these causal considerations has been used in USEPA guidance (USEPA 2000; Suter et al., in press).

Considerations related to site exposure and response observations

One of the most critical issues in a site assessment is the relationship between stressor exposure and biological responses. These can be addressed by strength of association, consistency, specificity, and patterns.

- Strength of association—The stronger the response to a potential cause, the more likely it is that the response indicates true causation. This means either a severe effect or a large proportion of organisms or species responding in the exposed areas relative to reference areas, and a large increase in response per unit increase in exposure. A weak response is more likely to result from sampling error or unrecognized confounding factors.
- Consistency of association—A more consistent association of an effect with a potential cause is more likely to indicate true causation. The case for causation is stronger if the number of instances of consistency is greater, if the systems in which consistency is observed are diverse, and if the methods of measurement are diverse. Consistent conjunction alone cannot prove causality, particularly when, as in studies of environmental impairments, the cause is not replicated. However, if the effects are replicated, a causal inference is stronger. For example, if fish kills repeatedly occur below a particular

outfall, there is a consistent association over time of those incidents with a potential cause.

- Specificity—More specific cause–effect relationships are more likely to be demonstrated to be causal. Regular association of cause and effect is more readily established if a specific effect is associated with a specific cause. For example, if a chemical is known to cause hepatic tumors in fish, a causal association is clearer than if a chemical causes a variety of nonspecific effects. Specificity of cause is even more persuasive. That is, if an effect has only one or a few known causes, then the occurrence of one of those causes in association with the effect is strong evidence of causation. In the extreme, causation is clear when both effects and causes are specific (*x* causes only *y*, and *y* is caused only by *x*). One implication of this consideration is that both effects and causes should be defined as specifically as possible in order to increase the specificity of the association. For example, a specific cause such as highly embedded substrate is more clearly associated with identified effects than is a general cause such as poor habitat quality. Similarly, associations of effects with localized pollution are more clearly causal than are associations with a regional pollutant.
- Temporality—A cause must always precede its effects. Episodic effects have episodic causes, which may include natural episodes such as events in the life history of the receptor organisms.
- Biological gradient—The effect should increase with increasing exposure. The classic requirement of toxicology is that effects must be shown to increase with dose, but it is applicable to other potential causes. For example, if low substrate texture is believed to cause reduced diversity of benthic invertebrates, then diversity should decline along a gradient of texture. This phenomenon, however, may not occur if interactions between stressors do occur. For example, varying levels of suspended solids, organic matter, or temperature may alter chemical bioavailability or metabolism, thereby altering the gradient effect.
- Complete exposure pathway—An exposure pathway is the physical course a stressor takes from the source to the receptors (e.g., organisms or community) of interest. If the exposure pathway is incomplete, then the stressor does not reach the receptor and cannot cause an effect. Evidence for a complete exposure pathway is case-specific and may include measurements such as body burdens of chemicals, presence of parasites or pathogens, or biomarkers of exposure. For stressors that do not leave internal evidence (e.g., siltation), measurements that show that the stressor co-occurs in space and time with the receptor may be useful.

Considerations related to analysis of preliminary assessment

The assessors must critically review their assessment to establish whether stressor–effect relationships are plausible and can be more strongly established through additional analyses.

- Plausibility—Given what is known about the biology, physics, and chemistry of the potential cause, the receiving environment, and the affected organisms, is it plausible that the effect resulted from the cause? Unless basic scientific laws are violated, plausibility is a weak criterion for rejecting a proposed cause because some true causes are implausible when first considered. For example, initially it was considered improbable that a pesticide could cause the collapse of peregrine falcon populations because they were not directly exposed.
- Experiments—Do toxicity tests (e.g., in situ exposures of caged organisms, Chapter 8) or other controlled experimental studies demonstrate that the potential cause can induce the observed effect? Experimental results may be taken from the literature, but case-specific studies are usually more relevant. Toxicity tests of effluents or of ambient waters (low and high flow) or sediments are particularly relevant.
- Analogy—Is the hypothesized relationship between cause and effect similar to any well-established cases? For example, the fish community in a channelized stream resembles communities in previously studied channelized streams. In some cases, studies of similar sites tend to refute the causal association. For example, high levels of clean suspended sediments at other sites did not cause tumors.
- Predictive performance—Does the potential cause have any initially unobserved properties that were predicted to occur, and was that prediction confirmed? The ability to make and confirm predictions is one of the hallmarks of a good scientific hypothesis. For example, if the proposed cause of a fish kill is spraying of the stream with an organophosphate insecticide, we could make the specific prediction that cholinesterase levels would be reduced or the more general prediction that insects and crustaceans would also be killed. Multiple predictions (e.g., plants and protozoa would not be harmed) would strengthen this criterion. These predictions could be confirmed by surveying the specified groups.

Considerations related to relationships among lines of evidence

In the final assessment steps, multiple lines of evidence (from the various assessment approaches) are reviewed together. Currently, this process is relatively nonquantitative and involves best professional judgment.

- Consistency of evidence—Is the hypothesized relationship between cause and effect consistent with all available evidence? The strength of this consideration increases with the number of lines of evidence (Yerushalmy and Palmer 1959).

- Coherence of evidence—Does a mechanistic conceptual model explain any apparent inconsistencies among the lines of evidence? For example, all the evidence may be coherent if reproduction is not occurring at a site, but juvenile or adult fish recolonize the site from unexposed locations. Another example would be cases in which the evidence is coherent if the measured total metal concentration is not 100% bioavailable. These explanations depend on the expertise and judgment of the assessors for their strength. However, they may lead to experiments or predictions in future iterations of the causal assessment that could support stronger inferences.

These causal considerations may be evaluated in terms of the format presented in Table 1-1. The considerations and Table 1-1 provide one means of weighing multiple lines of evidence. Others include the original criteria and considerations of Hill (1965), Susser (1986a, 1986b), Koch's postulates (Woodman and Cowling 1987; Kapustka 1996; Suter 1998), the State of Massachusetts method (Menzie et al. 1996), and expert judgment.

Minimum Levels of Confidence

After determining the cause of an effect, it is necessary to consider whether there is sufficient confidence in the results to proceed to a regulatory, remedial, or restorative activity. The level of confidence needed depends on the management goals. A high level of confidence is always desirable but not always necessary. A resource manager may perform a quick causal analysis to ascertain the major stressors in a geographical area. Existing data of undefined quality may be adequate for that purpose, particularly if the ranking is clear. However, if the causal analysis will support a decision concerning an important resource or a decision that is likely to lead to a legal challenge or to a particularly expensive solution, a greater level of certainty is needed. Ideally, the nature and quality of data to be obtained will be agreed upon in advance of the causal analysis.

Some tools for attaining greater levels of confidence include

- reanalyzing the data, using only the highest quality data;
- collecting more data; or
- emphasizing confirmation steps.

The results of any analysis are only as good as the quality of the data. Poor quality data are certainly not desirable, even for gross screening procedures. Sometimes using best professional judgment may be better than using low quality data. In other situations, even data of unknown or questionable quality can contain valuable information if that information is cautiously mined from the data. Degrees of certainty required can vary depending on the use of the results. Resource managers and investigators should carefully consider the level of certainty needed, along with the type of information needed.

Table 1-1 Format for a table to summarize results of an inference concerning causation in ecoepidemiology[a,b]

Consideration	Results	Effect on hypothesis[c]
Strength of association	Strong, moderate, weak, none	+++, ++, +, 0
Consistency of association	Invariate, regular, most of the time, seldom (preferably, present numeric results)	+++, ++, +, 0
Specificity of cause	High, moderate, low	+++, ++, 0
Specificity of effect	High, moderate, low	++, +, 0
Temporality	Compatible, incompatible, uncertain	++, ---, 0
Biological gradient	Clearly monotonic, weak or other than monotonic, none found, no evidence	+++, +, -, 0
Exposure pathway	Evidence for all steps, incomplete evidence, no evidence, some steps implausible	++,+,0,-
Plausibility	Plausible, implausible	+, -
Experimental studies	Concordant, ambiguous, inconcordant, absent	+++, +, ---, 0
Analogous cases	Many or few but clear, few or unclear, none found, similar sites show no relationship	++, +, 0, -
Predictive performance	Confirmed specific or multiple, confirmed general, failed, none	+++, ++, ---, 0
Consistency of evidence	All consistent, most consistent, many inconsistencies	+++, +, ---
Coherence of evidence	Inconsistency explained by a credible mechanism, no known explanation	+, 0

[a] Source: Based on Suter 1998. This approach has been improved and expanded in subsequent publications (USEPA 2000, Suter et al. in press).

[b] Note: In an application, 1 result and a corresponding effect on hypothesis rating would be selected for each consideration.

[c] Plus signs indicate strength of evidence in favor of hypothesized cause provided by a causal consideration. Minus signs indicate evidence against hypothesized cause. Zeros indicate no influence.

Next Steps

The conclusion of this assessment process may be that

- a cause is adequately identified and measures are taken to correct the problem;
- no cause is identified, and because the existence of impairment is doubtful, the biological surveys are repeated; or
- no cause is identified and the impairment is clear, so the analysis of causation is repeated with other data.

If the causal analysis is repeated, it is possible to take 1 or both of 2 fundamentally different approaches. First, the ecoepidemiological approach may be further pursued, including better characterization of the nature and distribution of potential causal agents or better characterization of the nature and distribution of biological responses. Second, it may be possible to design field or laboratory experiments that can clearly reject causal hypotheses.

The following chapters will provide detailed information on methods and considerations that should be applied to this process. In addition, Chapters 2 through 8 provide detailed case examples of alternative approaches that have been successfully used to identify and better define complex relationships of stressors in terrestrial and aquatic ecosystems.

CHAPTER 2

Linking Individual-level Responses and Population-level Consequences

Lorraine Maltby, Tim J. Kedwards, Valery E. Forbes, Keith Grasman, Jan E. Kammenga, Wayne R. Munns Jr., Amy H. Ringwood, Judith S. Weis, Simon N. Wood

The protection of populations is at the heart of ecological risk assessment, yet most studies measure effects on individuals. In this chapter, we outline the need to enhance our ability to project and interpret the effect of stressors on natural populations and to manage risk more effectively. We review the range of population-level endpoints that can be studied before discussing population growth rate in more detail. We discuss how individual-level information may be used to project population-level consequences of stress, and we discuss the advantages and limitations of each method. Finally, we consider models and approaches that have the potential to enhance ecological risk assessment by reducing the uncertainties associated with extrapolating from individuals to populations and by increasing the ecological significance of the projected effects.

Why Assess the Effects of Stress on Populations?

Ecotoxicological tests in the laboratory, micro- and mesocosms, and in situ chambers are aimed mainly at obtaining data on possible deleterious effects of stressors (predominantly chemical substances) upon groups of individual organisms. Based on these toxicity parameters, an ecological risk assessment then investigates how and to what extent populations might be affected by the observed stressor and, as a consequence, might lose their ecosystem function or even be at risk of becoming extinct. It is clearly evident that, while the focus of such risk assessments is on the protection of populations, most studies measure effects on individuals. In part, this is a result of the fact that ecotoxicology has developed from toxicology, where the individual is the focus of interest. However, this is also a consequence of trying to simplify complex ecological issues and to understand how stressors interact with biological systems.

Ecological Variability: Separating Natural from Anthropogenic Causes of Ecosystem Impairment.
D.J. Baird and G.A. Burton, Jr., editors.
ISBN 1-880611-43-0

The use of population-level endpoints as measures of stress can aid our understanding of the risk stressors pose to the structure and function of populations. They can provide insight into the effect of stressors on population dynamics, including persistence, as well as on the size and structure of populations. Whereas some of these attributes can be measured directly, others must be projected from individual-based measurements or estimated from models (Barnthouse 1993). In this chapter, we describe and evaluate a number of population-level endpoints but consider population growth rate in detail. We consider how information on the effects of stressors on individuals can be used to project population-level consequences and evaluate the uncertainties and ecological significance of these projected effects.

In the context of this chapter, a "population" is a group of individuals of the same species that live together in the same place and interbreed, therefore possessing an average set of properties such as birth rates and death rates. This definition recognizes that populations are made up of individual organisms but does not require knowledge of which individuals give birth or die. Instead, the population is characterized by mean population birth and death rates, and variability in these is treated as a statistical property of the population. In such risk assessment models, knowledge about the individual is sacrificed in order to have a practical theory that does not require information about the inclination or location of individual organisms.

What Population-level Endpoints Can Be Studied?

There are numerous potential population-level measures, which either may be considered suitable endpoints in themselves or may complement each other. Table 2-1 outlines the characteristics of each of the potential endpoints along with an indication of the applicability to experimentation, natural populations and modeling, conceptual tractability, ecological interpretation, and mathematical accessibility to ecotoxicologists. This section further outlines the potential population-level attributes provided in Table 2-1.

Measures of population size

Population size

The number of individuals (or their summed biomass) is the most basic measure of population size and one that is easy to understand. The most obvious impact of a stressor in field, laboratory, or model systems is often that the size of the population has decreased with respect to the control. But because most real populations fluctuate naturally, some sophistication is needed in order to establish whether an observed change in abundance is important (i.e., attributable to the stressor). This usually means that the size frequency distribution of the population should be considered. Ideally, the size distribution would be evaluated before and after impact and compared (e.g., in a 2-sample Kolmogorov-Smirnov test): Such a comparison is easy with a model, time consuming in the laboratory, and perhaps impractical in the field. In the latter situation, the best we can hope is to compare what population size data are available after exposure to some pre-stress reference distribution. It is worth

Table 2-1 Characteristics of potential population-level endpoints, indicating applicability to experimentation, natural populations and modeling, conceptual tractability, ecological interpretation, and mathematical accessibility to ecotoxicologists

Characteristic	Experimental applicability	Applicability to natural populations	Use in modeling	Conceptual tractability	Ecological interpretation	Mathematical accessibility
Population size	General	General	Descriptive, projective, predictive	Easy	Population size, energy transfer	High
Equilibrium abundance	Limited	General	Descriptive, projective	Moderately easy	Partial indicator of persistence, recovery	Medium
Stability	Limited	Limited	Descriptive, projective	Hard	Likelihood of persistence and recovery	Medium
Recovery time	General	Limited (exposure)	Descriptive, projective	Easy	Likelihood of persistence in response to repeated stressors	High
Time to extinction	Limited	Limited	Projective	Difficult for natural populations, easier in experimental populations	Likelihood of persistence	High
Minimum viable population	Limited	Limited	Projective	Difficult for natural populations, easier in experimental populations	Likelihood of persistence	Medium
Quasi-extinction	Limited	Limited	Projective	Hard	Likelihood of persistence	Low
Tolerance distribution	General	Limited	Descriptive	Easy	Response to past impacts, susceptibility to future impacts	Medium
Size and age structure	Limited	General	Descriptive, projective	Easy (unless age or size cannot be determined, e.g., birds), sampling bias		High
Sex ratio	General	General	Not widely used	Easy	Differential mortality, developmental problems	High
Genetic diversity	Limited but growing	General	Descriptive, projective	Easy	Response to past impacts, susceptibility to future impacts, adaptability	High
Population growth rate	High	High	Descriptive, projective, predictive	Easy	Population fitness	High

noting that, while population size may be the easiest population measure to assess, the statistical difficulties in analyzing such data are far from trivial. These statistical difficulties should not be ignored if the intention is to detect changes in population size, a fact attested to by the enormous statistical literature on animal abundance estimation and abundance trend estimation (Seber 1982; Buckland et al. 1993).

Equilibrium abundance

The equilibrium abundance of a population is a theoretical construct. It is the population abundance at which the population neither grows nor declines. In practice, it is often assumed that a population fluctuates around its equilibrium abundance, but this is an approximate rule of thumb; the equilibrium abundance is rarely the same as the long-term average abundance, although it may be close. The concept is easy. Suppose that we have a simple model for a single-species population,

$$\frac{dN}{dt} = f(N) \qquad \text{(Equation 2-1),}$$

where $f(N)$ is a function determining the rate of change of population abundance N as a function of N. The equilibrium of the population occurs when the N does not change through time, that is, when $dN/dt = 0$. This means that the equilibrium population abundance (N^*) is given by the solution of

$$f(N^*) = 0 \qquad \text{(Equation 2-2).}$$

This can also be expressed in discrete time as

$$N_{t+1} = g(N_t) \qquad \text{(Equation 2-3),}$$

where N_t is the abundance at time t. For this model, the condition that population size does not change says that $N_{t+1} = N_t$, implying that the equilibrium population size is the solution of the following (see Figure 2-1):

$$N^* = g(N^*) \qquad \text{(Equation 2-4).}$$

Begon and Mortimer (1986) stated that equilibrium population size, at which birth rate equals death rate, represents the population size that the resources of the environment can just maintain without a tendency to either increase or decrease. It is therefore clear that changes in equilibrium population size have a much greater ecological significance than do changes in population size. Indeed, a change in population size without a change in equilibrium population size indicates an effect that may well be temporary, whereas a change in a population's equilibrium population size probably indicates something more long term.

Given the theoretical nature of equilibrium population size, it requires careful experimentation to establish in the laboratory and is estimable only in the field with the aid of a population model. It can be calculated for all models that do not predict explosion or extinction and for most that do.

Stability

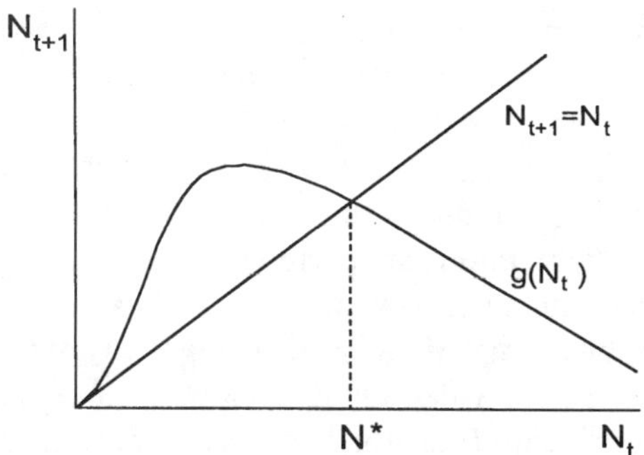

Figure 2-1 Equilibrium population abundance in discrete time. Relationship between N_{t+1} and N_t according to discrete model in Eq.2-3 is given by curve, and straight line indicates stable population size. Equilibrium abundance N^* occurs when N_{t+1} equals N_t, i.e., where the line intersects the curve.

A stable population is one that returns to an equilibrium after perturbation. The concept may be extended: "a stable limit cycle is a cycle to which a population's dynamics return after perturbation" (Nisbet and Gurney 1982; Gurney and Nisbet 1998). In theoretical models, the stability of an equilibrium is established by starting the model at equilibrium, introducing a small perturbation away from that equilibrium, and then determining whether the population returns to the equilibrium or tends to move away from it. Stable populations may return to equilibrium smoothly (monotonically) or by decaying oscillations around the equilibrium. Unstable populations will explode, become extinct, or oscillate around the equilibrium, displaying cycles or chaos.

Clearly, stability is of great ecological importance. Destabilization of population dynamics can lead to wild population swings. Peaks in abundance may negatively impact other species or aspects of system function (e.g., pest outbreaks may severely reduce the abundance of their prey or host; algal blooms can lead to oxygen depletion and alterations in nutrient cycling). Alternatively, severe decreases in population size may increase the likelihood of extinction. Unfortunately, stability is very hard to measure in a model-free way; it is only by constructing a population model that we can hope to measure changes in stability. Of course, increased variability gives circumstantial evidence of decreased stability, but because variability can also be the result of stochastic effects, it is rarely conclusive. Stability also has the difficulty of being poorly understood even in the ecological community and certainly in the wider world.

To ensure that meaningful estimates of stability are obtained, appropriate scales of time and space must be used (Connell and Sousa 1983). With respect to time scale, it is necessary to either observe or predict patterns of replacement over at least 1 complete generation of all individuals. To do this, we predict the transition probabilities of future replacements for all adults by estimating the number of juveniles in the population. However, this assumes that each juvenile has an equal chance of surviving to become a replacement and that this probability remains constant. Both of these assumptions are unlikely to be valid. Alternatively, we could measure juvenile survival probability directly and use this value to estimate replacement.

It is also necessary to estimate stability on a spatial scale at least equal to the minimum area required for the species to persist, defined by Connell and Sousa (1983) as the area providing the necessary environmental conditions for the development, growth, and survival of offspring. The minimum area can be measured directly by following the replacement of all adults on a number of different spatial scales for a minimum period of 1 generation. The smallest of the scales on which replacement occurs is defined as the "minimum area." Alternatively, minimum area may be estimated indirectly using the vital rates and age structure of the population in an area in which suitable habitats are expected to occur within the period of adult replacement. Individual vital rates are an integration of many biochemical and physiological processes. The life histories of individuals are determined by the total availability and allocation of energy and nutrients between maintenance, growth, and reproduction (see "Extrapolating from bioenergetic responses," p 63). Species that produce offspring that are widely dispersed or that require environments very different from the adult stage are likely to have larger minimum areas, and estimates of stability will need to take this into account.

Thus, both the temporal and spatial scales over which stability estimates are appropriate are functions of the characteristic vital rates of the population (i.e., fecundity, growth, development, survival). In addition, they are likely to be influenced by characteristics of the environment and by the temporal and spatial scales of environmental heterogeneity.

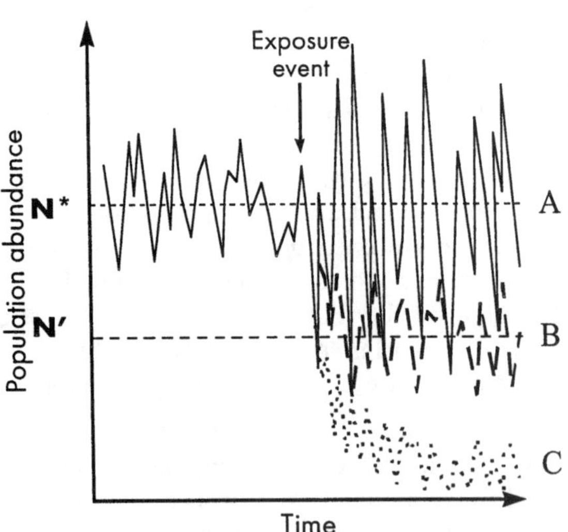

Figure 2-2 Theoretical population projections over time before and after exposure to a stressor. N^* = equilibrium population size; N' = predefined critical abundance. Population A exhibits increased frequency and magnitude of oscillations post-exposure but no change in equilibrium population size; B exhibits reduced equilibrium population size and C exhibits monotonic directional change in trajectory after exposure to stressor.

Recovery time

When a population experiences an insult or is exposed to stressors, population dynamics may be affected in a number of ways. Consider the case of a population that is characterized by fluctuations around some equilibrium abundance. Possible responses are no effects, alterations in overall mean population size, or changes in the frequency or magnitude of oscillations (Figure 2-2). One obvious concern is that these kinds of changes may make the population more susceptible to extinction (see "Measures of persistence," p 34). It is also possible that alterations in abundance of 1 species may affect the abundances of other species or other community dynamics.

If actions are taken to reduce or remove the stressors, then an important question is when has there been sufficient recovery? While it may be unrealistic in some circumstances to require that the population must return to the initial steady-state conditions, we need to define criteria for determining when the population has sufficiently recovered. Population recovery time is a highly attractive population-level endpoint because it is readily and easily comprehensible to regulators, managers, and the general public. Indeed, the applicability of recovery time as an insightful ecological risk assessment endpoint has recently been investigated and advocated by a number of European (Campbell et al. 1999; Giddings et al. 2002) and North American (U.S. Environmental Protection Agency [USEPA] 1999) initiatives.

However, defining population recovery and its attainment is far more complex and controversial in reality than in concept. Indeed, distinction between ecological and ecotoxicological recovery has also been made (Van Straalen and van Rijn 1998). "Ecological recovery" implies the restoration of the original species composition and ecological functions after disturbance, while "ecotoxicological recovery" is defined as the disappearance of the stressor to a level at which it no longer has any more adverse effects on the community. It is clear that in the context of this chapter we are concerned with the population-level endpoint of ecological recovery, which has been defined in many ways, including these:

- Number of days an affected population's growth rate lags behind an unaffected population (Kareiva et al. 1996)
- Time for the population to recover to 80% of the control (Thacker and Jepson 1993)
- Time after treatment at which mean numbers in the treatment approach a notional standard error (SE) value derived from the variation in numbers in the treatment and control on the occasion prior to treatment (Jepson and Thacker 1990)
- Return of the perturbed system to the window of natural variability (Weins 1996)
- First day when the numbers relative to the control return to an arbitrary 0.9× the number before the chemical was added (Sherratt et al. 1999).

Clearly the criteria used are very much dependent upon the species in question; for example, greater regulatory concern would be raised if the species were endangered compared to that of a rapidly reproducing cladoceran. The characteristics of potentially at-risk species are further discussed in the section titled "Evaluating effects in the context of natural variability" (p 69).

It is clear that populations of rapidly reproducing, highly fecund individuals have an inherently higher recovery potential than do those of individuals with opposite life-history characteristics. The influence of life-history traits in determining time to recovery for freshwater macroinvertebrates can be readily examined over the Internet on the Pond-*FX* website (http://www.ent3.orst.edu/PondFX). The site allows the online application and simulation of discrete time population models (see "Deterministic models," p 48) in assessing the recovery potential of aquatic invertebrate species to generic perturbations (Heneghan et al. 1999). It is particularly useful in illustrating the importance of life history and strategy in determining recovery potential of different species.

Measures of persistence

Other valuable endpoints for evaluating population-level effects are related to population persistence. Measures of population persistence provide information about the likelihood of a population existing through time. All the potential methods discussed here can be termed "population viability analyses" (PVAs). Such measures can be quantified as functions of the time estimated for a population to exist under a set of environmental conditions, the minimum population size required for the population to persist for some specified length of time (with a specified probability), and/or the probability that a given population will fall below some predefined, critical population size. PVA is often oriented toward the conservation and management of rare and threatened species, with the goal of applying the principles of population ecology to improve their chances of survival. It is worth noting that much of the work in this area has been pioneered by conservation biologists and has only relatively recently been considered within mainstream ecotoxicology.

Time to, and probability of, extinction

Time to extinction (TE) as a population-level endpoint is derived directly from the probability of extinction measure (Gillman and Hails 1997). It has been applied to a range of animal species, including the Bay checkerspot (Foley 1994), plant species (e.g., *Orchis morio*; Gillman and Silvertown 1997), and butterfly populations (Harrison et al. 1991). Foley (1994) describes an analytical technique for estimating probability of extinction applied to various taxa, which can then be converted into TE if extinction is viewed as a Poisson process (Foley 1994, Box 6.1). Therefore, the expected (or average) TE is approximated as

$$TE = \frac{-t}{Log_e(1-P)} \qquad \text{(Equation 2-5)},$$

where P is the probability of extinction over a time interval t.

Quantified using procedures that estimate extinction probabilities as a function of demographic, genetic, and environmental considerations, this measure has fairly limited application in both experimental and natural populations, largely because it requires detailed and complex assessments involving multiple, replicate populations that experience a similar set of environmental conditions. Such experiments are feasible only for species that display certain life-history characteristics; for example, only those species that have rapid turnover rates or small size are amenable to such analysis in the laboratory. Some of these limitations are relaxed somewhat in evaluations of TE in natural populations (for which issues of experimental holding conditions and manipulation are irrelevant), although issues of replication, as well as incorporation of significance levels of impact to cause extinction, need to be addressed. The effects of human-induced disturbance on population dynamics must be simulated (or realized) to effect extinction events or to result in trajectories that reasonably could lead to extinction. TE is perhaps best used in modeling analyses of population behavior, in which environmental stochasticity and potential influences of stressors can be reasonably simulated. Its utility in such exercises is limited to projecting expected population-level effects under specified conditions. TE is conceptually understandable to environmental managers and decision-makers as well as to the general public; indeed, any measure involving true extinction of a population is expected to resonate with nontechnical stakeholders. However, it is important to note that such measures do have technical definitions and as such should not be used without careful consideration and statement of all assumptions, limitations, and uncertainties.

Minimum viable population

Related to TE is the concept of "minimum viable population" (MVP), the smallest population size that will persist for some specified length of time with a specified probability. The currency of MVP is the number of animals needed in the population to meet a specified probability of persistence. MVP has been defined as the smallest population size that has a defined small probability (often 5%) of becoming extinct in a certain time interval (often 100 years). For example, Goldingay and Possingham (1995) conducted a PVA on the yellow-bellied glider in order to determine the minimum habitat areas needed to support populations with a 95% probability of persisting for 100 years. Using a model that incorporated life-history data from species and environmental uncertainty, they determined the number of family groups required for a viable population and found that adult mortality rate had the greatest influence on population viability.

The area required to support this MVP is called the "minimum area requirement" (Remmert 1994). Sometimes MVPs are quoted in terms of the populations with only a 1% probability of extinction per year; this sounds acceptable, but note that an extinction probability of 1% on a yearly basis leads to an extinction probability of ~ 63% in 100 years.

Theoretically, a negative relationship is expected to exist between population size and probability of extinction. This may be due to a number of causes, not the least of which is the expectation that stochastic environmental fluctuations can result in decreases in population size; smaller populations will have a greater likelihood of being driven to extinction than will larger populations (distribution of organisms among subpopulations, separated in space, also should reduce the risk of population extinction from random environmental fluctuations). Environmental stress is expected to exacerbate the effects of environmental fluctuations, as well as impact the ability of populations to recover from negative oscillations. The applicability of this measure is similar to that of TE, being limited in experimental settings and with respect to natural populations. As with TE, the use of MVP in modeling situations is limited primarily to projections. However, we suspect that MVP as a concept and measure is more difficult to interpret than TE, because it involves consideration of the relationships between population abundance and probability of extinction from stochastic variation and from stressor-related impacts to the population. Indeed although MVPs were among the earliest applications of PVA (Samson et al. 1985; Shaffer and Samson 1985), use declined when the application proved both biologically and politically complex (Soule 1987; Gilpin 1996).

Quasi-extinction

A third measure of persistence, so-called "quasi-extinction," is defined as the probability that the population will fall below some critical abundance (Ginzburg et al. 1982). Critical abundance can be thought of as specific population densities below which adverse effects are known or suspected to occur. For example, critical densities of whales in the North Atlantic may exist, below which individuals are no longer able to find mates for reproduction. Similarly, critical population sizes of commercially harvested finfish might be identified, through a cost–benefit (or similar) analysis, below which harvest by the fishing fleet is no longer economically feasible. True extinction, in which the critical population size is 0, can be considered a special case of quasi-extinction.

Figure 2-2 illustrates this concept. Assume that population size fluctuates in time with an average or equilibrium population $N^* = 1000$. At certain times, the population size may fall below the predefined critical abundance N' of, for example, 200. The amount of time spent at or below $N' = 200$ is a measure of the risk of becoming quasi-extinct at this critical abundance.

Stressors have the potential of affecting the probability of quasi-extinction or extinction in a number ways. Figure 2-2 illustrates 2 such ways in which a stressor might affect a stable population equilibrium by reducing the equilibrium population size or by causing monotonic directional changes in overall trajectories. Interestingly, it is possible to make conservative estimates of quasi-extinction risks and times without formal knowledge of the density-dependence in population dynamics (Ginzburg et al. 1990).

Quasi-extinction has limited application in both experimental and natural population situations (for reasons similar to the previous measures of persistence) and has primary value in modeling analyses that target projection of population-level effects. Indeed, estimates of the probability of decline, or quasi-extinction, are derived from Monte Carlo simulations that require estimates of variation in vital rates. In order to compare a control with a stressed population, 2 Monte Carlo simulations would have to be performed, providing 2 estimates of the probability of decline and subsequently finding either the difference or the ratio. Therefore, such estimates are not particularly tractable from an analysis standpoint. Furthermore, the concept of quasi-extinction may not be easily understood by end users and stakeholders.

Extinction probability curves could potentially be used to investigate significant impairment and subsequent population recovery. Using the values of N and N', the extinction probability curves can be constructed for the pre-insult conditions and the post-insult conditions (Figure 2-3). The difference between the 2 curves, indicated by shading in Figure 2-3, is a measure of the impact suffered by the population. Likewise, pre- and post-exposure extinction curves can be used to determine whether the population has sufficiently recovered. Goodness-of-fit analyses (e.g., Kolmogorov-Smirnov) could then be used to determine whether the curves are statistically different. Because these curves represent basic frequency distributions, it would also be possible to define 95% confidence limits. Therefore, using this approach, recovery would be deemed acceptable when the confidence limits of the 2 curves overlap. The generation and depiction of confidence limits would also be valuable for illustrative purposes, particularly for communicating the concept to stakeholders. Regardless of how the criteria for recovery are established, there are some real advantages to the application of quasi-extinction theory to recovery issues. The pre-impacted and post-impacted curves integrate a number of important population characteristics, including the magnitude and frequency of population fluctuations as well as the population size and variability.

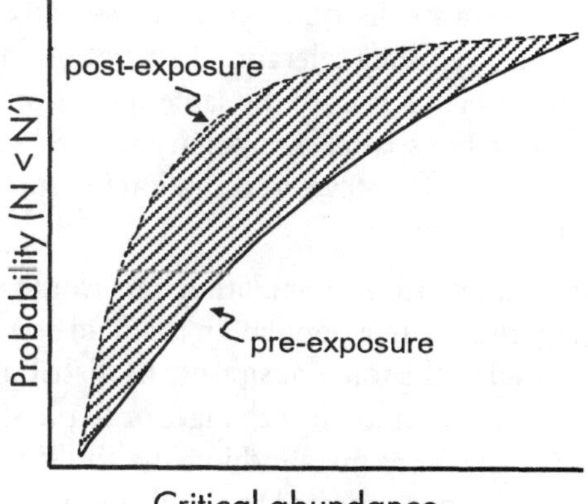

Figure 2-3 Theoretical construct illustrating risk of decline curves for pre-and post-exposed population. Shaded area is measure of impact and could be used to determine whether population has suffered ecologically significant impairment or indeed has recovered from effects of stressor.

In order to use extinction models for any risk assessment purpose, it is important to identify relevant parameters as suitable endpoints that can be incorporated in a decision framework. Quasi-extinction models implicitly incorporate population size, population growth rate, and hence all the vital rates that are related to population growth rate. However, it should be noted that, unlike life-cycle matrix analyses in which elasticity and sensitivity values represent the contribution of vital rates to population growth rate, it is not often deemed practical to assess the sensitivity of estimated probabilities of quasi-extinction to vital rates (see Goldingay and Possingham 1995). In order to do so, it is necessary to define a discrete change in one of the vital rates, re-run the Monte Carlo simulation, then find the change in probability of decline, and such computations can be expensive.

Measures of population structure

The previous sections discussed how stressors can affect the size and persistence of populations. In addition, stressors may alter the structure of populations. Population-level responses to stress include alterations in sex ratio and in size and age structure as well as changes in genetic diversity and tolerance distributions.

Tolerance distribution

The tolerance distribution is an essential component of toxicity and ecotoxicity testing. Typically tolerance distributions are used to compare the relative toxicity of different chemicals to the same species or strain (i.e., as bioassays). However, temporal or spatial changes in a species' tolerance distribution can provide important population-level information on the effects of a variety of natural and anthropogenic stressors.

In an unimpacted population, one would expect a bell-shaped curve of tolerance (Figure 2-4). In a population that had acquired tolerance, this distribution would be skewed to the tolerant side of the distribution. This alteration may be due to actual genetic (evolutionary) changes or to physiological acclimation through such mechanisms as metallothioneins and stress proteins. The likelihood of genetic adaptation is expected to positively relate to the amount of genetic variation for tolerance present in the population. Numerous studies (e.g., Nevo et al. 1984; Guttman 1994) have shown altered allozyme frequencies in populations of organisms that live in polluted environments and that are more tolerant of the toxicants present in their environment. When possible pollution-induced genetic shifts in natural populations are examined, it is important to take natural geographic gradients of genetic structure into consideration, as these can confound pollution-induced effects (Weis et al. 1999). There is little data to demonstrate clearly that the particular alleles which increase in such environments are themselves directly associated with tolerance (fitness). It is more likely that the allozymes are linked with the genes that convey tolerance (but see Klerks and Levinton 1989).

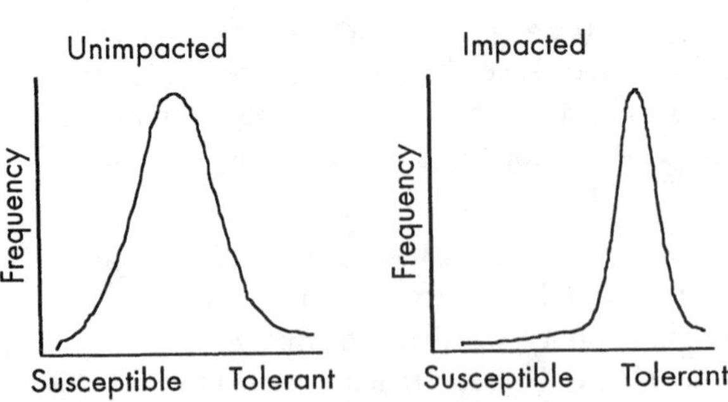

Figure 2-4 Theoretical tolerance distributions for unimpacted and impacted populations

The manifestation of tolerance may not be seen at all life-history stages, and the development of tolerance to 1 toxicant may allow the organisms to be more tolerant of other toxicants. However, it is doubtful whether the increase in tolerance toward a particular toxicant truly reflects overall increased fitness. Increased tolerance due to physiological acclimation is likely to have energy costs to the individual. Detoxification and tolerance to a particular stressor may come at the expense of tolerance to other environmental factors, both natural and anthropogenic (Bush and Weis 1983), or reduced fitness in other ways (Baker 1987). Antonovics et al. (1971), for example, showed that resistance in plants growing near smelters was associated with lower fitness and that tolerant strains were out-competed by nontolerant plants in unpolluted sites. Similarly, Postma and Groenendijk (1999) showed that Cd-tolerant midges had increased mortality and larval development time under control conditions.

Genetic diversity

Genetic diversity can be an important indicator of responses to past stressors and of susceptibility to future impacts. The bottleneck effect occurs when a disturbance or stressor reduces the population to a small size. The population may then grow or remain small, but in the absence of immigration, the genetic diversity in the population represents only those alleles found in survivors, that is, it can be much lower than the diversity before the stressor. Normal mutation rates, which are usually measured in thousands of generations, are too slow to cause an appreciable increase in genetic diversity following a bottleneck. Chronic or multiple episodic stressors may cause multiple bottlenecks, magnifying this effect of reduced genetic diversity. Genetic variation within populations translates directly to individuals with a greater adaptive ability. Changing environmental conditions may favor the survival and/or differential reproduction of individuals with certain adaptations or characteristics. Hence, a population with low genetic diversity is more likely to go extinct when confronted by disease, stressors, or disturbance if all or most members of the population are susceptible to the perturbation.

Conservation biology recognizes the importance of maintaining high genetic diversity for the long-term persistence of wild and captive populations. A variety of methods are available for measuring genetic traits, including random amplified polymorphic deoxyribonucleic acid (RAPD; based on amplification from short primers), restriction fragment length polymorphism (RFLP; based on cleavage sites for restriction enzymes), microsatellite DNA sequences, and isozymes. Individuals are scored for the presence of these genetically determined characteristics, and these scores can then be used to calculate indexes of genetic diversity or similarity. These methods are widely applicable to natural populations, often requiring small sample sizes ($n = 20$ to 30 per site) and allowing noninvasive tissue sampling. Emerging evidence shows reduced genetic diversity in nonmigratory populations at sites that have experienced pollution or disturbance events (Lavie and Nevo 1982; Lavie et al. 1984; Nevo et al. 1986; Benton and Guttman 1990). Genetic diversity can be incorporated into models that describe the current status of populations, and it can be used in projective models that simulate future status. PVA and MVP analyses are projective models that frequently include genetic diversity (see "Measures of persistence," p 34). The concept of genetic diversity should be readily accessible to biologists and risk assessors because the importance of genetic variation for adaptability of populations is a basic biological principle.

Size and age structure

A stable population shows a temporally consistent age distribution, with a somewhat larger number of younger, smaller individuals and progressively smaller numbers of older, larger individuals. If a population becomes dominated by larger, older individuals, this could indicate a problem in reproduction and hence a reduction in population growth rate. On the other hand, a population highly dominated by smaller, younger individuals, may be responding to a stressor that affects individual longevity and/or inhibits growth. It is not possible to age individuals of many species; investigators must rely on size and frequency data. For species that can be aged (e.g., fish), comparing size at different ages makes it possible to accurately calculate growth rates in field populations and to compare growth rates at different sites.

The exposure of an organism to a stressor may have an affect upon one or more of the vital rates and subsequently may affect the size or age structure of the population. For example, a reduction in feeding behavior, reproduction, or growth rate, or an altered predator avoidance behavior may lead to reduced population size and affect the size structure of the prey population (Hairston and Pastorok 1975; Hairston et al. 1982, 1983; Werner et al. 1983; Sih et al. 1985; Gleason and Bengtson 1996a, 1996b).

Sex ratio

Sex ratio, an estimate of the relative abundances of the 2 sexes in sexually reproducing dioecious populations (and sometimes seasonally parthenogenetic species such as daphnids, rotifers, and aphids), can provide information about the status of the population. Substantial theory exists concerning the evolutionary consequences of skewed sex ratios in populations (Maynard Smith 1978), although the use of this measure as an indicator of population-level effects caused by environmental stressors is limited primarily to understanding differential effects between the sexes. However, it also can provide information regarding the mechanisms of effect. For example, certain chemicals have been demonstrated to interfere with hormonal systems that control sexual differentiation in certain birds (Fry and Toone 1981), mammals (Jones and Hajek 1995; Gray and Kelce 1996), reptiles and amphibians (Bergeron et al. 1994; Guillette et al. 1994, 1995), and fish (Gimeno et al. 1996; Jobling et al. 1996; Gray and Metcalfe 1997). In such cases, ecological theory (if not empirical evidence) suggests that a change in sex ratio affects population dynamics by influencing total reproductive output of the population and by altering frequencies of encounter between the sexes during periods of mating (Kalmus and Smith

1960; Hamilton 1967). Stressor-induced imposex, such as that reported for certain invertebrates (e.g., Gibbs and Byran 1986; Moore and Stevenson 1991), should have population-level ramifications similar to those of changes in sex ratio.

Sex ratio as a population-level endpoint has general applicability in experimental evaluations of population impacts as well as in investigations involving natural populations. For example, Hairston and Munns (1984) and Hairston et al. (1985) used changes in sex ratio to identify the seasonal onset of predation by visually oriented predators on copepod populations in freshwater lakes. The change in copepod sex ratio is described as

$$d/dt(M/F) = (M/F)(d_f - d_m) \quad \text{(Equation 2-6)},$$

where M and F are male and female abundances, and d_m and d_f are mortality rates for males and females, and the slope of a line relating the natural logarithm of sex ratio to time is equal to the differences in mortality rate between the sexes:

$$\ln(M/F)_t = \ln(M/F)_0 + (d_f - d_m)\ t \quad \text{(Equation 2-7)}.$$

Simple regression techniques, which fitted a piecewise linear model by least squares, were used to partition the sex ratio data into 2 groups: 1 representing changing sex ratios and 1 reflecting constant sex ratio. The intersection of regression lines fit to these 2 groups was taken to indicate the onset of increased predation (illustrated conceptually by Figure 2-5). Although applied to a problem involving species interactions, this approach is potentially useful in evaluating the timing and magnitude of stressor insults.

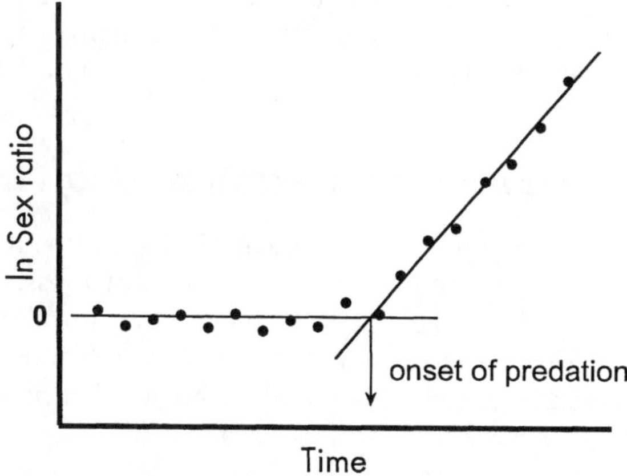

Figure 2-5 Conceptual approach to estimation of onset of changes in sex ratio

In models of population dynamics, sex ratio typically is incorporated either as an assumption (e.g., Kuhn et al. 2001) or as an influence on average birth rates. Sex ratio is a potentially useful endpoint in descriptions of the effects of stressors (particularly in situations where such effects might be diagnostic), and although not widely used, it can be part of either projective or predictive modeling assessments. It is both conceptually tractable and generally accessible and is particularly useful to understanding the effects of differential mortality and developmental effects.

Population growth rate

For the survival of any population, it is imperative that a certain species-specific population growth rate is reached to compensate for losses through natural death, predation, competition, and environmental variability. The species-specific population growth rate is the direct result of evolution and is a direct measure of population well-being, or "fitness." Population growth rate, and the extent to which it is impacted by stressors, can be viewed as a gauge of the demographic parameters that compose a population and is therefore of central ecological and evolutionary importance. It is generally acknowledged to be the key intervening variable that links individual effects to populations (Calow et al. 1997; Sibly 1999). It forms a fundamental part of population ecology underpinning population regulation (Sinclair 1996) and underlies fisheries and wildlife management (Sutherland and Reynolds 1998). However, the use of population growth rate in ecotoxicology has been limited (Sibly 1996), despite the fact that its usefulness as a measure of stress was demonstrated more than 35 years ago (Marshall 1962).

Population growth rate as a population-level endpoint has obvious attractions: It is readily and easily understood by all parties and can be viewed as an integrative measure of age-specific survivorship and age-specific fecundity (Table 2-1). It can be derived from knowledge of the whole life cycle of an organism, attainable through laboratory studies, and can also be derived from partial life-history information (see "Instantaneous population growth rate and partial life-table information," p 45). The population growth rate can be used not only as a descriptive function of a population but also as a projective tool to provide an indication of what may happen to a population, given a specific set of environmental conditions through time. Indeed, if parameterized correctly from a well-defined scenario, population growth rate can be used as a tool for short-term prediction. Because of its utilitarian nature, population growth rate is considered in more detail in the rest of this chapter.

Estimating Population Growth Rate from Life Tables

A recurring theme in the ecotoxicological literature is the need for greater insight into and understanding of the effects of stressors on populations and communities (e.g., Baird et al. 1996). Stressors may cause physiological or biochemical harm to individuals and so may adversely affect their birth, growth, or death rates (Walker et al. 1996). These adverse effects might cause population declines were it not for compensatory processes (such as density dependence) or tradeoff mechanisms (see "Compensation," p 75). There is a need for ecological studies that examine the effects of stressors on population growth rate, which may be viewed as an indicator of population fitness.

Population growth rate has a long history of use as an index of population fitness (Leslie 1945; reviewed in Caswell 1989) and as such has many ways of being expressed, including for example,

- r_i, the instantaneous rate of increase,
- r_m, the per capita rate of increase (also called "intrinsic rate of natural increase" or "malthusian parameter"), and
- λ, which is equal to e^r. λ and is variously referred to as "reproductive rate," "population multiplication rate," or "finite rate of increase."

In an increasing population $\lambda > 1$ (r_i or $r_m > 0$), in a declining population $\lambda < 1$ ($r < 0$), and in a constant-size population $\lambda = 1$ (r_i or $r_m = 0$). Whether one expresses population growth rate in terms of λ or r is partly a matter of preference. In some cases, use of λ allows simpler expression of equations, but r has the advantage of being symmetrical in its representation of population increase and decrease, which may facilitate statistical analyses (Sibly 1999). λ is the measure calculated in age-classified matrix models, whereas either λ or r have been used in stage-classified models.

Methods for Deriving Measures of Population Growth Rate

There are 2 alternative population growth rates that might concern us. One is the realized growth rate, the rate that a population achieves under a specific set of conditions, which may be affected by density-dependent or other factors. For many natural populations, the long-term average realized growth rate is 0. The other interesting population growth rate is the potential long-term growth rate in the absence of density-dependent factors. This is the rate of population growth a population would eventually achieve if all its age-dependent, per capita vital rates stayed fixed through time.

There are a number of alternative techniques for the measurement of population growth rate that differ in the manner in which they handle age structure and other assumptions. The simplest approach is to ignore age structure and derive r_i, the instantaneous rate of increase. We assume that population size N at time t is governed by the simple equation

$$\frac{dN}{dt} = r_i N \qquad \text{(Equation 2-8),}$$

which is appropriate for a population that has achieved an equilibrium age structure or for one in which age structure is dynamically unimportant. Rewriting this equation, we find it clear that

$$r_i = \frac{1}{N}\frac{dN}{dt} \qquad \text{(Equation 2-9).}$$

Integrating this from population N_0 at time 0 to N_t at time t yields

$$r_i = \frac{\ln(N_t / N_0)}{t} \qquad \text{(Equation 2-10).}$$

Solving for r_i yields a rate of population increase or decline: A positive r_i indicates a growing population, a zero r_i indicates an unchanging population, and a negative r_i indicates a population in decline.

Another method of estimating population growth rate, the Euler-Lotka equation, takes into account the age structure of the population. It is assumed that the population is governed by the age-structured model of Leslie (1945):

$$\mathbf{N}_{t+1} = \mathbf{L}\mathbf{N}_t \quad \text{(Equation 2-11)},$$

where $\mathbf{N}_t$ and $\mathbf{N}_{t+1}$ are vectors containing the numbers of organisms in each age class at times t and $t+1$, and $\mathbf{L}$ is the matrix defined by

$$\mathbf{L} = \begin{bmatrix} m_0 & m_1 & \cdot & \cdot & m_v \\ P_0 & 0 & & & \\ & P_1 & & & \\ & & \cdot & & \\ & & & P_{v-1} & 0 \end{bmatrix},$$

where m_i is the fecundity of age group i (the number of young produced per individual of age i during a time period that live to the next time period), and P_i is the proportion surviving from age class i at time t to age class $i+1$ at time $t+1$. A population governed by such an equation will eventually achieve a stable age distribution and will increase or decrease with instantaneous population growth rate r_m. It turns out that $\lambda = e^{r_m}$, where λ is the dominant eigenvalue of $\mathbf{L}$, also known as the "finite rate of increase" (when the population has achieved a stable age structure $\mathbf{N}_{t+1} = \lambda\mathbf{N}_t$). The Euler-Lotka equation is one way of finding λ (or r_m):

$$1 = \sum l_x m_x e^{-r_m x} \quad \text{(Equation 2-12)},$$

where x is age and l_x is the probability of surviving to age x, which can be calculated from the P_i's using the fact that $l_{x+1} = l_x P_x$ and $l_0 = 1$. Note that if a population can be accurately described by the Leslie model and has achieved a stable age structure, then r_i is mathematically equivalent to r_m, so that r_m can be estimated using r_i.

Population growth rate is widely used by population biologists, although its use has been criticized for the assumption of a stable age distribution (e.g., Laughlin 1965). In fact, the calculation of r_m does not assume stable age distribution; rather, r_m is the population growth rate that would be achieved once the population had achieved a stable age distribution (assuming that the population could be described by the Leslie model). It is only if we wish to use r_m as an estimate of the realized population growth rate for a particular population that we must assume the population concerned has achieved a stable age distribution.

Note also that a wide variety of population models can be derived from the Leslie matrix model by making the survival and reproduction parameters random variables or functions of environmental parameters or population size, but in these circumstances, simple population growth measures are harder to calculate and interpret.

Instantaneous Population Growth Rate and Partial Life-table Information

The methods for deriving population growth rate described in the previous section require detailed life-table information. While it may not always be practicable to produce a complete life table for a species, it may be possible to produce a partial life table. Many standard chronic toxicity tests are designed to provide information on reproduction and survival, and these data can potentially be used to estimate population growth rate. Van Leeuwen et al. (1985), for example, used information from a standard 21-d reproduction study to investigate the effect of Cd exposure on the intrinsic rate of increase of *Daphnia magna* populations. Daniels and Allan (1981) had previously demonstrated that there was almost a perfect correlation ($r = 0.99$) between r_m values calculated for a 21-d exposure and values based on entire life-cycle experiments with *Daphnia pulex*.

Partial life tables are likely to be most appropriate indicators of population growth rate for exponentially growing (density-independent) populations for which early reproduction is more important than late reproduction. Van Leeuwen et al. (1985) noted, however, that partial life-cycle studies may underestimate long-term toxicity. Because experiments are initiated with unexposed animals, compounds that show a slow increase in toxicity with time may not exhibit their effects until after one or more broods have been produced. These delayed effects may not be as marked in natural populations where neonates may be exposed prior to release from the brood pouch (i.e., during oogenesis and embryogenesis). Van Leeuwen et al. (1985) suggest that a more realistic estimate of chronic

Advantages and limitations of r_i estimates and r_m estimated from partial life tables.

Advantages

- Several existing standard toxicity tests are partial life-cycle studies, and the intrinsic rate of increase (r_m) can be determined from these studies with little extra effort and cost.
- Instantaneous growth rate (r_i) can be measured directly and does not require detailed life-table information.
- Instantaneous growth rate can be a good substitute for r_m and may be less time- and labor-intensive to measure.

Limitations

- Population growth rate estimates, based on the results of partial life-cycle studies may underestimate the long-term toxicity of slow-acting stressors or of stressor effects acting on populations that are at steady state.
- Measurement of r_i is viable only for species that have short life cycles and that can be cultured under laboratory conditions.
- Measurement of r_i provides no information on which vital rate changes are causing the change in population growth rate.

toxicity may be obtained by testing F_1-generation neonates from pre-exposed animals.

Another approach that has been suggested is to measure population growth through direct observation of a population over time using the instantaneous rate of increase (Kareiva and Sahakian 1990; Walthall and Stark 1997a; Sibly 1999). The instantaneous rate of increase (r_i) is given by the change in number of animals in a cohort over time, and like r_m, it integrates both survivorship and fecundity as well as density-dependent regulatory mechanisms. Walthall and Stark (1997a) found that r_i was not statistically different from r_m and thus concluded that r_i could be used as a substitute measure of population growth rate. However, because complete life-table information is not needed, the calculation of r_i is less time- and labor-intensive. Instantaneous growth rate is derived from a phenomenological model that excludes population structure and information on vital rates. It is therefore not possible to determine what is causing a change in r_i. Exposure of populations to 2 chemicals may result in the same r_i but for different reasons (i.e., 1 chemical may affect survival, whereas the other may affect fecundity).

Demographic Population Models

Our concern in this chapter is the impact of stressors on population dynamics. In an ideal world, we would address this by building a perfect quantitative model for each population of interest, linking effects of stressors on the individual to population dynamics. This model would give us a complete understanding of population dynamics. Since this is impractical, we must instead focus on measures that each give only partial insight into population dynamics but which are practical to measure or infer. In this section, we discuss how the measures of different population dynamic characteristics can be used in model projections and predictions.

It is important to note that the application and acceptance of demographic population modeling within risk assessment per se is a challenging issue. Risk assessors and regulators alike may not have been exposed previously to population biology and indeed may view modeling with a good degree of scepticism. It is only by increasing their knowledge and understanding of the underlying aims, premises, and assumptions of the models that acceptance will be generated (Ashby 1999). We hope that the following sections may shed some light on the topic, illustrate some of the strengths and limitations of population modeling, and further provide a springboard to the accessibility and application of such techniques in ecological risk assessment.

In population models, it is clear that the basic unit is the individual. It is therefore the task of model developers to translate their knowledge about the mechanisms at the individual level into models for the change in number of individuals. Clearly what is required is a modeling methodology, which in principle can accommodate as little or as much biological detail as is available, in a readily tractable mathematical framework. A plethora of mathematical techniques currently meet these requirements, and it is often difficult for the ecotoxicologist to distinguish easily among them. Mathematical models of ecological theory can be classified as deterministic or stochastic. Stochastic models contain a random component, as opposed to deterministic models in which the same inputs produce the same output no matter how many times the models are run. Both deterministic and stochastic models have important roles to play in the analysis of any particular system. Provided that population numbers never become too small, a deterministic model may enable sufficient biological understanding to be gained about the system; if at any time population numbers do become small, then a stochastic analysis is required (Renshaw 1990). However, as a general rule, stochastic models are analytically more intractable than their deterministic counterparts (see Table 2-2).

Table 2-2 Assumptions and limitations of deterministic and stochastic population models

Model type	Assumptions	Limitations
Population, deterministic	No random elements	No stochasticity
Discrete state, discrete time	No interactions between time steps; individuals aggregated within stages	If density-dependent, more difficult to calculate elasticity and sensitivities
Discrete state, continuous time	All individuals within stage are identical except with respect to single growth index	Difficult to simulate numerically
Continuous time, continuous state	All individuals born at same time, act same way through life	Computationally the most difficult and analytically difficult
Population, stochastic	Environment may be random; individuals may differ randomly	Analytically more intractable; outcomes very sensitive to noise models assumed
Discrete state, discrete time	Form and extent of variability known and quantified	Sensitive to form of noise model chosen
Individual-based models (IBMs)	Individual variability important enough to keep explicit track of; either event-based or discrete time; other assumptions tend to be case- and implementation-specific	No standard formulation; case-specific; almost always investigated only by simulation; hard to understand and therefore to evaluate critically: "Black box"
Stochastic differential equations in continuous time	Strong assumptions about formulation and solution; not very accessible to non-mathematicians	Mathematically very complex; lots of theory; limited practical value

Table 2-2 illustrates the different deterministic and stochastic population modeling approaches that have been reported, along with a short description of the respective assumptions and limitations. As a general rule of thumb, the mathematical complexity of the modeling framework and ecological realism increases as you move down the table. It is evident that a balance needs to be struck among the mathematical complexity of the model formulation, the ecological and toxicological information available to parameterize the model, and the resulting realism of the output. Where this balance is drawn is largely dependent upon the question being addressed. For example, Nisbet and Gurney (1982) considered mathematical models of 2 broad types:

1) Strategic population models are simple, mathematically tractable models constructed with the aim of identifying possible ecological principles, that is, to answer the question "Could it happen?" as opposed to "Will it happen?" As opposed to tactical models, strategic models are mathematically simple.
2) Tactical models are designed to yield short-term forecasts of population changes. They require precise and detailed information on the species involved. Often the life-history information required to parameterize such models simply is not available. However, if parameterized correctly, such simulation models yield good short-term predictions of population trends.

Deterministic models

The first category of models to consider are those termed "deterministic," in which there are no random elements, or stochasticity, within the model. Although such simulations are mathematically accurate, the absence of any individual or environmental variation is ecologically and biologically naïve.

Discrete time, discrete state

Models that are dependent upon discrete time and state have been extensively studied, and a number of methods are well documented (Leslie 1945; Lefkovitch 1965; Caswell 1989; Gurney and Nisbet 1998). As with any modeling technique, a number of important assumptions and limitations should be considered. The manner in which any organism ages is continuous; organisms do not jump from age 2 to age 3. However, in models of discrete state, the continuous age variable is divided into discrete states as a modeling convenience; the same is also done for time. The life cycle of the organism in question is essentially divided into different stages or ages, depending upon how best it is described. As organisms enter each stage or age class, they have a defined probability of either remaining in the stage until the next time step or moving into the next stage. Because of this, and individual may remain in a stage for only 2 or 3 time steps before it progresses to the next, while other individuals may reside in the same stage much longer before maturing to the next stage. Therefore, some models may exhibit some strange stage transit times as a consequence of individuals being aggregated into discrete stages or ages.

The matrix model, a discrete age, discrete time model, is described as follows:

$$\begin{bmatrix} N_0(t+1) \\ N_1(t+1) \\ . \\ . \\ N_v(t+1) \end{bmatrix} = \begin{bmatrix} m_0 & m_1 & . & . & m_v \\ P_0 & 0 & & & \\ & P_1 & & & \\ & & . & & \\ & & & P_{v-1} & 0 \end{bmatrix} \begin{bmatrix} N_0(t) \\ N_1(t) \\ . \\ . \\ N_v(t) \end{bmatrix},$$

where $N_i(t)$ is the population of the ith daily age class at time t, and the daily fecundity rates m_i and survival rates P_i are related to the continuous per capita birth and mortality rates by

$$m_i = \int_i^{i+1} \beta(x)dx \text{ and } P_i = e^{-\int_i^{i+1} \mu(x)dx} \quad \text{(Equation 2-13).}$$

Clearly, the population growth rate depends on the values of the vital rates; it is therefore important to understand how it would change if the vital rates were to change. This understanding is particularly crucial in ecotoxicology, in which interest lies in the consequences of changes in vital rates caused by stressors. By calculating the complex conjugate transpose (Caswell 1989), it is a relatively simple mathematical process to determine the sensitivity of the population growth rate to any of the matrix entries in models that do not contain density dependence. However, this calculation is made significantly more complex by the introduction of density dependence.

Continuous time, discrete state

In continuous time and discrete state models, such as those described by Nisbet (1996), the difficulty encountered in discrete time with transit times is overcome by allowing an individual to enter a given class with a development index of 0 and grow within that class until it attains an index of 1. In this way, an individual develops within a class and matures to the next only when sufficient growth has been attained. Continuous time models aim to predict values of state variables at all future times, not just at integer multiples of some time increment experienced in discrete time models.

The modeling approach is based upon delay-differential equations (DDEs), essentially a bookkeeping device that tracks the development of each individual. Let n_i be the size of the population in the ith stage at time t, and assume that time starts from $t = 0$, for a 2-stage model.

For juveniles,

$$\frac{dJ}{dt} = \beta.A(t) - \beta.A(t-\tau).e^{-\mu J.\tau} - \mu_J.J \quad \text{(Equation 2-14),}$$

and for adults,

$$\frac{dA}{dt} = \beta . A(t-\tau).e^{-\mu J.\tau} - \mu_A . A \qquad \text{(Equation 2-15)},$$

where β = birth rate, A = number of adults, J = number of juveniles, τ = development time, and μ = per capita mortality rate experienced by the stage.

Simply put, the rate of change of population size can be considered to be the difference between recruitment into and losses out of the population. Assuming that all individuals in a given stage at a particular time develop at the same rate, the stage behaves like a conveyor belt (Figure 2-6), where the growth rate is the rate at which the belt turns. Hence all the individuals that mature out of the stage at time *t* must have been recruited into it simultaneously. This approach therefore assumes that all individuals within a stage are identical except with respect to single growth index.

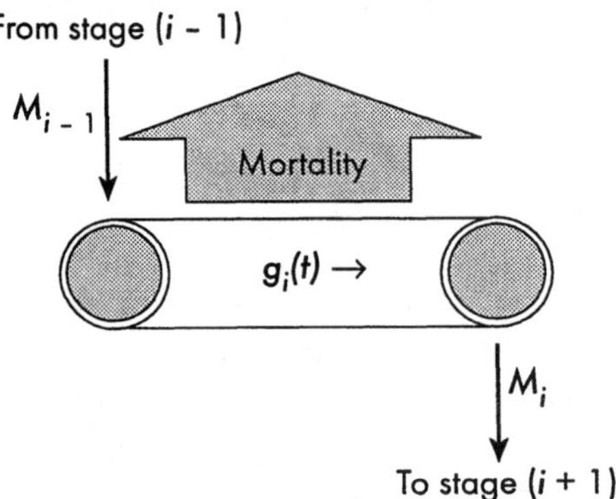

Figure 2-6 Schematic representation of development of individuals through i^{th} stage of stage-structured population model, where M_i = maturation rate into i^{th} stage, and g_i = development rate of individuals in i^{th} stage (From *Ecological Dynamics* by WSC Gurney and Roger Nisbet, ©1998 by Oxford University Press, Inc. Used by permission of Oxford University Press, Inc.)

Although such models lend themselves well to parameterization in terms of instantaneous rates of mortality, reproduction, and development, they are difficult to simulate numerically. However, there are a number of commercial and freely available software packages that can assist in the solution of the equations.

Continuous time, continuous state

Partial-differential equation models describe the population in terms of a continuous state (e.g., age or size) in continuous time. The state of the population is given by a function *f*, which gives the density of individuals in a stage *a* at time *t*. Because, after birth, individuals either get older or die,

$$\frac{\partial f}{\partial t} + \frac{\partial f}{\partial a} + \mu f = 0 \qquad \text{(Equation 2-16)},$$

where $\mu(a)$is the per capita death rate at age *a*. Birth yields the boundary condition

$$f(0,t) = \int f(x,t)\beta(x)dx \qquad \text{(Equation 2-17)},$$

where $\beta(a)$ is the per capita birth rate of individuals of age *a*.

Such models have advantages over discrete time models because they can include the most detailed information on the vital rates. This is because individuals are not lumped into discrete stages or divided into discrete time intervals. They permit coupling to the environment either through direct density dependence of the parameters or through dynamic feedback from resources. Again like DDE models, they can also take advantage of parametric form for the vital rates (Caswell et al. 1996).

However, continuous time and state models are the most complex to study numerically of all 3 types of deterministic models. Indeed, much of the analytical theory that exists for other modeling types simply does not exist for this type. The development of the Escalator Boxcar Train method (De Roos 1988; De Roos et al. 1992) has made their study easier, but they still remain out of the reach of all but the most mathematically proficient.

Stochastic models

All the models discussed up to now are deterministic, that is, the system state at any future time can be predicted from its present state. This is of course a rather untenable assumption; unpredictability or randomness enters all ecological dynamic systems, and stochastic models are useful tools to investigate these random events and their affects upon population projections. In contrast to the deterministic approach, stochastic models include more reality because they include variability in vital rates. Variability may be demographic, meaning that vital rate values vary among individuals. This is called "demographic stochasticity" and plays an important role primarily in small populations. Environmental stochasticity is imposed on each individual and plays a crucial role in both small and large populations. Therefore, in order to link individual vital rates to extinction probability, it is necessary to have detailed information on the environmental variability specific to each vital rate.

A large body of papers underline the need to include environmental stochasticity as an important ingredient of population dynamic models that are designed to estimate species extinction probabilities. Ginzburg et al. (1982) and Ferson et al. (1989) investigated both linear and nonlinear stochastic models for population growth rate, and Tuljapurkar and Orzack (1980) and Tuljapurkar (1982) explored the behavior of linear stochastic Leslie matrices, including variance and covariance matrices of vital rates. Tuljapurkar (1982) defined the overall long-term population growth rate of the average population projection matrices over time and showed that when the level of environmental stochasticity was small, it was approximately equal to the population growth rate of the mean matrix. Lande and Orzack (1988) elaborated on this concept and showed that the overall long-term population growth rate is a good estimator even if stochastic noise is up to 30%.

Unfortunately the inclusion of stochasticity into, for example, the deterministic models described above is not a simple process. An in-depth discussion of the technical details of stochastic population models is beyond the scope of this chapter, and therefore the following sections briefly outline each of the approaches and suggest suitable references for further reading.

Discrete time, discrete state

The introduction of randomly varying vital rates in discrete time and state models has the effect of increasing the ecological realism of the model, as such variation is ubiquitous and can strongly influence the dynamics of populations. Such randomness is introduced into the model by using a statistical description, such as probability distribution, of a given vital rate. The form of distribution and the degree of autocorrelation assumed can have quite a strong influence on the predictions of some population endpoints, and rarely are sufficient data available to avoid making quite bold assumptions in this respect. Nonetheless, the deterministic modeling assumption of no noise is probably about as brazen as is possible, and discrete time, discrete space models are certainly the most straightforward framework in which to examine the effects of environmental stochasticity.

Interested readers should consult Tuljapurkar (1996), who provides an excellent review of the main ideas and techniques for analyzing stochastic matrix models.

In the discrete time context, demographic stochasticity is usually best addressed using birth and death process models. Markov chains are quite popular and benefit from the existence of an enormous body of theory. Most introductory statistics books introduce the topic (e.g., De Groot 1986).

Individual-based models

Individual-based modeling (IBM) covers an enormous range of techniques (De Roos et al. 1991). Their common theme is that the basic unit of model formulation is the individual, with the population being nothing more than a collection of individuals and the population dynamics arising solely from the interaction of a collection of individuals (and some external environment). So, most of the formulations that we have already met can be used for IBM, given a restrictive enough description of the individuals within the modeled population. However, the term “IBM” is usually used to refer to models in which each individual is represented separately within a simulation model, and the model for each individual is stochastic (e.g., whether or not it dies today is a random event). This leads to a great deal of flexibility, in that differences between individuals can be modeled, as can quite complex interactions. The simulation tends to be nontrivial, however. The choice is usually between working in discrete time, in which case the ordering of events within a timestep can become important, or using event-based methods, in which time is viewed as continuous, but updating of the population state occurs every time an event occurs in the population (events might be anything from “A found a scrap of food” to “A gave birth to B”). The latter is not trivial to implement for nontrivial models.

Clearly, IBMs have not only the potential for great realism but also the potential to eat huge amounts of computer time without venturing near a datapoint. Similarly, analytical understanding is usually very difficult to achieve: Even if a very complicated model reproduces reality quite well, we may have no better feeling for how the model works than we had for how the original system worked. However, any of the population endpoints discussed in this chapter can potentially be investigated numerically using a suitable IBM.

Another serious problem is the difficulty in assessing the validity of IBMs. There is not a standard implementation or standard means of describing them, which can make it very difficult to objectively assess the merits or otherwise of any particular IBM. On the other hand, when well done, these are perhaps the models least likely to do serious abuse to the biology that they attempt to describe.

Stochastic differential equation models

Stochastic differential equation models are usually used in a discrete state, continuous time setting and are typically produced by adding a stochastic element to differential equation models. They are analytically horrendous (Øksendal 1989), partly because there is not even a standard calculus for solving them. Some analytical results are available, however (at least once you have defined your calculus), so that with enough persistence it is possible to use these models to predict or project average growth rates, recovery times, and extinction probabilities (Turelli 1997). But given the difficulties, it is probably not worth using these models in the current context, especially since inexpensive computer power now makes other approaches (like IBMs) more accessible.

Birth and death process models

There is a considerable literature on birth and death process models, which model the transition of a population between discrete population sizes as births and deaths occur within the population. Nisbet and Gurney (1982) provide a useful introduction. These models are particularly suitable for looking at extinction-related questions in small populations because there is a good deal of supporting mathematical theory, and unlike pure simulation approaches, the model assumptions can usually be stated quite clearly and explicitly.

Effects of Model Formulation on Estimates of Population Growth Rate

It is often difficult and confusing for nonmathematicians to fully understand the differences between each of the modeling types discussed above. The array of potential mathematical techniques is often bewildering and sometimes begs the question of how critical model choice is in the resulting estimates of population growth rate and projections or predictions. In order to determine the effect of different deterministic model formulations upon the estimation of population growth rate, the 3 types of deterministic models described above (discrete time and state; continuous time, discrete state; continuous time and state) were parameterized by using the same life-table data and the results compared and contrasted.

Methods

All 3 deterministic models were parameterized by fitting directly to the 28-d mysid life-table data reported in Kuhn et al. (2001). Data consisted of daily counts of the number of individuals surviving from an initial cohort of 60 neonates, along with cumulative egg production each day. These data were available for 9 *para*-nonylphenol concentrations, although only the lowest 6 are analyzed here because these are the most ecologically interesting. The model unknowns were the functions $\beta(a)$ and $\mu(a)$ for the matrix and PDE models, and the parameters β, μ_J, $N_i(t)$, and τ for the DDE model.

Given particular values for the unknowns, each model can predict the number of individuals surviving and the cumulative egg production in the 28-d experiments. Hence, the unknowns can be estimated by adjusting their values in order to achieve the best fit between model predictions and data (model fit was measured by sum of squares of differences between data and model predictions). Working with unknown functions requires some method of model selection in order to avoid fitting with overly complex functions; generalized cross validation was used here (Craven and Wahba 1979; Wood 2000b for the multidimensional generalization). The general method of model fitting used here is described in Wood (2000a), which builds on ideas introduced in Wood (1994) and Wood and Nisbet (1991).

When the model is age-truncated, the maximum age is limited to 28 days in the following manner:

$$\frac{dA}{dt} = \beta.A(t-\tau).e^{-\mu J.\tau} - \beta.A(t-28)e^{-\mu_J\tau-\mu_A(28-\tau)} - \mu_A.A \quad \text{(Equation 2-18).}$$

Each model was fitted to the 28-d data for each of the 6 *p*-nonylphenol concentrations. Each fitted model was used to predict the population after 55 days in a multigeneration experiment run at the relevant toxicant concentration. Each model was also used to estimate the asymptotic population growth rate r_m expected in the absence of density-dependent vital rates.

Results

The results are summarized in Figures 2-7 and 2-8. It is important to note that, following Kuhn et al. (2001), the PDE and matrix models assume that individuals are lost to the system at day 28 (in fact, mysids live substantially longer than that), an assumption that makes little difference to growth rate estimates at high growth rates but becomes very important as rates decline. The DDE model does not make this assumption, and it is this which almost entirely explains the differences in N_{55} and r_m estimates between it and the other models. Indeed, if the 2-stage DDE model is modified so that individuals are lost at day 28 (i.e., age-truncated), then the predictions are very similar to those of the other models.

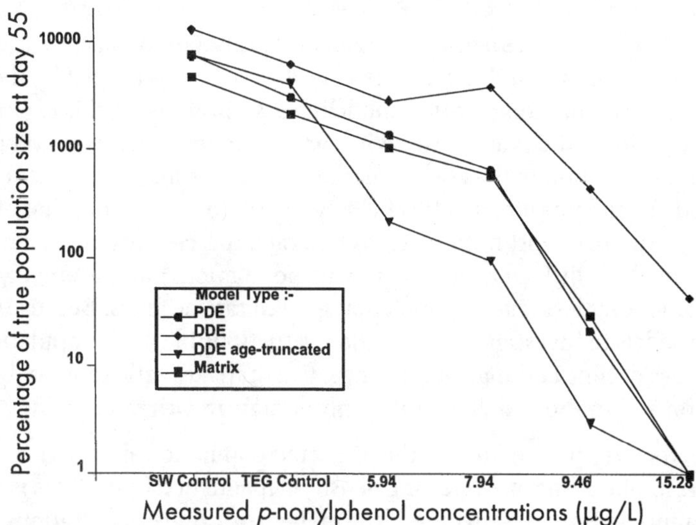

Figure 2-7 Estimation of percentage of true population size at day 55 for 4 different model formulations using data from Kuhn et al. (2001) (SW = seawater; TEG = triethylene glycol)

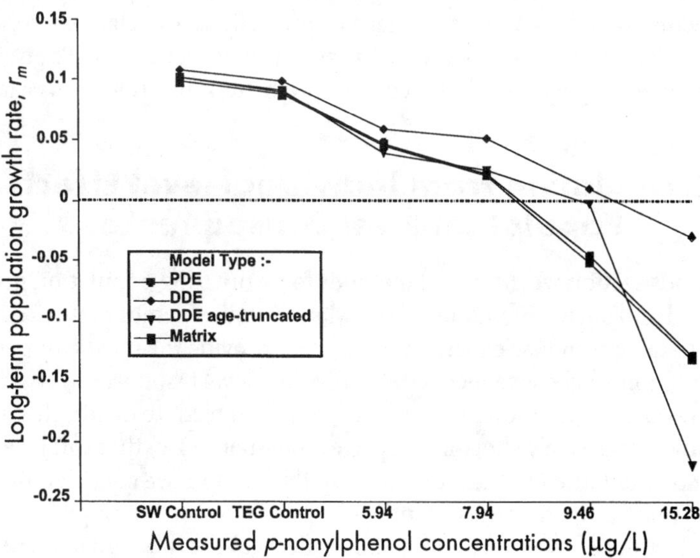

Figure 2-8 Estimated long-term population growth rate (r_m) from data reported by Kuhn et al. (2001), derived by 4 different model formulations in the control and 5 treatments (SW = seawater; TEG = triethylene glycol)

Discussion

It is clear that the choice of demographic modeling formulation makes some difference to estimates, but in this case, these differences are swamped by the differences that arise from more critical modeling assumptions. In this example, the decision to kill all individuals after day 28 in some of these models was crucial. The impact of this assumption in these examples reinforces the inappropriateness of justifying modeling assumptions in this field by appeal to the folklore that older individuals contribute almost nothing to growth rate. It is clearly true that older individuals contribute little in a rapidly growing population, but it is equally obvious that this becomes ever less true as population growth rate declines. Because we are formulating models to investigate exactly those situations in which population growth rate does decline and may become negative, it makes little sense to base this formulation on assumptions that are valid only at high growth rate.

Furthermore, the very poor nature of the predicted population sizes at day 55 highlights the implications of neglecting density-dependent factors. This is reinforced by examining Figure 2-7, where at the higher toxicant concentrations, predictions at day 55 are significantly improved because at lower densities the effects of density dependence are not pronounced. Examining Figure 2-8 clearly shows the age-truncated DDE model provides far and away the best predictive capacity of all the model formulations at low abundances when the density-independent nature of the model is not paramount.

Although the results presented here suggest that, provided the gross characterization of the biology is the same, predictions or projections are relatively robust to the details of model formulation, this is not always the case. Wood and Thomas (1999) provide a good example of just how sensitive predictions or projections can be.

Extrapolating from Individual-level Effects to Population-level Consequences

In the previous section we discussed methods for estimating population growth rate from detailed and partial life-table information. Detailed demographic analyses based on life-table response experiments (LTREs) provide the most complete picture of the population-level consequences of individual-level responses to environmental stress. However, because individuals are tracked from birth to death, this approach is viable only for relatively short-lived species (life span ~ 1 y) that can be easily cultured and maintained in the laboratory. In this section, we consider alternative approaches to investigating the potential effects of stressors on population growth rate, and we explore how information about effects at the individual or suborganismal level may be used to provide insight into population-level consequences.

Population size is determined by the vital rates of individuals within the population, as well as by the rates of emigration from and immigration into the population. Any factors that influence these processes, either directly or indirectly, can therefore have population-level consequences. Individual vital rates are influenced by population density acting through intraspecific competition for resources (e.g., food, space) as well as by species interactions such as predation, interspecific competition, and mutualism. The effect of species interactions on vital rates may depend upon the size and structure of the population as well as the type of habitat being occupied. Many predators, for example, are size selective, and therefore the predation risk of a particular individual will depend upon its size and the size structure of the population to which it belongs (Hairston and Pastorok 1975; Hairston et al. 1982, 1983; Gleason and Bengtson 1996a, 1996b). Contaminants often affect vital rates directly by decreasing survival or reproductive success, but they may also have indirect effects through changes in species interactions.

Individual vital rates are an integration of many biochemical and physiological processes. The life histories of individuals are determined by the total availability and allocation of energy and nutrients among maintenance, growth, and reproduction, and several physiologically based individual models have been developed to predict vital rates from energy budgets (see "Extrapolating from bioenergetic responses," p 63). Any toxicant-induced response that results in a decrease in energy or nutrient intake (e.g., decreased feeding rate) or an increase in energy expenditure (e.g., increased maintenance costs) could have consequences for an individual's vital rates. Feeding rate can be influenced by many factors, including food quality and quantity, physiological condition of the organism, and behavior of the organism. Maintenance costs may be increased by the induction of detoxification or repair processes.

The translation of responses observed at 1 level of biological organization to consequences at other levels of organization is particularly important in retrospective risk assessment, where the effects of stressors on natural populations are often investigated by measuring physiological (e.g., energetics) or biochemical (e.g., molecular biomarker) responses. Understanding linkages between levels of biological organization may also be important in prospective risk assessment, especially in situations where LTREs are not appropriate either because the study species has a long life span or because it is difficult to culture under laboratory conditions. Indeed, the vast majority of standard ecotoxicological tests employ individual-level responses as endpoints, assuming, implicitly if not explicitly, that these endpoints are linked to responses at higher biological levels. Finally, an understanding of how effects translate across levels of organization is essential if we are to gain mechanistic insight into how stressors effect populations.

Linkages can either be investigated by correlating effects observed at different levels of organization or by determining causal relationships and constructing mechanistic models. Here we focus on mechanistic models and in particular consider models that link physiological responses to vital rates and hence population growth. We also consider whether it is possible to use specific life-history traits as indexes of population growth rate and explore to what extent changes in behavior and biochemical responses can provide information of possible population-level effects. The conceptual approach that will be used is illustrated in Figure 2-9.

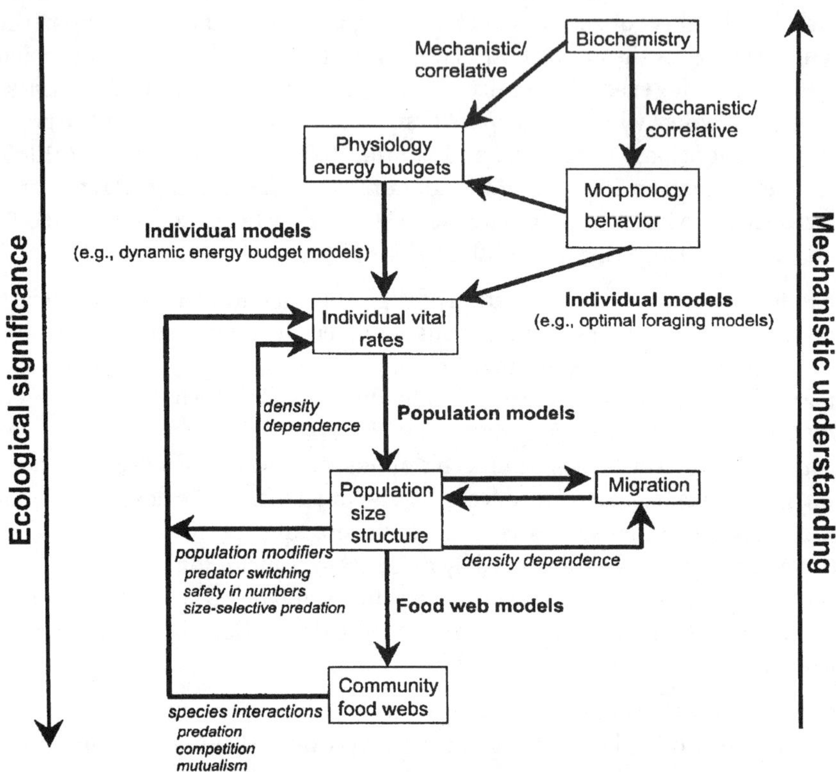

Figure 2-9 Conceptual framework linking effects at different levels of biological organization. Evidence of some linkages is based on correlations, whereas for others, there is evidence for causal relationships and mechanistic understanding. Mechanistic models can be used to project population- and community-level consequences of toxicant effects on individuals.

Using Individual Life-history Traits to Project Population-level Effects

Population growth rate is an integration of several vital rates. The question we address here is this: "Is it possible to identify those vital rates that are driving changes in population growth rate and to focus studies on those?"

Standard acute toxicity tests frequently focus on the effect of a chemical on survival rates, with results expressed as an LC50 value. There is some evidence that, within a species, acute LC50 may provide a good predictor of population-level consequences. Kuhn et al. (2000) compared data for *Americamysis bahia* (formerly *Mysidopsis bahia*) exposed to 20 different toxicants. Data collected during a 28-d life-cycle test were used to model the dynamics of mysid populations using an age-classified discrete state, discrete time matrix model. The exposure concentration predicted to produce $\lambda = 1$ (i.e., zero population growth) was estimated for each chemical and compared to the 96-h LC50. There was a very strong positive correlation between these 2 endpoints ($r = 0.96$) indicating that, for this species, acute LC50 values are a good predictor of projected population-level effects. A further study (Kuhn et al. 2001) investigated the relationship between population growth rates projected from 28-d life-cycle tests with those observed in 55-d multigenerational studies. There was a very good rank correlation ($r = 0.95$) between the mysid abundances projected from the 28-d study and those measured in the 55-d study. Moreover, the critical concentration (i.e., $\lambda = 1$) of 16 µg/L nonylphenol projected using an age-classified matrix model was very similar to the observed critical concentration of 19 µg/L nonylphenol.

In contrast to the above study, Stark et al. (1997) concluded that LC50 values were not good predictors of the population-level effect of the pesticides dicofol and Neemix on the mites *Tetranychus articae* and *Iphiseius degenerans.* In their study, the species-specific differences in susceptibility of population growth rate to pesticide exposure were related to species-specific differences in reproductive rate, not survival. The least susceptible species, *T. urticae,* had a higher reproductive rate and its reproduction was less adversely affected by pesticide exposure. Forbes and Calow (1999) reviewed 41 invertebrate studies (28 species and 44 toxicants) and evaluated 1) whether any single life-history trait was consistently the most sensitive to toxicant exposure and 2) the relationships between specific life-history traits and population growth rate. Across all species and toxicants, there was no single life-history trait that was the most sensitive to toxicant exposure, although reproductive endpoints (e.g., number of offspring, net reproductive rate) were the most sensitive in 69% of cases. Percentage reduction in population growth rate was significantly correlated with percentage impairments in number of offspring ($r = 0.56$) and time to reproduction ($r = 0.72$) but not with reduced survival ($r = 0.3$).

The proportional sensitivity of population growth rate (λ) to changes in life-history traits (a_{ij}) can be evaluated by calculating the elasticity (e_{ij}) of each trait (Kroon et al. 1986):

$$e_{ij} = \frac{a_{ij}}{\lambda} \cdot \frac{\partial \lambda}{\partial a_{ij}} \qquad \text{(Equation 2-19).}$$

The higher the elasticity of a trait, the greater its contribution to effects on population growth rate. The influence of different life-history traits on population growth rates of the polychaetes *Capitella* sp. I and *Streblospio benedicti* has been investigated by Levin et al. (1996) and Hansen et al. (1999a). These studies illustrate 2 important points:

1) Life-history traits may be relatively insensitive to stressor impacts but make a major contribution to effects on λ if they have high elasticity (Caswell 1996). In the study by Hansen et al. (1999a), time to first reproduction was relatively

insensitive to nonylphenol (declining by 20%) but, because of its high elasticity, explained more than half (55%) of the effect on λ. In contrast, fertility was substantially (by 75%) reduced by nonylphenol but, because it had relatively low elasticity, explained only 44% of the effect on λ. These results support an earlier study by Kammenga et al. (1996), who investigated the effect of Cd on the population growth rate of the nematode *Plectus acuminatus*. They found that a 45% reduction in reproductive period, the life-history trait most sensitive to Cd, had no significant effect on r_m. In contrast, a 7.5% reduction in the least sensitive trait, age at first reproduction, caused a 5% reduction in r_m.

2) The sensitivity of population growth rate to changes in life-history traits varies depending on species and treatment (Table 2-3). Levin et al. (1996) concluded that increased age at first reproduction and, to a lesser extent, decreased fertility were driving the reduction in the population growth rate of *Streblospio benedicti* exposed to hydrocarbons. A similar result was obtained for the *Capitella* sp. I exposed to nonylphenol (Hansen et al. 1999a). In contrast, decreased fertility and juvenile survival were the principal causes of a reduction in the population growth rate of *Streblospio benedicti* exposed to blue-green algae (Levin et al. 1996).

Table 2-3 Contribution of differences in life-history traits to treatment effects on λ in a 2-stage model

Species	Blue-green algae	Hydrocarbon
Streblospio benedicti[a]	Fertility > juvenile survival > age at 1st reproduction > adult survival	Age at 1st reproduction > fertility >> juvenile and adult survival
Capitella sp. I[a]	Age at 1st reproduction > fertility > juvenile survival > adult survival	No significant effect on λ
Capitella sp. I[b]	Not applicable	Age at 1st reproduction > fertility >>> juvenile and adult survival

[a] Source: Levin et al. 1996.
[b] Source: Hansen et al. 1999a.

The extent to which population growth rate responds to stressors depends on the severity of the effects on the individual life-history traits (which varies among stressors) and on the sensitivity of population growth rate to changes in each of the life-history traits contributing to it. The sensitivity of population growth rate to different life-history traits depends upon the status of the population (Stearns 1992). In populations that are growing rapidly, selection on age at first reproduction is strong, and selection on survival and fecundity is relaxed. In populations nearer equilibrium, selection on survival and fecundity rates is strong, and selection on age at maturity is relaxed.

Population growth rates projected from data generated under controlled laboratory conditions are likely to vary from those for organisms in natural populations. This is because many factors known to be important in natural populations (e.g., predation, competition, variation in environmental factors) are controlled or eliminated under laboratory conditions. Moreover, the calculation of population growth rates and the subsequent analyses of sensitivity assume density independence. Grant (1998) demonstrated how density dependence alters the sensitivity of population growth rate to specific life-history traits, and Hansen et al. (1999b) demonstrated how competition and predation could cause changes in elasticities and hence alter the effect that toxicant-induced reductions in specific life-history traits have on population growth rate.

Advantages and limitations of using single life-history traits to project population growth rate

Advantages

- Determining individual life-history traits is less labor- and time-intensive than the determining complete life tables.
- Within species, individual life-history traits may be good predictors of population growth rate.
- Sensitivity analysis of λ reveals which traits contribute most to population response.

Limitations

- Interspecific comparisons demonstrate that 1) no single life-history trait is most sensitive across a range of stressors and 2) the correlation between changes in any specific life-history trait and changes in population growth rate is weak.
- Large stress-induced effects on life-history traits may make small contributions to the effect on population growth rate and vice versa.
- Importance of specific life-history traits in determining population growth rate may be affected by competition, predation, and presence and form of density dependence.
- Population growth rates projected from data generated under controlled laboratory conditions are likely to vary from those for organisms in natural populations.

Extrapolating from Individual- and Suborganism-level Responses

There are 2 main sources of uncertainty when we use individual- or suborganism-level responses to project population-level effects: measurement uncertainties and extrapolation uncertainties.

1) Measurement uncertainties relate to sampling issues and the consequences of temporal and spatial variation in organism and environmental quality (see Chapter 5). The sources and magnitude of variation in biological responses need to be considered both in the design of sampling programs and in the interpretation of resulting data (Chapter 5; Depledge and Lundebye 1996; Luoma 1996). There is also a need for baseline data that describes how individual organisms respond to natural stressors so that true anthropogenic effects can be distinguished from natural variability (Chapter 6; Ringwood et al. 1999).

2) Extrapolation uncertainties are uncertainties associated with making the critical linkages across levels of biological organization. Experimental demonstration of causal relationships strengthens the confidence of extrapolations. Where this is not possible, extrapolation uncertainties may be reduced by establishing correlations between responses and by demonstrating that linkages are mechanistically plausible.

Extrapolating from behavioral responses

Stressors may have profound effects on many aspects of behavior affecting an organism's ability to obtain food, avoid predation, and reproduce. Toxicant-induced impairment of reproductive behavior has been reported for both vertebrates and invertebrates, including reduced parental care in fish (Weber 1993), reduced singing and displaying in birds (Grue and Shipley 1984), and reduced precopular pair formation in amphipod crustaceans (Pascoe et al. 1994). Studies on a variety of stressors and species have demonstrated that exposure to toxicants often results in a decrease in foraging behavior and food intake (Cairns and Loos 1967; Little et al. 1990; Morgan and Kiceniuk 1990; Scherer et al. 1997; Maltby 1999). Moreover, short-term exposures to toxicants may result in long-term impairment of feeding behavior. Fjeld et al. (1998) report impaired feeding efficiencies and reduced competitive abilities of grayling (*Thymallus thymallus*) 3 years after they were exposed to methylmercury for 10 days as embryos. Stressors may also influence food consumption by affecting the behavior of prey, making them more susceptible to predation (Farr 1977; Kraus and Kraus 1986; Little et al. 1990; Jung and Jagoe 1995).

Behavior is an individual-level response that potentially can be linked to the population level. By altering feeding and predator avoidance, stressors can cause changes in predator–prey interactions, resulting in population changes in the predator, the prey, or both. Changes in feeding behavior can result in reduced feeding rate, which can be directly related to impaired growth and population declines (see "Extrapolating from bioenergetic responses," p 63). However, behavioral responses observed under experimental conditions may not be manifested in natural populations because of differences in either exposure or response. For example, birds provided with pesticide-contaminated food have reduced food consumption. However, in the field, not all food is contaminated, and birds maintain normal levels of food consumption by switching to uncontaminated food (Bennett 1994; Grue et al. 1997).

There have been numerous studies of contaminant effects on behavior, many of which focused on ecologically relevant behaviors, but clear demonstrations of the population-level consequences of these effects have been less frequent. In a recent review of the effects of pollutants on the reproductive behavior of fishes, Jones and Reynolds (1997) concluded that, whereas effects on parental care and courtship behavior have been reported, few studies have measured reproductive success or extrapolated results to effects on populations. Peakall (1996) examined behavioral changes in natural populations of wildlife associated with pollution. He states that, "although changes in behaviour have been observed, they are generally more difficult to quantify and are less reproducible than biochemical changes." Moreover, he concludes that "there is no clear evidence in wildlife that behavioural changes caused by pollutants are a serious threat to populations." A similar conclusion was reached by Grue et al. (1997) in a review of the neurophysiological and behavioral effects of organophosphate and carbamate pesticides on nontarget wildlife.

An example of a possible link between contaminant-induced behavioral changes and population-level effects is provided by a study of *Fundulus heteroclitus*. These fish are significant predators of grass shrimp (*Palaemonetes pugio*; Knieb 1986), but they also consume detritus, even though it is of little nutritional value to them (Prinslow et al. 1974). *F. heteroclitus* collected from a contaminated site in northern New Jersey, USA, had poorer prey capture ability than did those from nearby reference sites, due primarily to fewer attempts to capture prey (Smith and Weis 1997). Moreover, fish collected from the contaminated field sites had less animal material in their guts and twice as much detritus as did fish collected from reference sites. These differences in feeding behavior were associated with differences in population size and structure. *F. heteroclitus* that inhabited contaminated sites were present at lower densities and were smaller than *F. heteroclitus* at reference sites (Weis and Weis 1989). They also exhibited reduced growth and condition index (CI), earlier reproduction, and reduced longevity (Toppin et al. 1987). Similar shifts in size structure have been reported for another population of *F. heteroclitus* that inhabits polychlorinated biphenyl (PCB)-polluted New Bedford Harbor, Massachusetts, USA, but in this case, the population size of the fish does not seem to be affected (Black et al. 1998).

> ***Advantages and limitations of using behavioral responses to project population-level effects***
>
> ***Advantages***
> - Integrates biochemical and physiological effects
> - Sensitive and therefore may provide early-warning indicator
>
> ***Limitations***
> - May be difficult to quantify, especially in natural populations
> - Link to population-level is often unclear

Extrapolating from bioenergetic responses

Bioenergetics provides a useful tool for linking individual-level effects and population-level consequences and can be described as an energy budget that balances energy inputs (consumption, C) with energetic outputs (metabolism, R; fecal egestion, F; urinary excretion, U; production, P):

$$C = R + F + U + P \qquad \text{(Equation 2-20).}$$

Individual organisms have a finite amount of energy to allocate to maintenance, growth, and reproduction. Increasing energy allocation to 1 component of the energy budget (e.g., growth) will therefore result in less energy being available for other components (e.g., maintenance, reproduction) and will consequently have vital rate implications (Calow and Sibly 1990). Reducing energy allocated to maintenance will result in a reduction in survival rates, reducing energy allocated to growth will result in increased developmental times, and reducing energy allocated to reproduction will result in reduced offspring number and/or size. Stressors may affect energy budgets by reducing energy acquisition (i.e., reduction in *C*) and increasing energy expenditure (i.e., increased maintenance costs of defense and repair processes, increase in *R*). The effect of stressors on energy budgets and on the amount of energy available for growth and reproduction can be determined by calculating an individual's "scope for growth" (SfG). SfG is defined as the difference between energy absorbed from food and that lost through metabolism and excretion (Warren and Davis 1967). A positive SfG indicates that energy is available for growth and reproduction, while a negative SfG indicates that reserves must be used to maintain the organism. Short-term measures of SfG correlate well with long-term measures of vital rates (e.g., Bayne et al. 1985; Maltby 1994).

Scope for growth provides a powerful tool for the study of stress (Donkin and Widdows 1986). It has been used successfully in field and laboratory studies with a variety of aquatic species, although it is most commonly measured using marine bivalves (e.g., Widdows, Donkin, Brinsley, Evans, Salkeld et al. 1995; Sobral and Widdows 1997). Resident or transplanted bivalves are collected from the field, and components of their energy budgets are determined in the laboratory. This approach introduces uncertainties, as it is not known to what extent energy budgets are influenced by the collection procedure and laboratory conditions. The feeding rate of many species can be measured in situ, thus reducing some of these uncertainties. Reductions in energy intake are often the main cause of a reduction in SfG (Maltby 1999), and an in situ feeding assay has been developed for the freshwater amphipod *Gammarus pulex* (Maltby et al. 1990; Crane and Maltby 1991).

Dynamic energy budget (DEB) models describe the rules by which individual organisms assimilate and use energy from food (Kooijman 1993; Gurney and Nisbet 1998). They predict growth and reproduction in a known food environment and can therefore be used to extrapolate from toxicant-induced reductions in feeding rate to vital rates. DEB models have been applied to a number of vertebrate and invertebrate species, although *Daphnia* is the most common species used (Noonburg et al. 1998). The application of DEB models to ecotoxicology is discussed by Kooijman and Bedaux (1996) and by Nisbet et al. (1996), and Gurney and Nisbet (1998) describe how individual-based energetics models can be used to explicitly link SfG to long-term consequences for individual organisms.

By combining DEB models with demographic models, it is possible to use information about the effects of toxicants on the energy budget components of individuals to predict population-level effects. Kooijman and Metz (1984), for example, used an age-structured model to predict the dynamics of chemically stressed *Daphnia* populations from information about the effect of toxicants on individual energy budgets. A similar approach was adopted by Klok and de Roos (1996). They used an individual-based physiological model in conjunction with a stage-structured matrix model to predict the effect of Cu-induced changes in feeding and metabolism on the growth and reproduction of the worm *Lumbricus rubellus*. Using this approach, they projected a critical Cu concentration (i.e., zero population growth rate) of between 200 and 300 mg/kg. This corresponded well with the observed steep decline in the size of field populations at Cu concentrations exceeding 300 mg/kg.

Baveco and de Roos (1996) used a simplified Kooijman-Metz model combined with either a deterministic partial differential equation model or a stochastic individual-by-individual model to project the effect of toxicant stress on lumbricid populations. Both population models incorporated density-dependent regulation in the form of predation. Under constant stress, population density responded most sharply to a decrease in energy intake and, to a lesser extent, to an increase in maintenance costs. A 20% reduction in energy intake or a 20% increase in maintenance costs was sufficient to cause *Lumbricus terrestris* populations to go extinct. Baveco and de Roos (1996) demonstrate how this modeling approach can be used as a risk assessment tool to project the effect of toxicants on population size and the ability of populations to recover. The population-level consequences of changes in initial toxicant dose, toxicant half-life, and frequency of exposure can also be explored.

Advantages and limitation of a bioenergetics approach

Advantages

- Energy-budget components (i.e., consumption, egestion, excretion, and metabolism) can be measured for a wide range of organisms and can be used to calculate SfG.
- Short-term measures of SfG correlate well with long-term changes in vital rates. These links can be made explicit using individual-based physiological models.
- Physiological models can be used in conjunction with demographic models to project population-level consequences and to gain insight into the physiological mechanisms responsible for stressor-induced changes in vital rates.
- SfG is a useful tool in retrospective risk assessment, as it can be determined for field deployed animals or individuals from natural populations.
- Energy intake (i.e., feeding rate) is reduced by a range of stressors and is a major determinant of SfG. This can be measured in situ for several species.

Limitations

- Not all population-level effects are a consequence of energy budget changes.
- Energy-budget models can only evaluate reduced survival due to starvation. They do not take account of the specific modes of action of chemicals.
- Energetics models generally ignore interspecific interactions and often ignore density-dependent processes.
- Energy budget components are affected by a variety of intrinsic (e.g., body size, physiological condition) and extrinsic (e.g., temperature, salinity) factors, and consequently they may exhibit considerable natural variability.

One of the main advantages of models based on individual energy budgets is that they mechanistically link effects at the individual level to consequences at the population level and can be used to provide insight into the physiological mechanisms responsible for stressor-induced changes in vital rates. Moreover, they can be incorporated into food web and carbon–nutrient cycling models (De Ruiter et al. 1994) and thereby provide a link between population-level effects and ecosystem function (Chapters 3 and 4).

Extrapolating from molecular and cellular responses

Within the past decade, there has been an increasing emphasis on the potential use of biomarkers as indicators of anthropogenic impacts. Molecular and cellular responses (i.e., biomarkers) are believed to be among the most sensitive and earliest detectable responses. The underlying premise of biomarker tools is that effects at higher levels of organization (i.e., populations and communities) represent the net sum of effects on individuals that resulted from alterations in cellular and molecular responses.

There is a conceptual basis for linking biochemical responses to population growth rates (Figure 2-9), and there are several examples in the literature in which biochemical responses have been linked mechanistically to individual- or population-level effects; rarely, however, are these links quantitative. For example, tributyltin (TBT), a compound used in antifouling paints, has been shown to cause imposex (i.e., masculinization of females) in gastropod mollusks, resulting in significant declines of gastropod populations near harbors and marinas polluted with TBT (Bryan et al. 1986, 1988; Gibbs and Bryan 1986; Gibbs et al. 1988). Experimental studies demonstrated that TBT inhibited the P450 aromatase system, which catalyzes the aromatization of androgens to estrogens, causing elevated androgen levels and androgen–estrogen imbalances and resulting imposex (Betin et al. 1996). There is, therefore, a mechanistic link between TBT-induced biochemical changes, individual-level effects, and population-level consequences.

The effect of dichlorodiphenyldichloroethylene (DDE) on avian eggshells provides another example of a link between biochemical and population-level effects. During the 1950s and 1960s, reduced population sizes were associated with DDE exposure, primarily in the orders Pelicaniformes and Falconiformes. Reduced egg hatchability was associated with thinned eggshells, and laboratory studies demonstrated that DDE caused thin eggshells by reducing the activity of Ca–Mg pumps in the shell gland epithelium. Ca–Mg pumps are responsible for driving the movement of Ca from the blood to the site of shell deposition (Lundholm 1987, 1993). A final example is provided by the work of Köhler et al. (1998), who studied the induction of stress gene transcription (including messenger ribonucleic acid [mRNA] stability) and the accumulation of the corresponding stress protein hsp70 in slugs exposed to Cd or Zn-enriched food. To validate the suitability of these 2 aspects of the cellular stress response to act as early-warning markers for metal effects on life-history parameters, the fecundity, offspring number, longevity, and mortality of slugs were recorded in life-cycle experiments. They found that the elevation of the hsp70-mRNA level caused by short-term (14-d) metal exposure coincided with both diminished fecundity and reduced offspring production due to chronic metal exposure in terms of threshold concentrations for Cd and Zn effects.

However, in many cases, the linkages from biochemical changes to population-level responses have not been demonstrated. Some molecular and biochemical responses to environmental contaminants are part of an organism's normal metabolic processes (e.g., gluathione, oxidative enzymes such as catalase, superoxide dismutase). The potential higher-order effects of large alterations in responses that are essential for normal metabolic and physiological processes may be readily appreciated. However, the potential effects of small alterations in biochemical responses for individuals and populations are less clear. Many biomarker responses (e.g., metallothioneins, heat shock proteins, P450 pathways) represent detoxification mechanisms. In these cases, simple changes may not represent significant adverse effects at the individual or population level, particularly if they effectively ameliorate potential adverse effects of toxic chemicals. Others, such as lysosomal destabilization, lipid peroxidation, and genotoxicity, function as valuable damage indicators.

Advantages and limitations of a biomarker approach

Advantages

- Sensitivity.
- Valuable for identifying potential causative factors.
- Amenable to biotechnological development.
- Many cellular responses are conserved to some degree across species, so insights with 1 species frequently apply to other species.
- Can be used in a field or laboratory setting.

Limitations

- Mechanistic links between biochemical or molecular response and population-level effects are often unclear.
- Distinguishing transient responses from more definitive, longer-term responses may be difficult.
- Many responses are seasonally variable and must be interpreted in the light of such variation.

Detoxification mechanisms such as heat shock proteins, metallothioneins, and P450 pathways are regarded as energetically expensive, so that the energy required for detoxification processes may be diverted from other metabolic processes, including growth and reproduction. For example, Krebs and Loeschcke (1994) demonstrated that when female *Drosophila melanogaster* were exposed to elevated temperature, fecundity was reduced, although survival to subsequent high temperatures was enhanced. The response was believed to be due to the diversion of energy from reproduction to the production of heat shock proteins. However, it can be difficult to distinguish between energetic effects and more direct effects. For instance, increased catabolism of proteins has also been associated with increased temperature, so reduced reproduction of temperature-treated *Drosophila* could be related to more direct effects of temperature than to changes in energy allocation. There are few examples in which relationships between biochemical responses and SfG (see "Extrapolating from bioenergetic responses," p 63) have been investigated. The relationships between SfG and heat shock proteins were investigated in mussels exposed to Cu (Sanders et al. 1991). The highest heat shock protein levels and significant adverse effects on SfG were observed with the highest Cu concentrations (32 and 100 mg/L), and significant elevations of heat shock protein were observed at concentrations > 3 mg/L.

Biomarkers have a number of appealing characteristics. There is generally some understanding of underlying mechanisms and potential consequences. They are particularly valuable for verifying specific chemicals or classes of toxins as putative causes. They are amenable to technological adaptations, which effectively increase potential sample size and standardized methods for validation. However, several important issues require careful consideration if biomarkers are to be used as risk assessment tools:

1) It is important to establish linkages to population-level changes or the potential for population-level changes.
2) Because cellular and physiological responses may vary seasonally because of changes in metabolic rates and reproductive status, seasonal variation must be considered when native populations are evaluated (Viarengo et al. 1991; Bordin et al. 1997).
3) It must also be appreciated that when organisms encounter stressors, there are frequently alterations in responses that may be transient. If homeostatic and detoxification mechanisms effectively ameliorate the response, there can be a return to the initial state; but significant stress will occur if sustained changes alter homeostasis and reduce fitness.
4) Stress may elicit a cascade of responses that can only be characterized by using a suite of biomarkers. For example, when oysters were exposed to Cu in the laboratory, there were initially severe increases in membrane damage as indicated by increased lipid peroxidation. However, with continued Cu exposures, metallothioneins were induced so that by day 7, lipid peroxidation levels returned to baseline (Ringwood et al. 1998). Toxicity would be expected when the compensatory capacities of metallothioneins are overwhelmed. Therefore, when applying these tools as stress indicators, simple elevation of lipid peroxides would not be sufficient alone, but simultaneous elevation of metallothionein and lipid peroxides implies that detoxification capacities have been overwhelmed, causing significant tissue damage.

Generally, the use of multiple biomarker measures is advocated because they are believed to provide a more robust indication of significant alterations and a higher degree of certainty. However, simply knowing the number of altered responses is not as important as having an appreciation of the potential ramifications. Recently, there have been attempts to develop integrated models of biomarker responses that could be used to summarize complex multiparameter data and to develop integrated indexes (Narbonne et al. 1999). These approaches facilitate the development of criteria for ranking stress levels. As they are developed, it will be worthwhile to consider how integrated biomarker indexes can be incorporated into physiological and population models. Ultimately, the critical question is "Can we define distinct levels of integrated stress responses that are more relatable to population-level effects?"

Evaluating Projected Population-level Effects

In the previous sections, we discussed how population-level effects of stressors may be measured or projected from individual-level effects. Projections are usually based on information obtained from relatively short-term laboratory experiments on closed populations, whereas measured effects on natural populations usually are restricted to limited observations at a single location. However, natural populations are rarely closed and may exhibit considerable temporal and spatial variability (Chapters 5 and 6). Evaluating the ecological significance of projected or measured changes in population-level endpoints in such dynamic systems is a major challenge. In this section, we discuss factors that must be considered when evaluating population-level endpoints against a background of natural variability. We also consider the extrapolation of effects projected using laboratory data for surrogate species and the extrapolation across spatial and temporal scales. Because the ecological significance of a projected or measured change in the size or structure of a population will depend upon the magnitude of the change and on the ability of the population to compensate for, or recover from, the change, we discuss compensation (recovery was discussed in "Measures of population size," p 28).

Evaluating Effects in the Context of Natural Variability

The significance of any perturbation in population dynamics must be evaluated in the context of the variability that is a natural component of most populations and in the context of other stressors that affect the population. Giving specific recommendations for evaluating ecological significance is difficult because significance varies with the biology of the species or ecosystem under consideration. For example, it may be important to maintain a minimum population size for the long-term survival of a population (e.g., MVP analyses) or for the maintenance of some ecosystem function (e.g., prey population size necessary to maintain a predator population; see Chapters 3 and 4 for discussions of this type of indirect effect). Here we discuss general principles, not species-specific factors, for evaluating ecological significance. It is important to note that risk assessment decisions must also incorporate social and political values, which may be based on considerations other than ecological significance and are beyond the scope of this discussion.

The ecological significance of a population-level change caused by a stressor is a function of

- the pre-impact population trajectory (state of population, natural variability);
- the magnitude, timing, frequency, and spatial extent of the impact; and
- the presence of other stressors, natural or human-induced.

The pre-impact population trajectory is often represented by changes in average population density and fluctuations about that average over time. However, the state of the population can be followed over time using other population-level endpoints such as age or size structure, population growth rates, recruitment rates, and genetic diversity (Table 2-4).

Table 2-4 Characteristics of populations that increase the ecological significance of stressors

High-risk characteristics	Low-risk characteristics
Low density	High density
Unstable age or stage structure	Stable age or stage structure
High natural stochasticity	Low natural stochasticity
Low genetic diversity	High genetic diversity
High natural mortality	Low natural mortality
Isolated population	Connected to other subpopulations
Low dispersal ability	High dispersal ability

Timing of impact is particularly important for populations with wide fluctuations in density (or other characteristics). A stressor that occurs when the population is at or near a minimum may cause extinction or at least reduce the population far below its long-term average. A stressor that occurs when the population is at a maximum may simply reduce population size to a level that is within the range of its normal fluctuation. A frequent or chronic stressor may inhibit or prevent population recovery, which is more ecologically significant than a short-term stressor from which the population recovers. The significance of a particular stressor may be affected by other stressors that occur before, after, or simultaneously. These other stressors may influence the pre-impact population and/or the recovery trajectories, as well as the susceptibility of the population to perturbations. The spatial extent of stressors also influences the ecological significance and persistence of the impact (see "Extrapolating across spatial scales," p 72).

Specific characteristics of some populations may increase the risk that a stressor will cause an ecologically significant change (Table 2-4). These characteristics may influence the population's pre-impact trajectory, susceptibility to stress, and recovery potential. These characteristics may be influenced by the life-history traits of the species (e.g., *r*- or K-selected species) and by natural and human-induced stressors in the environment. Hence, changes in the environment, including other stressors, may make a population more susceptible to a significant ecological impact. It is important to note that a species may possess a mixture of high- and low-risk characteristics. If a population has a number of the high-risk characteristics, then the impact of a stressor is more likely to cause a perturbation with long-term ecological significance. A population with few high-risk characteristics is more likely to experience a perturbation of smaller magnitude or shorter duration, which would carry less ecological significance. The mechanisms by which these high-risk characteristics may lead to significant changes by stressors are discussed throughout this chapter.

Extrapolating Across Species, Spatial Scales, and Temporal Scales

A major source of uncertainty in ecological risk assessment is the extrapolation of results from studies on a limited number of species conducted over short spatial and temporal scales, often in the laboratory, to effects on natural populations. Here we consider these uncertainties and how they may be addressed.

Extrapolating across species

In the section titled "Using individual life-history traits to project population level effects" (p 58), we noted that whereas the relationships between changes in individual life-history traits and changes in population growth rate within a single species may be strong, the relationships across species are much weaker. This is not to say that we cannot derive some rules of thumb that allow us to identify which vital rates are likely to be more or less important for population growth rate, given knowledge about the state of the population and the life-cycle type. There is a considerable body of life-history theory that considers the sensitivity of r_m to specific life-history traits (e.g., Cole 1954; Stearns 1992; Caswell 1996) and that can be used to inform ecological risk assessment. The main conclusion from life-history theory is that sensitivity is situational (Stearns 1992), that is, r_m is sensitive to different life-history traits under different sets of conditions. It is therefore important to consider a species' life history when we extrapolate results from 1 species to another.

Recent discussions of the application of life-history theory to ecotoxicology are provided by Calow et al. (1997), Kammenga et al. (1997), and Van Straalen and Kammenga (1998). Calow et al. (1997), for example, developed a theoretical framework based on simplified life-cycle models to make general predictions about how stressor effects on individual vital rates translate into changes in population growth rate for different broad categories of life-cycle type. They considered 3 life-cycle types, semelparous, moderately iteroparous, and strongly iteroparous, and explored how impacts on λ would vary among life-cycle types given similar changes in individual vital rates. The analysis also considered variations in vital rates within each of the broad life-cycle categories (e.g., time to first reproduction less than or greater than time between broods). This kind of analysis can provide a useful starting point for extrapolating stressor effects across species on the basis of their life-cycle type and for developing simple rules of thumb that would help to identify key vital rates for different species and population starting points (e.g., growing, shrinking, or at steady state).

Extrapolating across spatial scales

Our discussions so far have considered populations and their environment to be spatially homogeneous. The spatially independent models used to project and/or interpret impacts to populations have been based on datasets collected from controlled laboratory experiments or relatively confined field studies. However, natural populations exist in a spatial context. More often than not, a mosaic of habitat qualities and stressor concentrations exists in the landscape and across the range of the population. Such spatial heterogeneity influences the vital rates of individuals belonging to local groups (or "subpopulations"), and it is the summation of local subpopulation dynamics that determines overall population (or "metapopulation") dynamics. Further, unexposed patches in the spatial mosaic, as well as other forms of spatial refugia, can act as sources for immigration into areas where subpopulations have been impacted by stressors, potentially ameliorating the effects of those stressors. For example, the number of breeding adult Caspian terns (*Sterna caspia*) is relatively stable at sites in the Great Lakes that are highly contaminated with PCBs. However, the percentage of young raised at these colonies that return there to breed is very low compared to the percentage in unexposed colonies. Site-specific recruitment is low in these contaminated colonies, and the population of breeding adults appears to be supported by the immigration of birds raised at less contaminated colonies (Mora et al. 1993). Conversely, impacted subpopulations can act as sinks within the larger metapopulation, depressing overall abundance to an extent greater than projected. How, then, can effects projections that have been derived from experimental manipulations, confined field studies, and modeling analyses be extrapolated to natural populations?

Recognition that spatial context exists within populations has led to the development of theory and modeling approaches that accommodate spatial heterogeneity and its effects on population dynamics. These approaches have been applied to questions of population viability, conservation biology, pest dispersal and invasion, and other population management issues. The use of models incorporating spatial context to address population-level effects of chemical stressors is relatively new, although examples do exist (e.g., Thomas et al. 1990; Sherratt and Jepson 1993; Jepson and Sherratt 1996). Discussions of modeling population dynamics in a spatial context are given by Okubo (1980), Hanski (1991), Hanski and Gilpin (1991), Dunning et al. (1995), and Tilman and Kareiva (1997).

Population models that incorporate a spatial component offer a means for extrapolating projected population-level effects to real-world situations. Two broad classes of modeling approaches have emerged: metapopulation models and spatially explicit models. Some of the specific model formulations developed to date are identified in Table 2-5.

Models in the first broad class, metapopulation models, consider the existence of subpopulations that communicate with each other through immigration and emigration. Conceptually, subpopulations are treated almost as individuals, with

Table 2-5 Population modeling approaches that incorporate spatial context

Model	Description	Comments	References
Levins' model (and variants)	Continuous time, discrete space	Breaks habitat into an infinite number of distinct sites; assumes global random dispersal of recruits	Levins (1969) Hanksi (1991, 1996, 1997) Lamberson et al. (1992)
Incidence function	Continuous time, discrete space	Breaks habitat into a finite number of distinct sites of differing size; assumes localized dispersal of recruits	Hanski (1994)
Reaction–diffusion	Continuous time, continuous space	Local populations grow ("reaction") and diffuse through space ("diffusion")	Reviewed in Okubo (1980)
Cellular automaton-like	Discrete time, discrete space	Divides habitat into contiguous sites ("cells"); assumes dispersal among adjacent sites	Schumaker (1998) Akcakaya and Atwood (1997)

local extinction rates and recolonization driving local dynamics. Such models need not incorporate spatial structure explicitly; rather, the environment is idealized as a series of more or less identical patches. The influence on metapopulation persistence and spatial distribution of 1) the size, spacing, and density of patches; 2) the rates and forms of movement among patches; and 3) the rates of extinction within patches are easily evaluated using metapopulation models. Pioneering work in the development of metapopulation models was conducted by Levins (1969).

The second broad class encompasses spatially explicit models, which increase ecological realism by incorporating landscape structure and habitat quality explicitly. These models differ from metapopulation models in that the spatial context of patches is important not only to migration among patches but also to the internal dynamics of subpopulations within patches. Internal subpopulation dynamics can be modeled by relating aspects of habitat quality to average vital rates, or even by allowing environmental parameters to affect individual physiological and behavioral responses using an IBM approach. Variation in habitat quality can be incorporated using spatially referenced datasets of land use patterns, landscape characteristics, and stressor distributions. For example, Akçakaya and Atwood (1997) developed a spatially explicit model of the threatened California gnatcatcher (*Polioptila californica californica*) dynamics to support protection of this subspecies and its habitat from land development in southern California, USA. The habitat requirements of *P. c. californica* were determined by relating habitat characteristics to patterns of habitat use. These relationships were used to develop habitat suitability functions, which in turn were used to score the quality of grid cells modeled in the landscape, based on geographic information system (GIS) data on land cover and topography. A 2-stage stochastic matrix model with density dependence was used to model gnatcatcher dynamics within patches. Population-wide vital rates in this model were estimated from existing datasets, independent of habitat suitability, and were allowed to vary randomly to reflect environmental and demographic stochasticity. Akçakaya and Atwood (1997) used their model to relate probabilities of population decline (quasi-extinction) to a variety of environmental and demographic parameters, with an intent of supporting evaluation of various resource management options. Recently, Schumaker (1998) developed a spatially explicit modeling platform that permits incorporation of GIS-referenced habitat data, specification of functional relationships between habitat quality and demographic response, and specification of rules to describe movement of individuals among patches.

When properly constructed and employed, metapopulation and spatially explicit models can be used to extrapolate the population-level effects projected from limited information to total population dynamics in nature. Pulliam (1994) describes incorporation and evaluation of chemical stressors and their effects in spatially explicit modeling approaches. Such approaches have large potential in assessing risk from environmental stressors to populations at a landscape level of spatial scale and in contributing to our understanding of the susceptibility of populations to these stressors.

Extrapolating across time

Population growth rate is subjected to environmental variation on both a spatial and a temporal scale. The influence of spatiality was outlined in the previous section, and this section is dedicated to the influence of temporal variation with specific emphasis on the estimation of population extinctions in contaminated habitats. The estimation of extinction probabilities or TE requires data on temporal variation or stochasticity of the various vital rates. These data parameterize diffusion approximations, which are commonly used to estimate population extinction probabilities. The application of diffusion models is reliable only when the population size can be estimated by a continuous function of time, that is, if the change of population size per time unit is small (Halley and Iwasa 1998). Under the diffusion approximation, 3 different types of stochasticity can be discerned (Lande 1993):

1) demographic stochasticity, which originates from random changes in individual vital rates in finite population size;
2) environmental stochasticity, which arises from continuous, small ambient changes that affect the vital rates of all individuals equally; and
3) catastrophes, which are relatively large perturbations that cause sudden changes in population size.

Demographic stochasticity is important only in small populations because random variations in vital rates level out in populations larger than 100 individuals (Lande and Orzack 1988). Environmental stochasticity and catastrophes are important for both large and small populations. Here we will focus on environmental stochasticity because it plays a major role in extinction dynamics in many natural populations. Recently Lande (1998b), an authority on stochastic extinction modeling, reviewed the way anthropogenic factors constitute the primary deterministic causes of species declines, endangerment, and extinction. He pointed at land development, overexploitation, species translocations and introductions, and pollution. These primary anthropogenic factors produce ecological and genetic effects that contribute to extinction risk. Ecological factors include environmental stochasticity, random catastrophes, and metapopulation dynamics (local extinction and colonization) that are intensified by habitat destruction and fragmentation. Genetic factors include hybridization with nonadapted gene pools and selective breeding and harvesting. Lande stressed that all factors affecting extinction risk are expressed, and can be evaluated, through their operation on population dynamics.

The remarks made by Lande (1998b) imply that, in order to conduct an ecologically sound risk assessment at the population level, we should take into account the environmental variation. At present, we are not aware of any ecotoxicological reports that include this aspect, despite the fact that there is a huge time series dataset available, waiting to be opened up and analyzed.

Compensation

The fact that some populations persist in stressed environments, even though adverse effects on individuals and the potential to cause population-level effects can be demonstrated, raises the possibility of compensatory mechanisms that ameliorate predicted adverse effects. Compensation for contaminant exposure can occur at different levels of biological organization.

Individual and biochemical levels

Some biomarker responses represent a type of compensatory response at the level of biochemistry. For example, metal exposure triggers the synthesis of metallothioneins, which can bind metals and prevent them from causing toxic effects. Toxic responses at the individual level will be manifested only when metal concentration exceeds the metallothionein binding capacity. Another phenomenon that has been observed in organisms exposed to low levels of stressors is a compensatory response resulting in U-shaped or inverse U-shaped dose response (Calabrese and Baldwin 1999). This response is referred to as "hormesis" and is thought to be the result of overcompensation by homeostatic regulatory mechanisms. Hormesis has been observed for a variety of endpoints in a variety of organisms responding to a variety of stressors (including toxicants, radiation, and others). However, the precise mechanism that results in hormesis is unclear, and the degree to which it can be considered a general biological phenomenon is controversial (Forbes 2000).

Compensation at the population level

When individual-level effects are manifested that are of sufficient magnitude and type as would be expected to result in population-level effects, such effects are not always seen because of other types of compensation. One type of compensation that has already been discussed is the development of tolerance (see "Measures of population structure," p 38), which confounds the ability to project population-level effects. The development of tolerance permits populations to persist in stressful habitats. However, populations that have developed genetic tolerance may be at a disadvantage when the toxicants are removed (e.g., through cleanup or natural dissipation) because they may have reduced genetic diversity compared to unaffected populations and may have lost the ability to adapt well to other environmental factors. The populations that have adapted to stressful habitats may also have developed altered metabolism to deal with the contaminants in their environment (Nacci et al. 1998, 1999). In contaminated environments, the energetic costs of this altered metabolism are outweighed by the benefit of reduced susceptibility. However, in uncontaminated environments, the costs may outweigh the benefits (Shaw 1999). In addition, adapted populations may have experienced demographic shifts such as smaller size and age structure, earlier reproduction and reduced life-span, or production of fewer, larger offspring (Weis and Weis 1989; Maltby 1991). It would appear likely that these demographic alterations would be more plastic and respond more rapidly to environmental improvements.

Compensation may also occur as a result of density-dependent factors. A classic example of the importance of density-dependent effects in ameliorating individual-level impairments is provided by eggshell thinning in birds in response to DDE exposure. In populations that are below their carrying capacity, eggshell thinning causes a decrease in population size and was responsible for the serious declines in populations of predatory birds (including pelicans, cormorants, peregrine falcons, ospreys, and bald eagles) during the 1950s, 1960s, and 1970s. However, for populations at the carrying capacity, no population effects were seen. In these populations, a number of individual birds would have died naturally of density-dependent effects. The additional mortalities from eggshell thinning reduced the density and hence reduced the number of birds dying of density-dependent effects. The overall effect was therefore no change in population size. The susceptibility of natural populations to the adverse effects of stressors therefore depends on the state of the population, which in turn may depend upon the presence of additional stressors. In Great Britain, for example, the use of DDE in the late 1940s and early 1950s caused significant eggshell thinning and reduced reproductive success in peregrine falcons. However, these populations were near their carrying capacity, and therefore the adverse effects on reproduction were not manifested at the population level. The use of cyclodiene insecticides during the late 1950s killed many adult peregrine falcons and reduced the population to below the carrying capacity (Walker et al. 1996). Under these circumstances, DDE-induced eggshell thinning did result in a reduction in the population size and slowed population recovery. Thus, a factor that is compensatory under some circumstances may impact population size significantly under other circumstances (i.e., additional stressors or altered environment).

The ecological significance of individual-level effects will therefore be more severe when density dependence is weak or absent. However, even if a population is at its carrying capacity and under strong density dependence, contaminant-induced reductions in reproductive success are a concern because if other stressors, natural or human-induced, reduce population size in the future, the reproductive effects of contaminants may slow or prevent population recovery. In the Great Lakes, for example, double-crested cormorants experienced severe persecution by humans until the 1950s. However, the small population did not begin to recover immediately after persecution diminished, despite the presence of an abundant food supply (alewives). DDE-induced eggshell thinning prevented population recovery until DDE was banned in the early 1970s. Great Lakes cormorant populations exploded during the late 1970s and 1980s after DDE contamination and associated eggshell thinning decreased (Weseloh, Ewins et al. 1995). Similarly, many marine invertebrates and fishes normally produce a great excess of offspring and normally only a very small fraction ever survive to be recruited into the population. For such organisms, a toxicant that affected reproduction might produce a very major decline in reproduction, without producing an impact at the population level. If a population is subjected to heavy fishing pressure, however, it may be more vulnerable to additional reductions caused by toxicants.

Grant (1998) explored the influence of 2 types of density dependence, acting either by reducing juvenile survivorship or reducing fertility, on dieldrin effects in *Eurytemora affinis*. An important conclusion was that substantial reductions in some life-history traits by toxicant exposure had little impact on the population if they were compensated for by reductions in the intensity of density dependence. Other recent examples of the role of density dependence in mitigating the effects of toxicants include Walthall and Stark (1997b) and Linke-Gamenick et al. (1999).

For a population to persist in an environment, population growth rate must increase at low density and decline at high density. Sibly (1994) considered the interaction between population density and population growth rate and speculated on the mechanisms that prevent populations from going extinct in contaminated environments. He discussed 2 possibilities, one in which the main effect of the contaminant was to increase mortality and the other in which it caused a reduction in energy for growth or reproduction (i.e., decreased SfG). In both cases, the populations were assumed to be food limited. In the first example, the increased mortality would result in decreased intraspecific competition for food. The surviving individuals would therefore increase their energy intake and consequently have increased growth and reproductive rates. This would result in an increased population growth rate, and hence there would be no overall change in population size. The possibility of such population compensation has long been recognized. Nicholson (1954) and Slobodkin and Richman (1956) found that losses even as great as 25% of the population may have no long-term impact on a population, while losses of 50% may result in only a slight change. In Sibly's second example, reduced consumption rates would also reduce intraspecific competition for food. Those individuals that were not affected or were least affected by the toxicant may therefore increase their feeding rate and hence their rates of growth and reproduction. However, if the feeding rate of all individuals is affected, the remaining individuals freed from competition and food limitation still will not thrive, since their ability to consume the surplus food is impaired. This type of compensatory response is also restricted to situations in which the density dependence acts on food rather than on other limited resources such as space.

The extent to which species can compensate for effects at the individual level, and the precise form of such compensatory mechanisms, requires further investigation if we are to fully understand the ecological significance of projected population-level effects.

Summary

- The focus of ecological risk assessment is on the protection of populations; however, most ecotoxicological studies heavily rely upon the response of a group of individuals being indicative of a population-level response. The use and application of population-level endpoints as indicative measures of stressors can undoubtedly assist in our understanding of the risks stressors pose to the structure and function of populations. The chapter authors evaluated a range of population-level endpoints and discussed how the uncertainties associated with extrapolating between an individual- and a population-level response can be reduced.
- A variety of measures of population structure and dynamics can be used to investigate stress effects (see "What population-level endpoints can be studied?," p 28); however, population growth rate is probably the most readily understandable of these measures. Population growth rate is an integrated measure of age-specific survival and age-specific fecundity and is of central ecological and evolutionary importance. It can be used as a descriptive function of a population or as a tool to project the population-level consequences of stressor effects. Under certain circumstances, it may also be used for short-term prediction.
- There are numerous published methodologies for measuring population growth rate, some of which have been used in branches of ecology for several decades. However, the application to ecotoxicology has been relatively limited. Demographic population models provide a mathematical framework that encompasses the life-cycle information of a given organism. They offer the opportunity to explore potential stressor effects and recovery of species across a wide range of environmental scenarios. Until now, the examination

of such ecological processes has generally been limited to field studies or environmental monitoring. These approaches are logistically difficult, produce vast amounts of data that are often equivocal, and are difficult to extrapolate to other scenarios.

- The rate-limiting step in the research and development of many demographic models, and the resulting choice of a suitable modeling framework, is most commonly the lack of life-history information on the organism in question. More often than not, the basic population ecology of the organism has not been reported in the open literature or is simply not known. Although age-specific fecundity and reproduction are sufficient measures in themselves to parameterize the simplest of strategic population models (see "Estimating population growth rate from life tables," p 42), and hence to derive an estimate of population growth rate, tactical models require much more detailed life-history information. The collation of such information on publicly accessible Internet databases has started to improve this situation; however, tactical model parameterization and application are often felled by our basic lack of ecological knowledge.
- The choice of modeling framework used to derive measures of population growth rate has been shown to be far outweighed by experimental artifacts, such as not running the study for long enough to capture all the required life-history information (see "Effects of model formulation on estimates of population growth rate," p 53). Indeed, the rudimentary matrix and complex PDE approaches produced remarkably similar estimates of population growth. The exercise did highlight 1 critical factor: the intrinsic importance of density dependence in model formulation. The comparison of population growth rate estimates from stressed and control populations in an unconstrained, density-independent environment is of course very valuable. The subsequent use of these estimates to make long-term estimates of population size without including the density-dependence process not surprisingly leads to very unrealistic results.
- The most readily accessible and comprehensible approaches to demographic modeling are deterministic matrix and stage-structured approaches (see "Demographic population models," p 46). Although the absence of stochasticity within such approaches is biologically naïve, the move to stochastic modeling takes them away from all but the very mathematically and statistically proficient. The balance between developing a model with sufficient ecological realism and the resulting complexity of the mathematical framework is critical. Indeed, endeavoring to increase the ecological realism of population models does not necessarily improve prediction accuracy. It is important that demographic models should not be used in isolation but as a complementary tools to those already available. They can be used very successfully, not only to increase our understanding of the underlying

ecological processes that laboratory or field-based experiments failed to address but also to sharpen, rather than directly answer, the question posed.

- Demographic population models require detailed information, in particular age-specific survival and age-specific fecundity schedules. However, it is possible to estimate population growth rate from partial life tables, and several standard toxicity tests can provide the necessary information with minimal extra effort and cost (see "Instantaneous population growth rate and partial life-table information," p 45). Indeed, only recently the Organization for Economic Cooperation and Development (OECD) has introduced population growth rate as a reportable endpoint in the *Daphnia magna* reproduction test (Guideline 211). Approaches based on full or partial life tables are limited to species that have short generation times and that are easy to culture and maintain in the laboratory (e.g., small invertebrates). In the absence of full or partial life-table information, it may be possible to use a single life-history trait (e.g., fecundity, survival, age at first reproduction) as a surrogate of population growth rate. However, whereas single life-history traits may be good predictors of population growth rate within species, this is may not be true across species. To date, no single life-history trait has been demonstrated to be the most sensitive across species and toxicants, and therefore focusing studies on a single trait may be misleading and may introduce uncertainty into the risk assessment (see "Using individual life-history traits to project population-level effects," p 58).
- It may not be possible to measure or calculate population-level endpoints for certain species, and therefore it may be necessary to use information about stressor-induced effects at the individual or suborganism level to provide insight into population-level consequences. Projecting from individual-level effects to population-level consequences is important in retrospective risk assessment where effects are investigated by measuring behavioral, physiological, or biochemical responses. It is also important in prospective risk assessment where full or partial life-cycle studies are not appropriate.
- Stressors may elicit a wide variety of individual or suborganism responses. For example, stressors may have profound effects on many aspects of behavior that affect an organism's ability to obtain food, avoid predation, and reproduce (see "Extrapolating from behavioral responses," p 62). Behavioral responses to stressors may provide early-warning indicators of potential population-level effects, but they are difficult to quantify, especially in natural populations. Moreover, the precise relationship between stressor-induced behavioral changes and population-level effects has rarely been clearly demonstrated.
- Bioenergetics provides a useful tool for linking individual-level effects and population-level consequences (see "Extrapolating from bioenergetic responses," p 63). Stressors may affect energy budgets by reducing energy acquisition (e.g., reducing feeding rates) and/or by increasing energy expen-

diture (e.g., increasing maintenance costs of defense and repair processes), and these effects can be measured by calculating an organism's SfG. SfG has been used successfully with a wide range of species and can be measured in situ. Short-term measures of SfG, or its components, can be linked mechanistically to long-term consequences for individual growth, reproduction, and survival using DEB models. By combining DEB models with demographic models, it is possible to use information about the effects of stressors on the energy budget component of individuals to predict population-level effects. Moreover, individual-based energy budget models can be incorporated into food web and carbon–nutrient cycling models and thereby provide a link between population-level effects and ecosystem function (see examples in Chapters 3 and 4).

- It has been argued that molecular and cellular responses (i.e., biomarkers) are among the most sensitive and earliest detectable responses to stressors. The underlying premise of biomarker tools is that effects at higher levels of organization (i.e., populations and communities) represent the net sum of effects on individuals, which results from alterations in cellular and molecular responses. Although the mechanistic link between biochemical responses and population-level effects has been made, this is often not the case. Many biomarker responses are part of an organism's normal metabolic processes, and in these cases, changes may not represent adverse effects at the individual or population level. Other biomarkers, however, function as valuable indicators of damage. Biomarkers are particularly useful as diagnostic tools and provide insight into the mode of action of stressors. However, the mechanistic links between biomarker responses and population-level effects are often unclear (see "Extrapolating from molecular and cellular responses," p 66).
- In prospective risk assessment, projections of the population-level consequences of stressors usually are derived from short-term, laboratory or semi-field studies on closed populations. Conversely, in retrospective risk assessment, measurements are made on natural populations at a single location over a limited time period. However, natural populations are rarely closed, and they exhibit considerable temporal and spatial variability (see Chapters 5 and 6). Evaluating the ecological significance of projected or measured changes in population-level endpoints in dynamic systems is a major challenge. The significance of a stressor-induced change in population-level endpoints will depend upon 1) the pre-impact trajectory of the population and 2) the magnitude, timing, frequency, and spatial extent of the impact and the presence of other stressors, natural or anthropogenic. The probability that a stressor will cause an ecologically significant change will also depend upon the characteristics of the species. In general, stressors are more likely to have an ecologically significant impact on species with low dispersal ability in isolated, low-density populations with high natural mortality, high natural stochasticity, low genetic diversity, and unstable age

or stage structure (see "Evaluating effects in the context of natural variability," p 69).

- Another source of uncertainty in ecological risk assessments is extrapolating results from studies of a limited number of species over short temporal and spatial scales to effects on natural population. Recent developments that may reduce the uncertainties include the application of life-history theory to extrapolating across species and the use of spatially explicit models and metapopulation models to incorporate landscape structure and habitat quality explicitly into population-level projections (see "Extrapolating across species, spatial scales, and temporal scales," p 71).
- Even though adverse effects on individuals and the potential to cause population-level effects can be demonstrated, populations may still persist in stressed environments. This is because compensatory mechanisms may ameliorate the predicted adverse effects (see "Compensation," p 75). These compensatory mechanisms may operate at different levels of biological organization and may, for example, occur as a result of density-dependent factors or the evolution of tolerance. Understanding these compensatory mechanisms is fundamental to understanding the ecological significance of projected population-level effects.
- The models and approaches outlined in this chapter have very real potential to address many of the issues in ecotoxicology and environmental risk assessment. The focus on population-level endpoints will enhance our ability to project and interpret the effects of stressors on natural populations and hence to assess and manage risk more effectively. Although this chapter has emphasized methods for projecting population growth rate from measurements of effects on individuals, population-level endpoints other than population growth rate may be important in evaluating the impact and mechanisms of effect of certain stressors. Measures such as recovery time and PVA also have potential uses in risk assessment. The concept of population recovery time, like growth rate, has obvious attractions in being readily comprehensible. The advent of new Web-based technologies will undoubtedly assist in the communication and accessibility of such population endpoints.

CHAPTER 3

The Food Web Approach in the Environmental Management of Toxic Substances

Donald J. Baird, Theo C.M. Brock, Peter C. de Ruiter, Alistair B.A. Boxall, Joseph M. Culp, Peter Eldridge, Udo Hommen, Robbert G. Jak, Karen A. Kidd, Ted DeWitt

Public perceptions of the value of nature are based on 2 key concepts: 1) the need to safeguard biodiversity and 2) the desire to preserve life-supporting functions within natural systems (e.g., maintenance of safe drinking water, clean air), both of which depend on community- and ecosystem-level phenomena. Thus, it is surprising to note that regulatory structures and approaches to protecting nature from the effects of toxic substances are still largely based on the outcome of single-species laboratory tests, which cannot fully represent the risks prevalent in real ecosystems. Critics of the extrapolation from laboratory single-species toxicity tests to an ecosystem effects approach state that toxicity tests do not consider bioaccumulation of contaminants and ignore both temporal changes and multiple stressor effects. There is a growing awareness among policymakers and scientists that assessment studies should adopt a systems approach. The U.S. Environmental Protection Agency (USEPA 1992) has proposed and defined endpoints for the evaluation of contaminant effects at the population, the community, and the ecosystem levels, and more recently, the Health Council of the Netherlands (HCN 1997) has recommended using food webs as a starting point when formulating assessment programs.

What is the Food Web Approach?

The arguments presented in this chapter state that management tools for ecological phenomena at population, community, and ecosystem levels should not be developed in isolation because the behaviors of populations, communities, and ecosystems under chemical stress are inextricably interrelated. Furthermore, it is argued that the adoption of a food web approach, that is, gaining an understanding of the food web at the beginning of the community and ecosystem management process, is

Ecological Variability: Separating Natural from Anthropogenic Causes of Ecosystem Impairment.
D.J. Baird and G.A. Burton, Jr., editors.
ISBN 1-880611-43-0

the only approach that considers risks to long-term ecosystem function based on an understanding of the trophic structures and mechanisms behind ecosystem processes.

Should Indirect Effects Be Ignored in Ecosystem Assessment?

Food webs link population dynamics to community dynamics. This linkage makes it possible to establish the causal effects of stressors on species by simultaneously analyzing direct and indirect trophic interactions within a community of organisms. Direct effects occur, not surprisingly, through the direct action of a stressing agent on a component of the ecosystem. In the case of toxic substances, this can occur through increased mortality because of direct uptake of a substance from the ambient medium or through the transfer and accumulation of contaminants, especially within species at the top of food chains, such as predatory birds and mammals (e.g., Klok et al. 2000). In contrast, indirect effects appear as an indirect consequence of direct effects and involve alterations in the dynamics of interacting species (DeAngelis 1996), even when the organisms affected are not directly exposed to the stressing agent. Many studies have revealed that indirect effects are important; in some circumstances, they may be more pronounced than direct effects (e.g., Paine 1980, 1992; Yodzis 1988; Wootton 1994; Abrams et al. 1996). Indirect effects may also become apparent through changed ecosystem attributes such as nutrient cycling and oxygen metabolism (Kersting 1994; Vanni 1996). Indirect effects thus demonstrate that species that are themselves insensitive to a particular stressor (as measured, for example, in a toxicity test) might be seriously affected or even become extinct indirectly as a result of its actions. Conversely, the absence of a particular species in an ecosystem is not necessarily the direct result of exposure to a particular stressing agent.

Assessing Stressor Effects on Ecosystem Structure and Function

The food web approach integrates multispecies data, providing the opportunity to assess the stability of communities through the analysis of networks of trophic interactions. This makes it possible to describe the consequences of chemical stress on biological diversity within communities. For this purpose, food web models can serve as powerful tools for the analysis of community stability, in terms of response to perturbations. Both experimental and theoretical investigations of the effects of perturbations on community structure have demonstrated that interfering with particular trophic interactions can alter community stability (Paine 1980, 1992; Yodzis 1988; Wootton 1994; De Ruiter et al. 1995). These observations have given rise to the now familiar keystone species concept and the importance of safeguarding such species in conservation programs (Paine 1980). Furthermore, a consequent reduction in community stability can ultimately cause extinction of species that were not affected by the initial perturbation. In the context of risk assessment, this implies that even small effects of a toxic substance on a key species can have disproportionate effects on community stability, which, in the end, might lead to the extinction of many species or even, in extreme situations, to monocultures (Paine 1992). Food webs link communities to ecosystems through an understanding of mechanism, making it possible to translate changes in population and community attributes directly to effects on ecosystem processes. For example, the cycling of energy and nutrients is the direct result of food web interactions in which many species play a role. For many ecosystems, species that contribute little biomass still may have a large influence on nutrient cycling and energy flow and thus affect the functioning of other species. Examples include bacterial grazers, which stimulate microbial activity through nutrient recycling, and algal grazers, which stimulate the productivity of submerged macrophytes by providing better light conditions through the grazing of periphytic algae. Clearly, the extinction of such species could have a disproportionately large influence on ecosystem function.

Hence, food web approaches can be used to analyze the effects of stressors on key target species, on the biological diversity in communities, and on the functioning of ecosystems. In this way, food webs are the "wiring behind the circuit board of the ecosystem," spanning different levels of ecological organization.

Chapter Aims

This chapter outlines the benefits of adopting a food web approach to the management of multiple stressors in ecosystems by describing a number of real and hypothetical case studies in which the food web approach is applied. These cases demonstrate how the food web approach can integrate multispecies data in an ecologically realistic way, yielding a new understanding of effects that can prove critical for the practice of sound environmental management. The examples focus primarily, but not exclusively, on the adverse effects of toxic substances on food webs, yet the conclusions drawn have relevance for the study and assessment of other stressing agents.

The text is divided into 4 major sections:

1) Functional classification—How we obtain the building blocks of food web models through the process of functional classification into trophic categories, and how this approach could be extended to encompass life-cycle patterns, offering the possibility of future use as a diagnostic tool in monitoring and assessment programs.
2) Trophic transfer—The most widely adopted application of the food web approach: the description of chemical fate by trophic transfer and illustration of the application of the food web approach in the protection of populations. In this section, the application of new techniques such as stable isotope analysis in trophic transfer studies is discussed, together with the limitations of existing approaches in bioaccumulation modeling and the difficult area of linking fate and effects in ecotoxicological studies.

3) Assessment of single and multiple stress effects—The variety of techniques available for the study and prediction of the effects of contaminants within food webs. Examples from field and mesocosm studies in marine, freshwater, and terrestrial environments, and a consideration of mathematical modeling approaches illustrate how the food web approach can be employed in the protection of communities and ecosystems.
4) Application to environmental management—The problems and prospects for the food web approach, including a plea for its inclusion at the very beginning of the risk assessment process, rather than its adoption at a later stage, as is the currently prevailing view.

Spinning Food Webs using Functional Classification

In order to construct a food web, it is necessary to know who eats whom. Classifying species by trophic category within food webs is an established method, yet not without its pitfalls, and some examples of this are given below. After discussing these approaches, we offer suggestions concerning how some new ecological concepts can be incorporated to improve the application of this important food web tool.

Functional Classification in Terrestrial and Aquatic Ecosystems

Yodzis (1996) has argued that food webs must be taxonomically well resolved so that subunit compartments are composed of consistent aggregations of species. Functional classification of food webs thus requires clustering of species, and this objective can be accomplished by a number of approaches. Traditional approaches to classification of community interactions for food web analysis have focused on trophic level groupings (Lindeman 1942) and functional feeding groups (Moore et al. 1988; Hawkins and MacMahon 1989). This classification of species into units with similar feeding habits is, in part, a utilitarian approach to reduce the often bewildering complexity in food web structure. Ecologists also argue that functional feeding groups or "guilds" (sensu Root 1967) are natural ecological units that are important arenas for interspecific competition (Hawkins and MacMahon 1989).

In risk assessment, the application of the functional feeding group concept has been useful for constructing food web models. Furthermore, this concept has been useful in food web models that simulate energy and nutrient cycling in ecosystems (Hunt et al. 1987; De Ruiter et al. 1994). Food web models in risk assessment (see "Extending the scope of functional classification," p 89) have been used to simulate exchanges of contaminants between the abiotic and biotic environments (Traas et al. 1998), to simulate the transfer of contaminants from prey to predator (Klok et al. 2000), and for the analysis of food web interactions in terms of energy flow and stability properties (De Ruiter et al. 1995).

Functional classification of food webs for risk assessment could also consider other species characteristics, including how habitat preferences and life-history characteristics affect exposure scenarios. For example, the exposure of springtails to Cd in soils was found to depend upon vertical separation in the soil horizons, with individuals at the surface receiving a much higher contaminant dose than those below (Faber 1991). The following discussion presents the advantages and limitations of functional classification in food web analysis and the current methods for establishing functional feeding groups. We also discuss the possibility of improving functional classification for ecological risk assessments of food webs through the inclusion of a suite of ecological characteristics, including feeding relationships, the effects of habitat and life cycle as they affect exposure to contaminants, and the relative strengths of interactions, or "connectance," among web members.

When ecological risk assessments incorporate food web approaches, feeding groups should characterize specific functions in the food web, for example, a subgroup of herbivores directly affecting a subgroup of primary producers such as periphytic algae. This definition emphasizes how components (e.g., herbivores) in the food web participate in energy transfer processes (e.g., consumption of primary producers) rather than simply describing the arrangement of species into trophic groups in the web. In addition, ecosystem models often use the functional group as the basic unit to link species interactions with energy flow (Hunt et al. 1987) or to simulate the processes of energy flow and nutrient cycling (De Ruiter et al. 1994). This is because an organism's importance is not necessarily related to its abundance (i.e., density or biomass) or rate of energy use, but rather to its interaction strength based upon per capita feeding energetics and population size (De Ruiter et al. 1994). Finally, the need to aggregate food webs into meaningful functional categories for simulation models also reduces food web complexity, thereby simplifying the analysis (Pimm 1982; Paine 1988; Yodzis 1996).

A key limitation in the functional classification of feeding groups is that ecologists seldom use precise criteria to delineate them. Hawkins and MacMahon (1989) note that most ecologists determine functional feeding group membership using either previously defined, resource-based feeding groups or a posteriori descriptions of feeding guilds determined through multivariate clustering of food resource usage among species. Often it is assumed that close taxonomic relationships can give insight into functional categorization. This is argued from the hypothesis that morphologically similar feeding mechanisms can in some way constrain the trophic habits of congeneric species within food webs (Cousins 1996). However, polyphagy and ontogenetic shifts in the diet of species can confound attempts to restrict species within functional feeding categories (e.g., Hawkins and MacMahon 1989). Thus, we cannot ignore the fact that lessons from natural history often blur our ability to generalize about food web relationships; essentially, the unique aspects of individual species and their dynamics may limit our ability to generalize (Paine 1988; Bengtsson and Martinez 1996). For food web approaches, then, it is important that functional feeding groups include all species in the community that use the same food resource at a specified time and location. Note, however, that Yodzis (1996) warns that after an entire food web has been aggregated in a consistent and objective procedure, we must resist the temptation to focus attention only on those species interactions that are of particular interest.

Functional positioning of species can vary between ecosystems because of differences in trophic interactions (e.g., competition, predation) with other species. This type of shift in realized niche (Hutchinson 1957) can affect the uptake of contaminants. For example, Kaag et al. (1997) demonstrated that the uptake of polychlorinated biphenyls (PCBs) varied because of a shift in feeding mode induced by competitive interaction with another species. In this example, as a result of competition from the filter-feeding mussel *Mytilus edulis*, the bivalve *Macoma balthica* altered its preferred habit of filter feeding and shifted to deposit feeding. Because sediments at the study site contained higher levels of PCBs than seston, PCB levels in *Macoma* increased as a consequence of this competition-induced shift in feeding strategy.

Commonly used functional feeding group classifications include those for freshwater invertebrates (Cummins 1973), marine benthos (Maurer et al. 1999), and soil arthropods (Moore and Hunt 1988). Cummins and co-workers (but see also Hawkins and MacMahon 1989) recognized 6 functional feeding groups based on food type (nature and particle size) and mode of feeding. It is interesting to note that this classification includes both food resources and location of the food in the habitat. In the terrestrial environment, Moore and Hunt (1988) identified 8 functional feeding groups through a combination of principal food source, mode of feeding, reproductive rate, defenses against predation, and distribution within the soil profile, based on information from laboratory observations, gut content analyses, and experimentation. Gomes and Haedrich (1992), in their study of food webs on the Grand Bank of Newfoundland, combined functionality (feeding relationships) with abundance of "regulars" (species present all year) to develop species guilds. They aggregated lower trophic-level species to the same taxonomic label (trophic species) while including compartments for individual abundant species at higher trophic levels. This trophocentric aggregation procedure focused on the dynamics of commercially important species at the expense of the supporting food web resource species.

Stable Isotope Analysis As a Tool In Trophic Categorization

Many techniques are available to aid the determination of feeding relationships, including direct behavioral observation, gut content analysis, biochemical methods such as fatty acid identification (Smith et al. 1996), and the analysis of stable isotopes in body tissues. In systems where direct observation is difficult or where dietary habits vary over seasons or the life cycle, stable isotopes of C and N can be particularly useful for defining the source of food resources (using C isotopes) and the trophic position (N isotope ratios) of organisms (see "Trophic transfer within food webs," p 91). Biotic ratios of the stable isotopes of carbon (^{13}C, ^{12}C) and nitrogen (^{15}N, ^{14}N) are being used by ecologists to determine the fate of C and the trophic positioning of organisms in food webs (Peterson and Fry 1987). Isotope ratios are generally conserved as C is transferred up the food web from prey to predator (Peterson and Fry 1987), with only limited fractionation observed. Thus, it is possible to assess the relative importance of different C sources to upper trophic level organisms, for example, aquatic versus terrestrial sources (Hobson and Sealey 1991) or benthic versus pelagic sources in lakes (Hecky and Hesslein 1995). In contrast, the heavier isotope of nitrogen (^{15}N) becomes more strongly enriched from prey to predator and can thus be used as a measure of trophic positioning. Such stable isotope techniques can thus provide a time-integrated measure of the dietary habits of the organisms over months to years in slow-growing animals, in contrast to more traditional methods of gut content analysis (Hesslein et al. 1993). Primary producers and consumers may vary considerably in their isotopic composition within and across seasons (Kidd et al. 1999) because of the dynamics of their physical and chemical environment, such as changes in nutrient and C availability, and high turnover rates at lower trophic levels (Gu et al. 1996). For this reason, it may be necessary to adequately characterize organic and inorganic C and N sources in the system of interest and evaluate temporal changes in both the prey and predators. The additional information provided by this technique may prove to be an important addition to food web assessments, since it would enable interpretation of feeding interrelationships and thus of the indirect food web effects of contaminants.

Although the utility of stable isotopes for analyzing the effects of contaminants on food webs has not been thoroughly investigated, these methods are useful for interpreting the bioaccumulation of contaminants through food webs (see "Use of stable isotopes in trophic transfer studies," p 95). Studies have also used the unique isotopic composition of contaminants (including nutrient amendments) to trace their uptake and incorporation into the food web (e.g., Wassenaar and Culp [1996] used stable isotope methods to infer the trophic importance of complex pulp mill effluents into a riverine food web).

Extending the Scope of Functional Classification

When we assess the effects of stressors, it can be argued that the fundamental classification scheme should include a suite of attributes for a species. Use of resources along principal niche axes of food, habitat, and time (sensu Schoener 1974) clearly are important criteria. For example, organisms that feed on similar food resources in similar habitats during the same time are more likely to interact among themselves than are species that do not share food resources in space and time (Moore and De Ruiter 1997a; Maurer et al. 1999). Thus, the suite of attributes would include identified trophic connections (i.e., the functional feeding group), microhabitat preference, life-cycle characteristics, and dynamic behavior of the species within the food web. The aim of this type of functional grouping would be to provide subsets of species with similar sensitivities to stressors (HCN 1997; Traas et al. 1995 for a contaminant example), and an example of how this might be achieved is given in Table 3-1.

Table 3-1 Example of how traditional approach to functional classification of species within food webs could be extended by including additional ecological characteristics

Functional feeding group	Microhabitat preferences	Life-cycle characteristics	Contribution to food web dynamics
Dietary preferences	Location within predefined microhabitat	Number of life stages (e.g., instars)	Energy channel membership
Feeding mechanism	Resident or transient occupancy of habitat (e.g., aquatic insect larva with terrestrial adult phase)	Number of reproductive bouts	Interaction strength with other species groups
		Population dynamics subject to density dependence	
		Life cycle duration	
		Body size (order of magnitude length or biomass scale)	

Because trophic connections are of overriding importance to food web approaches, functional feeding groups must remain the primary basis for classification within the food web.

Habitat use provides a secondary attribute in functional classification schemes for retrospective risk assessment. Its value lies in the fact that habitat preferences determine toxic exposure scenarios, in terms of rate of uptake of a substance and its exposure route. In addition, Yodzis (1996) suggests that subgroups of food webs suffering long-term perturbations could be based on population dynamics characteristics, for example, for groups of species that have minor effects on others or whose dynamics are much slower (or faster) than others'. Moore and De Ruiter (1997b) also suggest that association of a species with a particular energy channel pathway within a food web may affect its recovery from perturbation. For example, Moore and Hunt (1988) found that energy channels based on bacteria (i.e., bacteria → nematodes, protozoa → omnivores; nematodes → predacious collembola → predatory mites) recovered more rapidly after perturbation than did fungal based energy channels. Finally, life-cycle characteristics should be included in the functional classification because of the importance of this biological information for risk assessment procedures (Suter 1993b). Many aspects of a species life cycle will affect its sensitivity to stressors, including life stage, body size, and presence of invulnerable stages. Clearly, our proposal to develop a suite of characteristics for functional classification of species will require further research and assessment to determine its value for food web approaches.

Conclusions about Functional Classification

- Classification of food webs into meaningful functional categories reduces food web complexity and improves our ability to produce food web models that can be used to make quantitative predictions.
- Improved functional classification of food webs for ecological risk assessments should be based upon a suite of ecological characteristics, including feeding relationships, habitat use, life cycle, and dynamics of food web interactions, to provide subsets of species with similar sensitivities to stressors.
- Stable isotope techniques provide promising alternative approaches to traditional techniques for determining food web relationships and functional feeding groups (see also "Use of stable isotopes in trophic transfer studies," p 95).

Trophic Transfer within Food Webs

Organisms are exposed to contaminants in their environment through water, air, soil, sediments, and food. The importance of these exposure routes will vary, depending on the habitat preference and feeding strategies of the organisms and on the distribution of the contaminant in the environment. Trophic transfer, the direct uptake of contaminants from the tissues of prey organism into predators, is an important route of exposure for many aquatic and terrestrial species and has been examined and modeled extensively in laboratory and field studies (Thomann and Connolly 1984; Munger and Hare 1997).

The following section describes the factors that affect the exposure and uptake of contaminants into organisms, the subsequent movement of contaminants through food webs, and some new approaches to understanding the fate of pollutants in terrestrial and aquatic ecosystems.

Bioaccumulation of Contaminants in Food Webs

Previous studies on the fate of persistent organochlorines in aquatic and terrestrial food webs have found tissue concentrations of certain substances such as PCBs to be 200-fold higher in top predators than in primary consumers within the same food web (e.g., Oliver and Niimi 1988), a phenomenon termed "biomagnification." In aquatic systems, however, primary producers and many organisms inhabiting lower trophic levels are believed to take up contaminants largely from water, sediments, and air (e.g., Swackhamer and Skoglund 1993; McLachlan 1996; Berglund 1999). However, the trophic transfer of contaminants is generally accepted as a more important source of exposure for animals at higher trophic levels in aquatic systems (Thomann and Connolly 1984; Oliver and Niimi 1988). For example, in the case of persistent organochlorines, 99% of the body burden of a predatory fish can be attributed to dietary uptake (Thomann and Connolly 1984). Further evidence for this can be gained from studies that show the variability of contaminant concentrations in predator species is closely related to variable contaminant concentrations in their prey (Madenjian et al. 1994). It is through this trophic transfer process that the concentrations of persistent contaminants can be magnified within food webs (see Table 3-2; Weseloh, Hamr et al. 1995) and can often approach toxic levels for many piscivorous species, resulting in reproductive failures and population declines, as demonstrated in the case of dichlorodiphenyltrichloroethane (DDT) accumulation in peregrine falcons and ducks (Kolaja 1977). Concentrations of PCBs in the Lake Ontario food web increase several fold, from zooplankton to the top predators lake trout, on both a wet and lipid weight basis (Table 3-2; Kiriluk et al. 1995). The biomagnification of contaminants has also been observed in terrestrial food webs. Table 3-2 describes the example of an agricultural food chain in Germany (grass → cow → humans) in which concentrations of a PCB were 140 times higher in the top predator, humans, than in the lowest trophic level. Similarly, Larsson et al. (1990) observed approximately 10-fold higher concentrations of PCBs and DDT in predatory shrews than in herbivorous voles from similar locations. Organochlorines in wildlife are ultimately linked to environmental concentrations (Larsson et al. 1990), and reductions in their usage have resulted in lower contaminant burdens in fish (Borgmann and Whittle 1991) and fish-eating birds (Hebert et al. 1997) in the Laurentian Great Lakes. An understanding of the factors affecting the movement of contaminants through food webs is important for assessing and, through remedial action, reducing the exposure of top consumers to pollutants in food, thus addressing public concern about risks to human health and wildlife.

Table 3-2 Concentrations of PCB within organism compartments of Lake Ontario food web and of hexachlorinated PCB in terrestrial food web in southern Germany[a,b]

	Aquatic food web[c]		Terrestrial food web[d]	
Trophic level	Organism	ΣPCB (μg/g)	Organism	PCB 153 (pg/g)
Primary producers	—	—	Grass	44
Primary consumers	Zooplankton	0.09	Cow's milk[d]	207
Secondary consumers	*Mysis*[c]	0.22	Human milk[d]	6125
	Alewife	0.52	—	—
Tertiary consumers	Lake trout	4.18	—	—

[a] Source: Kiriluk et al. 1995.
[b] Source: McLachlan 1996.
[c] An invertebrate zooplankton predator.
[d] Lipid weight concentrations converted to wet weight assuming 3.5% lipid.

A range of factors influence the magnitude and rates of trophic transfer of contaminants, including the physical and chemical properties of the contaminants and their fate in the environment (surface adsorption), the biochemical composition of prey and predators, the gut physiology, the metabolism of the chemical, the environmental properties, and the structure of the food web. The possible influence of a range of these factors on biomagnification within a food web is summarized in Table 3-3. Further research is required to determine the significance of different routes of uptake for different classes of chemicals. Such research would enable the suggested classification system to be validated and further refined.

Measuring Trophic Transfer of Contaminants

A range of approaches is available for assessing the accumulation of a contaminant within a food web, including

- comparison of results with those of previous studies on contaminants with similar properties,
- performance of experimental studies to assess uptake through feeding,
- performance of stable isotope studies, and
- use of trophic transfer models.

The uptake of a contaminant within a food web is likely to be highly dependent on contaminant physical–chemical properties. For neutral organics, the main factor affecting bioaccumulation will be the octanol–water partition coefficient, K_{ow}. By using concentration data derived from previous experiments on food web components for a given substance with similar chemical properties, it should be possible to obtain a semiquantitative estimate of body burdens within the web for a substance for which no data are available. However, it should be stressed that this simple approach does not consider the effects of other factors such as metabolism, making extrapolation to other systems more difficult and illustrating the need for more research effort in this important area.

Dietary Uptake of Contaminants in Aquatic Species

For compounds for which ingestion is a quantitatively significant uptake route, biomagnification within a food web is likely. Therefore, by assessing uptake through ingestion, the likelihood of a contaminant becoming biomagnified through a food chain can be assessed. The importance of food in the accumulation of contaminants in biota has been assessed using laboratory studies. One approach has been to expose an organism to a contaminant in its diet and subsequently to contrast body burdens in those animals to others that have been exposed to a contaminant in both food and water. In Munger and Hare's study (1997), a 3-link planktonic food chain was used and included algae, a grazing zooplankton species, and its predator *Chaoborus*. Cd was introduced into the algae at the base of this experimental food chain, and each subsequent step in the food chain was exposed to Cd in its food and in the surrounding water or only in its food. Results from this study demonstrated

Table 3-3 Key factors affecting the uptake of contaminants by organisms within food webs

Factor	General rules	Risk of bioconcentration or biomagnification	Explanation	Examples	Reference
Physical–chemical properties	Log K_{ow} <3	Low	Compounds are hydrophilic.	Phenanthrene	EC 1996
	Log K_{ow} 3 to 5	High	Hydrophobic compounds		EC 1996
Contaminant type	Log K_{ow} 5 to 7	High	Hydrophobic compounds + feeding may be a significant route for contaminant uptake.	Benzo(*a*)-pyrene	
	Log K_{ow} > 7 Molecular size > 0.95 nm	Low	Effects of solubility and molecular size may restrict transport across biological membranes. Effects of molecular size may vary, so this is not a strict cut off.		Opperhuizen et al. 1995
	Ionic organics	Low	Compounds likely to be dissociated at environmental pH values and therefore not available for uptake	Penta-chlorophenol	Howe et al. 1994.
		High (depending on pK_a)	Feeding may be important uptake route within organisms with acidic gut		
	Metals	High	Uptake by feeding is important		Munger and Hare 1997
Food chain length	Short food chain	Low	—	Hg, PCBs	Rasmussen et al. 1990 Bentzen et al. 1996
	Long food chain	High	—	Hg, PCBs	
Metabolism	Compound metabolized in organism to polar excretable metabolites	Low	Polar metabolites formed that are excreted		Carey et al. 1998
	Compound not metabolized or metabolized to more hydrophobic unmetabolizable compound	High	Compound not metabolized and not excreted or metabolized to a more hydrophobic and non-excretable compound		Carey et al. 1998
Organism lipid content	High lipid content of prey or predator	High	High lipophilicity of certain contaminants	Organo-chlorines	Bentzen et al. 1996

that food, not water, was the important route of Cd exposure for the aquatic invertebrate predator *Chaoborus*.

Another approach to quantifying the uptake of chemicals from food and examining the factors that affect trophic transfer is to conduct feeding studies that expose an organism to known quantities of a contaminant in its food. The tissue burdens of the chemical are examined through both the exposure and depuration phases of 1 species at the same life stage, and the results are related to the physical–chemical properties of the contaminant (e.g., Fisk et al. 1998).

Laboratory studies such as those described above have unequivocally demonstrated the importance of food in the uptake and exposure of organisms to contaminants.

Use of Stable Isotopes in Trophic Transfer Studies

An alternative method to assessing the movement of contaminants through food webs uses naturally occurring stable isotopes of C and N to characterize trophic interrelationships (see Chapter 2, "Extrapolating from individual-level effects to population-level consequences," p 56). Stable isotope ratios of C and N in food web organisms are currently being used to examine the source of C to biota and their relative trophic position, respectively. These techniques have been used in studies of marine and freshwater systems, and they facilitate comparisons across food webs that differ considerably in community composition (Kidd 1998). Continuous measures of trophic position are better predictors of contaminant concentrations in upper trophic-level organisms (Van der Zande and Rasmussen 1996), and the application of stable isotope techniques in such studies is invaluable in improving our understanding of contaminant movement through food webs.

For example, concentrations of organochlorines and Hg in freshwater and marine biota have been shown to be significantly related to trophic position, as determined by stable N isotope ratios (Broman et al. 1992; Kidd et al. 1995; Atwell et al. 1998; Kucklick and Baker 1998). This technique has been used across systems that differ considerably in productivity, species composition, and physical–chemical environment. By using a continuous, nondiscrete measure of trophic positioning, it is possible to determine biomagnification rates of persistent contaminants through a food web (Kidd et al. 1998). Figure 3-1 demonstrates that the accumulation of organochlorines through food webs (as shown by the slope of the organochlorine–N isotope regression) is related to their lipophilicity. The highest slopes of these relationships are observed for the most lipophilic organochlorines after the effects of lipid have been removed (Kidd et al. 1998). These relationships can be compared and contrasted across systems to determine whether the movement of toxicants varies through food webs. For example, Kidd et al. (1995) demonstrated that the food web accumulation of the pesticide toxaphene was not significantly different in 3 subarctic lakes. Use of stable isotope analyses in contaminant studies will facilitate comparisons between systems that may differ in communities and will improve our understanding of the processes underlying contaminant biomagnification.

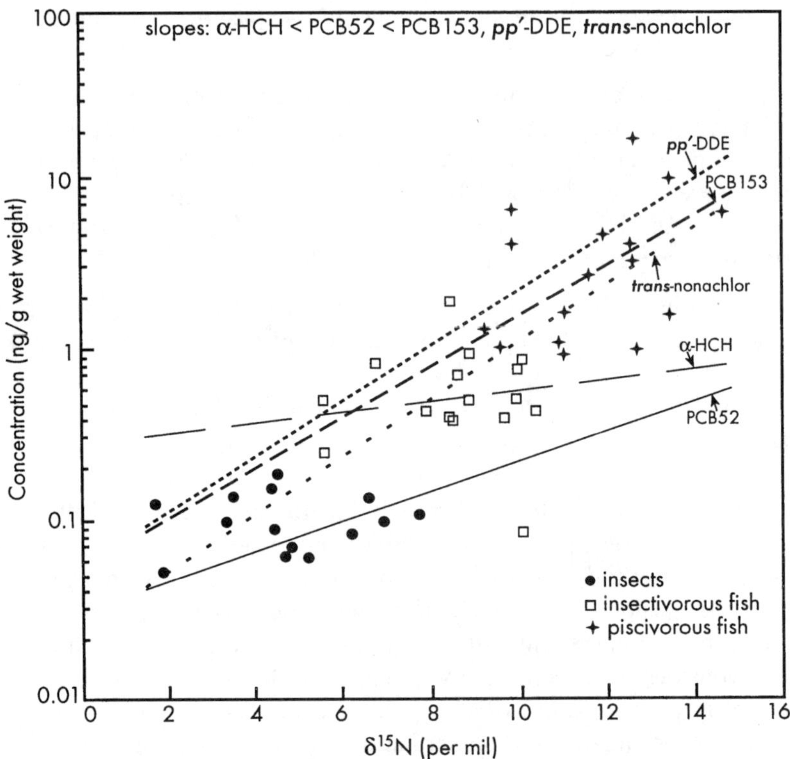

Figure 3-1 Relationships between lipid-adjusted concentrations of *p, p′*-DDE (log K_{ow} = 7.0), PCB153 (log K_{ow} = 6.9), PCB52 (log K_{ow} = 6.1), *trans*-nonachlor (log K_{ow} = 5.9), and α-HCH (log K_{ow} = 3.8) and trophic position ($\delta^{15}N$) in the food web of Peter Lake in the Canadian Arctic. (See Kidd et al. 1998b for further details.)

Modeling Chemical Fate in Food Webs

A number of biotic fate models have been developed to assess the fate of contaminants within a food web. These models generally include a trophic structure and a description of the chemical fate of a contaminant among the biotic components (e.g., Thomann et al. 1992; Gobas 1993; Mackay, DiGuardo, Paterson, Cowan 1996; Mackay, DiGuardo, Paterson, Kicsi, Cowan 1996). However, such models are greatly simplified descriptions of the ecosystem of concern, and they generally include a description of functional feeding groups, that is, detritivores, herbivores, and carnivores. The models have successfully been applied to the problem of describing PCBs and other hydrophobic organic compounds in large-scale ecosystems (e.g., Thomann and Connolly 1984; Connolly and Tonelli 1985; Campfens and Mackay 1997). Two currently available biotic fate models are summarized below:

1) The fate model developed by Mackay, DiGuardo, Paterson, and Cowan (1996) and Mackay, DiGuardo, Paterson, Kicsi, and Cowan (1996) is a mass balance

model describing contaminant flux through an aquatic food web. The model can deal with a range of organism types and requires the following inputs: compound mass, Henry's Law Constant, octanol–water partition coefficient, and concentration in water and sediment. Properties of the environment, that is, suspended solids concentrations, volume fraction of suspended solids, and sediment organic C (SOC) content can be varied within the model. Organism properties, namely volume, lipid volume, metabolic half-life, feeding rate, fractional respiration rate, gut absorption efficiency, and diet, can also be varied. Outputs of the model include fluxes of contaminant from prey to consumer, steady-state concentrations, and estimates of fugacity (i.e., escaping tendency).

2) The food chain bioaccumulation model, which is based on the equations of Gobas (1993), is a steady-state model. The model predicts bioconcentration, bioaccumulation, and biomagnification factors as well as contaminant concentrations in phytoplankton, zooplankton, benthic invertebrates, and fish, using a defined food chain. The model considers intake of a contaminant through the gills and the diet, along with elimination through the gills and feces, growth, and metabolism. A Monte Carlo simulation can be used to perform an uncertainty analysis. The approach has been linked to traditional environmental fate models to predict distributions at steady state or time-dependent distributions that follow concentrations through multiple age classes and generations.

Generally, currently available biotic fate models are limited because they are not linked to the dynamics of the food web itself. For example, they are unable to address the effects of changes in forage composition and trophic status that may occur in response to nutrients or exotic species invasion.

Models for the terrestrial environment have been developed to a lesser extent. In order to predict distribution in terrestrial organisms, multimedia approaches are required that consider the air, soil, and plant compartments as well as the transfer to animals and humans.

Linking Fate and Effects

In this section, we discuss the linking of exposure and effects and the approaches to predicting ambient exposure concentrations and body burdens of contaminants for populations within functional groups. Linking of exposure and effects has always proved elusive in ecotoxicological investigations, but some developing approaches offer hope in this area, and they are discussed below.

A wide range of models is available to predict the fate of contaminants in the environment. Although some of these models contain biotic compartments (see "Dietary uptake of contaminants in aquatic species," p 93), the rest are concerned solely with contaminant distribution in abiotic compartments such as air, water, and soil. They predict concentrations in each compartment, either at a steady state or at a specific point in time, and more recently have been used to produce probability distributions of concentrations (e.g., Mackay 1994). Environmental fate models can initially be used as screening tools to identify which food webs are most likely to be exposed. For example, if a contaminant accumulates in sediment, then a benthic food web should be considered.

When the food web at risk has been identified, it is then necessary to link the fate and food web models in order to predict contaminant effects on organisms within food webs. In order for a compound in a particular abiotic environmental compartment to enter a food web, it must be bioavailable. Bioavailability will be affected by a range of factors, including sorption to organic C in the water phase, dissociation of the chemical, biotransformation within food web organisms (including microbial degradation), diffusion of the substances from ingested material, and transfer of the chemical from 1 abiotic environmental compartment to another. Some common relationships for predicting accumulation of contaminants by organism and compartment type are summarized in Table 3-4.

Table 3-4 Equations for predicting contaminant uptake from different environmental compartments

System	Equation	References
Water →biota	$BCF = FL \times K_{ow}$	Hawker and Connell 1989 Gobas et al. 1991 Veith et al. 1979
Sediment →biota	$BSAF = C_s \times DOC / SOC$	DiToro et al. 1991
Air →plant	$BCFV = 0.2 + 0.7K_{wa} + 0.05K_{oa}$	Paterson et al. 1991
Soil or water →plant	$K_{pw} = (W_p + FL \times K_{ow}) \times pp / pw$	Briggs et al. 1982, 1983

BCF = bioconcentration factor; BCFV = plant–air volumetric BCF; BSAF = biota–sediment accumulation factor
C_s = concentration in sediment
DOC = dissolved organic C concentration
FL = lipid content in organism
K_{oa} = octanol–air partition coefficient; K_{ow} = octanol–water partition coefficient
K_{pw} = plant–water partition coefficient; K_{wa} = dimensionless Henry' s Law constant
SOC = sediment organic C concentration
pp = density of plant tissue; pw = density of water

Using concentration inputs from the fate models, we can predict body burdens in the components of a food web. This prediction can be compared directly with toxicant–dose information to obtain an indication of effect. However, dose–response relationships are probably available for only a few contaminants and even fewer species.

By using information on bioconcentration factors (BCFs) for an organism (obtained either experimentally or by using the equations described in Table 3-4) together with information from laboratory experiments, estimates of the lethal body burden (LBB) for specific toxic substances can be obtained, where

$$LBB = BCF \times LC50 \quad \text{(Equation 3-1).}$$

Predicted LBBs can then be compared with body burdens predicted from food web models to obtain an indication of potential effects.

This approach could also be applied to mixtures using so-called "toxic equivalence factors" or toxic units (TUs; e.g., Konneman 1981; Ahlborg et al. 1992), although these methods do not consider synergistic and antagonistic interactions of toxicants.

To date, several models have been used to predict the combined fate and effects of nutrients and contaminants in food webs. These include Integrated Fate and Effects Model (IFEM; Bartell et al. 1992), AQUATOX (Park et al. 1995), and Contaminants in Aquatic and Terrestrial ecoSystems (CATS 5; Janse and Traas 1996). However, there are still a number of limitations with these approaches, which are discussed below.

Generally, the models for predicting uptake from soil and sediment assume that uptake occurs through the pore water; such approaches do not consider uptake from ingestion of soil, sediment, and suspended sediment, which may lead to an underestimation of exposure. However, ingestion is an important route of uptake for certain contaminants, and measuring uptake between different trophic levels in a food web requires information on the assimilation efficiencies of each organism (ideally this information should be measured experimentally). In addition, the models do not explicitly account for the effects of either contaminant physical–chemical properties or environmental properties (pH, soil type, etc.) on uptake, leading to either over- or underestimation of exposure. A better understanding of factors that affect the uptake of contaminants from ingested material is therefore required.

Although the body burden approach may work well for predicting effects of certain classes of toxicant (e.g., narcotics), many studies indicate that body burdens may not always be clearly related to effects. Contaminants can exert effects through a number of mechanisms. Most contaminants need to cross biological membranes and enter the tissue of an organism; for these compounds, the effect will be dependent on dose. However, certain contaminants, such as polymeric substances or irritants, can exert an effect without the need to cross a biological membrane (e.g., Cd may exert an effect within an animal's gut [Taylor et al. 1998]). In such cases, body burden as predicted from current trophic transfer models would not be predictive of effects.

External factors may also modify the effect of a contaminant, making the use of LBBs problematic. For example, the toxicity of polycyclic aromatic hydrocarbons (PAHs) has been shown to increase in the presence of sunlight (Bowling et al. 1983;

Allred and Giesy 1985; Ren et al. 1994). For these compounds, it is likely that toxicity to benthic organisms can be predicted using LBBs; however, the effects of PAHs on water-column species that are exposed to sunlight will be much more pronounced and not easily predictable. It is therefore important to consider the mode of action of the compound being investigated when we link biotic fate models to effects. Better design of laboratory ecotoxicity studies may assist in identifying modes of toxic action and contaminants that are able to exert an effect externally.

For many species in a food web, it is unlikely that ecotoxicity data will be available for the chemical being investigated. In these cases, information on relative species sensitivity coupled with information on toxic modes of action may be useful in obtaining a qualitative indication of effect. For certain species, information on quantitative structure–activity relationships may be available, and it can be used to predict toxicity on the basis of a compound's physical–chemical properties (e.g., Turner et al. 1987; Organization for Economic Cooperation and Development [OECD] 1995).

Biotransformation may result in the breakdown of a contaminant into either less toxic or more toxic species. Biotransformation is considered in the currently available biotic fate models. While information may be available on the metabolism of a contaminant in a few species, it is unlikely that information will be available for all of the species in a particular food web. If a compound is readily metabolized to a less toxic compound, an overestimation of body burden, and hence of effects, will be obtained. Also, it may be possible to use information on detoxification mechanisms in individual species to obtain qualitative information.

Finally, it should again be emphasized that although there are a range of promising developments in predicting toxic substance fate and effects in food webs, these models and approaches urgently require careful verification under natural or seminatural conditions, (i.e., mesocosm studies; see "Assessing single and multiple stress effects within food webs," p 101). In addition, the development of new field-based bioassays to assess in situ bioavailability of substances to a range of organisms (see approaches outlined in Chapter 1, "Obtaining site information," p 15) could be coupled with more traditional mesocosm studies to improve our understanding of causal links between the fate and effects of toxic substances within food webs.

Trophic Transfer Within Food Webs: Research Gaps

- Considerable efforts have been made to study and quantify the movement of contaminants through aquatic food webs, yet there has been much less research on terrestrial food webs. There is an urgent need to redress this balance.
- In aquatic ecosystems in general, and temperate freshwater ecosystems in particular, there is now a good understanding of the factors and processes that influence the uptake of nonpolar organic compounds, allowing accurate

estimation of bioconcentration processes. The focus of research now needs to shift toward understanding similar processes for polar organics and inorganic substances such as heavy metals.

- There is a need to develop food web model scenarios that can be used by regulators to assess biomagnification. This can be greatly aided by new techniques being developed in the field of stable isotope analysis.

Assessing Single and Multiple Stress Effects within Food Webs

In this section, we present examples of how the food web approach can be used to investigate single or combined effects of stressors at the community or ecosystem level. These examples include studies of food webs through observation, experimentation, and modeling. In observational science, temporal and spatial recordings of ecosystem structure and function are monitored in a comparative manner to gain insight into the normal operating range and possible disturbed states of the ecosystem under study. The field observations may be used to generate hypotheses on food web interactions. In a subsequent phase, these hypotheses can be tested experimentally. Both observation and experimentation at the community and ecosystem levels are considered important to further develop and verify food web models.

Whole Ecosystem Experiments

Insight into the impact of stressors on food webs can be obtained by observing polluted ecosystems and comparing them with suitable reference sites. In practice, however, it is often difficult to ascribe the differences observed to a specific stressor because usually several stress factors interact in an uncontrolled way in the polluted ecosystem. For this reason, experimental approaches are adopted by stressing the ecosystem in a controlled way and by measuring its response. Whole-lake manipulation has been used to assess the ecological impacts of a selected stressor (Schindler 1998), for example, nutrient enrichments (Schindler 1977), acidification (Rudd et al. 1988), and contamination with heavy metals (Malley et al. 1996). Similar experiments have been performed in terrestrial agroecosystems (e.g., Hendrix et al. 1986; Brussaard et al. 1990). These experiments can be used to examine impacts of stressors on real agroecosystems without ignoring their temporal and spatial complexity and allowing for natural processes such as migration of organisms (including larger grazers and predators) and input and output of matter, energy, and nutrients. A disadvantage of whole-ecosystem studies is that often only 1 lake or field plot is manipulated. This lack of replication reduces our understanding of natural variability in response. In addition, experiments at this scale can be costly, and it may be difficult to find suitable reference locations. Furthermore, due to the complexity of the possible feedback mechanisms involved and the inevitable pars pro toto approach in systems ecological research, it can be difficult to unravel the cause–effect chains within the overall responses observed. Despite these problems, ecosystem-scale experiments have provided invaluable information on the direct and indirect effects of stressors on communities and ecosystems.

Model Ecosystem Experiments

Model ecosystems, which depending on their size also may be termed "microcosms" or "mesocosms," are bounded systems that are constructed with source material from natural ecosystems or by enclosing parts of real ecosystems. Model ecosystems are usually characterized by reduced size and complexity, when compared to whole real-world ecosystems. Nevertheless, they normally include an assemblage of organisms representing several trophic levels, in accordance with the previously established carrying capacity of the test system (Giddings 1980; Gearing 1989). A major advantage of model ecosystem studies is that they are controlled and replicated so that statistical evaluation of treatment-related effects is facilitated (Hurlbert 1984). In addition, they can be more or less standardized, which makes it easier to compare and interpret studies with similar or different stressors over time. Numerous examples of such studies are available for freshwater and marine experimental ecosystems (e.g., Giddings 1981; Gearing 1989; Brock and Budde 1994; Culp et al. 1996, 2000; Van den Brink et al. 1996; Belanger 1997; Jak 1997). However, such studies in terrestrial experimental ecosystems are rare (e.g., Morgan and Knacker 1994; Korthals et al. 1996).

The food web approach can greatly improve interpretation of micro- and mesocosm studies. To date in the ecotoxicological literature, numerous experiments in aquatic microcosms and mesocosms address the impact of contaminants on ecosystem structure and functioning. These studies are often aimed at deriving critical threshold levels for the most sensitive ecological endpoints for regulatory purposes. However, such detailed information on biological endpoints is rarely used to analyze food web interactions such as changes in pool sizes and/or process rates of functional groups. At best, suggestions are provided of possible causal links between exposure and observed indirect effects. However, it is clear that the results of experiments in model ecosystems and whole ecosystems can provide a wealth of information on possible interactions between populations of species within a community (including different types of indirect effects) and on functional redundancy within communities. To illustrate this, in the first case study we have applied a food web approach, reanalyzing data from studies on nutrient and insecticide manipulations in freshwater microcosms.

Case Study I: Combined Nutrient Addition and Insecticide Application in Freshwater Microcosms

The following example illustrates how food web analysis based on functional groups can help to identify indirect and combined effects of multiple stressors on an experimental freshwater ecosystem. Data are used from 2 experiments, designed to study the impact of nutrient and/or insecticide treatments on the ecology of macrophyte-dominated freshwater microcosms. In the first experiment, the impact of a single application of the insecticide chlorpyrifos was studied (Brock, Crum et al. 1992; Brock, Van den Bogaert 1992; Brock et al. 1993). The second experiment focused on the impact of additions of inorganic N and P on ecosystem structure and function, and on the combined effects of nutrient and chlorpyrifos treatments (Brock et al. 1995; Cuppen et al. 1995; Van Donk et al. 1995). Each of the 12 test systems used in each experiment contained a 10-cm sediment layer and had a water depth of 50 cm and a total volume of ~ 1 m^2. The systems were dominated by the aquatic vascular plant *Elodea nuttallii* and contained several populations of algae and invertebrates characteristic of drainage ditches in the Netherlands.

For the purpose of the food web analysis, the species studied in the microcosms were classified into 15 functional groups (see legend Figure 3-2A). The functional groups were ordered into the main trophic groups of the food web and interconnected by arrows showing the relationships between resource and consumer. Furthermore, the pool size of each functional group (expressed as g dry biomass) was quantitatively represented in the diagrams by box size, adopting 5 biomass (box) classes on a logarithmic scale.

The structure of the food web in the control systems (not treated with nutrients or insecticide) is presented in Figure 3-2A. The left part of the diagram represents the detrital food chain. The grazer food chain, based on periphytic and planktonic algae, is presented in the right part of the diagram.

The response of the functional groups to extra input of inorganic N and P is presented in Figure 3-2B. The addition of nutrients caused a significant increase in macrophytes (functional group 1) and periphytic and planktonic algae (functional groups 4 and 5). The increase in macrophytes caused a small increase in herbivorous shredders (group 6), most probably through the detritus-microorganism complex (group 15). This relatively small increase is explained by the fact that, at the end of the experiment, the majority of the macrophyte biomass was not yet in a senescent state. The increase in edible periphytic and planktonic algae (groups 4 and 5), however, resulted in a pronounced increase in grazers of algae, particularly herbivores of group 12 and to a lesser extent those of groups 9 and 10. In turn, the pronounced increase in herbivores of group 12 resulted in an increase of predators of these herbivores (group 14). In conclusion, in these experimental aquatic ecosystems, all trophic levels seem to be regulated by resource limitation, and the addition of extra nutrients resulted in pronounced bottom–up effects through food web interactions.

The response of the community in the microcosms to the addition of a single dose of the insecticide chlorpyrifos is presented in Figure 3-2C. In line with the results of single-species toxicity tests, the addition of chlorpyrifos caused a reduction of populations of crustaceans and insects within the functional groups comprising detritivores (groups 6 and 7), herbivores (groups 10 and 12), and carnivores (group 14). However, the functional groups of detritivores (group 8), herbivores (group 11),

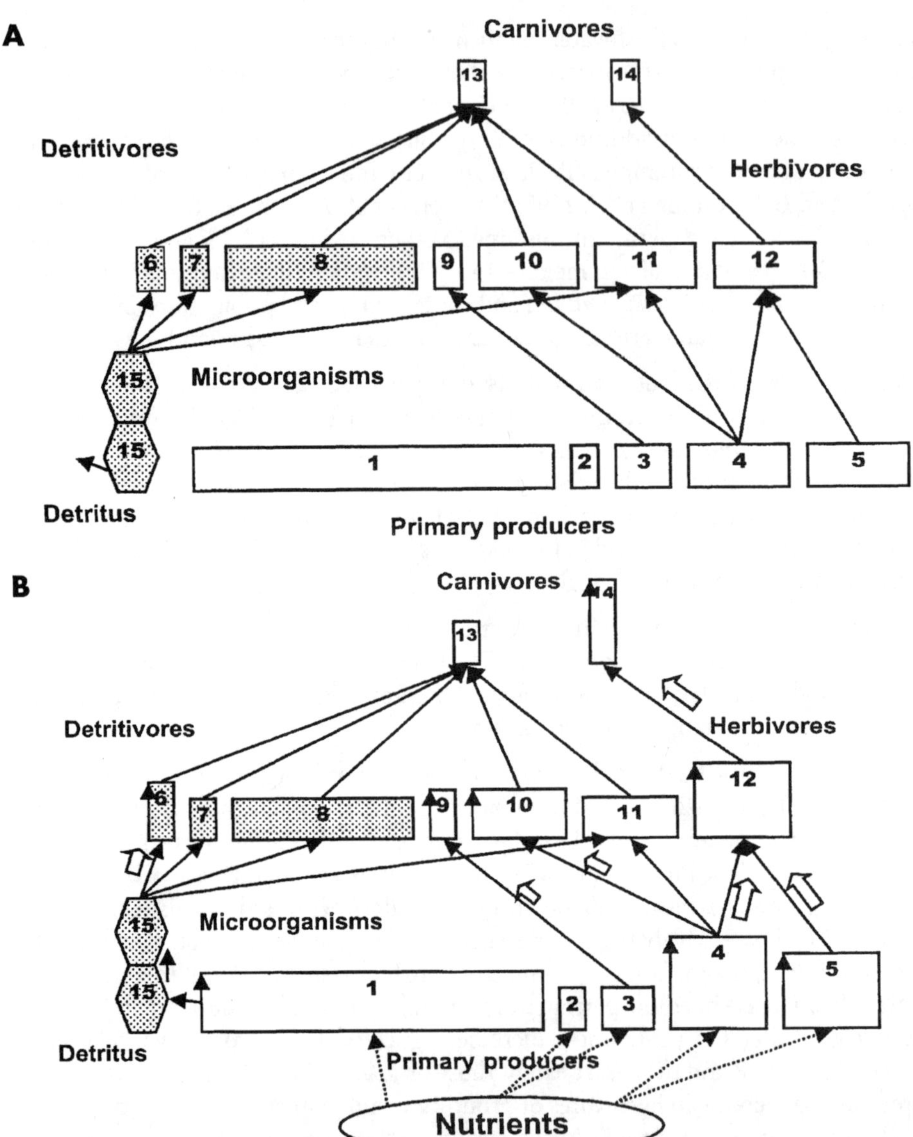

Figure 3-2 Trophic relations between functional groups (indicated by arrows) within the food web of freshwater microcosms that A) served as control test system, B) received extra doses of inorganic N and P each week, C) were treated with a single dose of the insecticide chlorpyrifos or D) were treated with a combination of insecticide and weekly input of extra nutrients. Pool size of each functional group is represented by box size (5 size classes in logarithmic scale). Classification of functional groups is as follows: [1] macrophytes (vascular plants and filamentous algae; biomass in control microcosms ~ 100 g dry weight [dw]/m^2); [2] less edible periphyton (blue green algae; biomass pool in controls ~ 0.1 g dw/m^2); [3] less edible phytoplankton (*Volvox*, and blue greens; biomass in controls ~ 1 g dw/m^2); [4] edible periphyton (mainly diatoms and green algae; biomass in controls ~ 3 g dw/ m^2); [5] edible phytoplankton (mainly cryptophytes, green algae, diatoms; biomass in controls ~ 2 g dw/m^2); [6] detriti-herbivorous shredders (Isopoda and some snails; biomass in controls ~ 0.4 g dw/

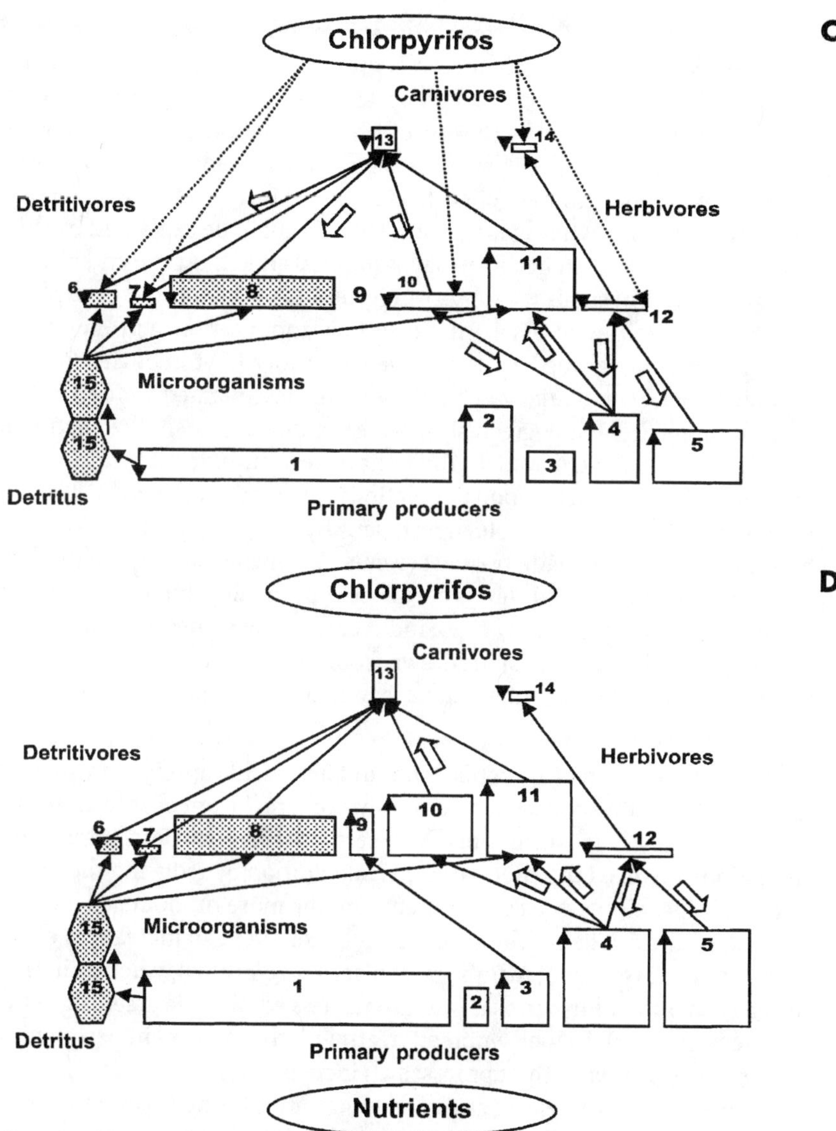

m^2; [7] detriti-omnivorous shredders (*Gammarus pulex*; biomass in controls ~ 0.4 g dw/m^2); [8] deposit feeders (sediment dwelling, detritus-associated Oligochaeta; biomass in controls ~ 5 g dw/m^2); [9] herbivores of less edible phytoplankton (rotifer *Hertwigia volvocicola* that is able to infest colony forming alga *Volvox*; biomass in controls ~ 0.3 mg dw/m^2); [10] herbivores of edible periphyton (several snails, oligochaet *Stylaria lacustris*, and insect *Cloeon dipterum*; biomass in controls ~ 2 g dw/m^2); [11] grazers of edible periphyton and macro-detritus (snails *Lymnaea* and *Bithynia*; biomass in controls ~ 3 g dw/m^2); [12] herbivores of edible phytoplankton and periphyton (Cladocera, most Copepoda and Rotifera; biomass in controls ~ 3 g dw/m^2); [13] predators of groups 8, 9, 10, 11, 12 (Turbellaria and Hirudinea; biomass in controls ~ 0.5 g dw/m^2); [14] predators of functional group 6 (insects *Plea*, *Chaoborus*; biomass in controls ~ 0.5 g dw/m^2); [15] microorganisms (not studied). **Note**: Where a number is shown without an accompanying box, this indicates complete extinction of the functional group.

and carnivores (group 13) not containing arthropod populations were also significantly affected, as were several functional groups of primary producers (groups 1, 2, 4, and 5). The decline in functional group 13 (carnivores comprising flatworms and leeches) could be explained by the loss of prey populations in the form of insects and crustaceans. The food web analysis suggests that predation by flatworms and leeches on functional group 8 (oligochaete worms) increased. The increase in edible planktonic and periphytic algae (groups 4 and 5) could be explained by the reduced grazing of the crustacean herbivores in groups 10 and 12. At the end of the experiment, an increase in snails was observed (group 11), which could be explained by the increase in periphytic algae. After insecticide application, a pronounced bloom of blue-green algae (group 2) was observed in the periphyton of the dominant macrophyte *Elodea nuttallii*, which cannot easily be explained by food web interactions. Although these blue-green algae were not directly eaten, their settlement in the periphyton of treated systems might have been stimulated by a decrease in the browsing activities of arthropod populations (physical interaction). By shading the macrophytes, the increase in blue-green periphytic algae caused a decline in macrophyte biomass in insecticide-treated systems. In conclusion, the results show that various groups at all trophic levels are affected by grazing and predation and that the addition of an insecticide may cause indirect top–down effects. However, a further important conclusion that can be drawn from this microcosm experiment is that factors other than trophic interactions may also play an important role in structuring the community.

Effects of the addition of both nutrients and insecticide application are presented in Figure 3-2D. Most responses can be readily explained from observations in systems treated with only nutrients or only the insecticide. Nevertheless, insecticide contamination enhanced the effects of nutrient addition by reducing the top–down control of algal biomass, as can be seen from the more pronounced increase of groups 3, 4, and 5. The combined treatment resulted in an increase in edible periphytic algae (group 4; mainly green algae and diatoms), whereas in the systems treated with only chlorpyrifos it was also the less edible blue green algae that increased (group 2). In the combined treatment, the increase in periphytic diatoms and green algae apparently suppressed an increase of blue-greens. Diatoms and green algae seem also to serve as a good food source for herbivores of functional groups 10 (particularly *Stylaria lacustris*) and 11 (particularly *Lymnaea stagnalis*). Hence, food web analysis can yield insight into the various cause–effect relationships, including direct and indirect effects, in both single and combined effects of stressors.

Classifying Food Web Models

In modeling the ecotoxicological effects of contaminants in food webs, a distinction can be made between food web accumulation models and food web interaction models. Food web accumulation models, or secondary poisoning models, have been developed to assess the transfer of toxic substances within the food web to specific target species of concern (e.g., invertebrates → fish → otter; earthworms → mole). While approaches to studying the bioaccumulation of contaminants in food webs have been described in the previous section, this section focuses on food web interaction models that describe and predict both the direct and indirect effects of the toxicant or stressor through population dynamical effects. These models may or may not incorporate bioaccumulation. Food web interaction models without bioaccumulation ignore the exposure of organisms to contaminants through the diet; the only concentration–effect relations that are used are those concerned with exposure through the ambient environment. This latter approach may be valid for nonpersistent toxicants that do not bioaccumulate, such as many modern pesticides.

The HCN (1997) distinguished between 3 types of food web interaction models: descriptive, analytical, and simulation (see also Paine 1980; Crowder 1988; DeAngelis 1992).

1) Descriptive (connectivity) food web models are based on the observations of diets of species or groups of species in the web. Such models provide qualitative information on the trophic relations between various populations within a community, but they cannot normally provide quantitative estimates of the direct and indirect effects of stressors (e.g., Lavigne 1995). The linking of such models with population modeling approaches (see Chapter 2) is an area that is worthy of further consideration.
2) Analytical (food web) models are used to gain insight into a reduced set of major ecosystem processes and interactions. A limited number of variables are introduced to enable comparisons to be worked out analytically or numerically. Analytical models are based on the assumption that, within ecosystems, it is possible to distinguish major regulating mechanisms and that interactions can be delineated. The major achievement of these models is that they contribute to the development of ecological theory and an understanding of mechanisms rather than having direct application to ecosystem management or ecological risk assessment (e.g., DeAngelis 1992; Scheffer 1998).
3) Simulation (complex food web) models are used if the degree of detail leads to complex model formulations because of a relatively large number of processes and interactions that are considered important (e.g., interactions between various functional groups). Chapter 4 provides an example of a simulation model of a soil food web. Other examples are the food web models of lake ecosystems, developed within the context of the International Biological Program to understand eutrophication processes (e.g., Park et al. 1974; Janse et al. 1995).

Of these 3 model types, simulation models offer the greatest promise in analyzing and predicting effects of stressors. According to their basic assumptions, these models can be further divided into steady-state models and dynamic models. The fundamental assumption of steady-state models is that C and nutrient inputs balance the system outputs. Hence, compartments in the steady-state model can consist of aggregates of species within functional groups (see "Spinning food webs using functional classification," p 86) or an individual population of a single species (see Chapter 2). In addition to the pools of living populations, there is a requirement in these models for detrital and dissolved organic material compartments because they form the nexus from which many of the biological compartments receive or deliver substrate. Steady-state models consist of a series of linear equations that describe the network of flows between all compartments and ecosystem processes (e.g., respiration, excretion). Input data include assimilation efficiency, production efficiency, and respiration per unit biomass, obtained directly from experiments or from the literature. The steady-state model can be spatially explicit but not time dependent. The results of the analysis are usually represented with flow diagrams. Examples of steady-state models include the Oregon marine mud-flat ecosystem and the Lovinkhoeve experimental farm case studies presented below.

In contrast with the steady-state approach, a fundamental assumption of dynamic food-web models is that C and nutrient flows that enter and leave compartments or that become incorporated within compartments need not balance. Hence, compartment sizes can vary. Dynamic models simulate how food webs respond to environmental factors (e.g., stressors), and density-dependent feedback structures that limit flows between compartments are taken into account in most cases. The feedback structures are of 2 forms: those that respond to the density of a resource and those that respond to the density of the compartment itself. The Monod, or Michaelis-Menten and Droop relationships (Droop 1973, 1974; Droop et al. 1982) are the best known of the resource feedback mechanisms, while Wiegert functions are generally used to simulate the effect of crowding (Wiegert and Wetzel 1979). Similar to the steady-state model, physiological processes and chemical interactions can be incorporated into dynamic models. Such models can be solved as a set of ordinary differential equations when the model is simply time dependent or with a system of partial-differential equations (PDEs) when the model is both time dependent and spatially explicit. The results of such dynamic models are normally represented in terms of concentration or biomass as a function of time. Some examples of dynamic food web models to assess ecological risks of stressors are provided by O'Neill et al. (1982), Hommen (1998; but see also the example below), and Traas et al. (1998).

A special type of dynamic simulation model is the hybrid model, characterized by a greater level of detail describing the population dynamics of the species that is central to the web. This species may be modeled with a stage, an age, or even an individual structured model. An example is provided by Swartzman and Rose (1984), who analyzed the effects of a toxicant in an aquatic microcosm by describing the response of the dominating *Daphnia* population with a stage-structured model and the response of the other populations by differential equations. Thus, the food web approach can also explicitly incorporate population dynamics into its toolkit (see Chapter 2 for a more comprehensive coverage of population modeling approaches).

Calibrating and Verifying Food Web Models

Model calibration requires that model outputs generally fit observations of a single mesocosm experiment or ecosystem study. Model verification has the additional requirement that the outputs are predictive of new conditions when new input data are applied. For this reason, the combination of experimentation in model ecosystems and construction of food web models is considered a promising approach when carried out in an iterative manner (e.g., Hommen 1998; Traas et al. 1998). Improving the validation status of food web models is especially important when environmental managers require a modeling tool that will predict ecosystem responses to multiple stressors. Optimally, the food web model developed for this purpose should predict the state of the stressed ecosystem and detail the ecological or biogeochemical processes responsible for alterations in the system. Quantitative prediction of response and determination of causality are equally important for the management of an ecosystem. Response prediction helps to determine when a potential stressor requires remedial action, while an understanding of causation provides information on those remedial actions likely to prove effective in returning the ecosystem to its prior state.

Case Study II: Oregon Mud Flat Ecosystem

The Oregon mud flat case study is an example of a steady-state food web model that illustrates the potential ecological risks of insecticide application in an estuarine environment.

The mud flats of the Yaquina River, Oregon, USA have an extensive intertidal zone due to a 2-m tidal variation that encompasses up to two-thirds of the estuary area at mean low-low tide. The river has dissolved inorganic nitrogen (DIN) concentrations of 10 to 20 mmolar N but relatively low phytoplankton concentrations. This probably is due to the rapid flushing of the estuary by the semidiurnal tidal cycle. The 2 major riverine sources of N in this system are the town of Toledo, located in the tidal freshwater reach of the river, and a fisheries cannery close to the coast in Newport, Oregon.

The mud flat community consists of dense colonies of *Zostera marina* in the lower intertidal to subtidal zones. The mid-intertidal areas are occupied by colonies of the mud shrimp *Upogebia* sp. and the upper intertidal by the sand shrimp *Neotrypaea* sp. There appears to be very little overlap of the seagrass and shrimp habitats: Seagrass transplants into the mud shrimp habitats are quickly covered by unconsolidated sediment, thus excluding *Zostera* from shrimp habitats. During much of the summer and autumn, green algae (*Ulva* and *Entermorpha)* invade all levels of the intertidal zone, either as drift algae or attached algae. Drift algae that land on the shrimp habitat quickly disappear from the sediment surface. Feeding studies suggest that although neither shrimp species feeds on these algae, they rapidly incorporate algal material into the sediments through bioturbation. Often whole sheets of *Ulva* are found in good condition more than 15 cm below the sediment surface in both shrimp habitats. The mud flat has a thin layer of benthic microalgae over the consolidated sediments, while no benthic algae are evident over the unconsolidated sediments. However, sediment chlorophyll measurements suggest that microalgae growing in the shrimp habitat are quickly incorporated into the sediment and may provide an important resource for the infaunal deposit feeders.

In the initial food web model, many more functional groups were represented, including seagrass epiphytes, crangonid shrimp, Dungeness crab, and several fish species. However, when biomass data, isotope constraints, and flux data were added to the analysis, the actual N flows involving these groups were trivial relative to the remaining groups. Therefore, all the fish, crabs, and pelagic shrimp species were aggregated into a single trophic species "fish," while seagrass epiphytes were ignored. The mollusk and annelid deposit- and filter-feeding organisms were aggregated into the infaunal group, which was then allowed to obtain resources from both sediment deposit and filter feeding.

To develop an appropriate database for the analysis, geographic information system (GIS) overlays of habitat area and total mud flat area were used to scale production and biomass per unit area of habitat. This provides a uniform dataset, scaled to equivalent units. All data including the range of ^{13}C values in each trophic species (functional group) were incorporated into food web constraints. Allometric, temperature, and physiological relationships from Eldridge and Jackson (1993) were also used in the model constraint system. A schematic representation of the model outputs is presented in Figures 3-3A and 3-3B.

The results of the model suggest that macroalgae and phytoplankton are the major primary producers on the mud flat. The macroalgae must be processed through the detritus compartment before the major consumer groups (*Upogebia*, fish, and *Neotrypaea*) can consume it. Little of the system's C or N is processed through the microbial compartment, even though approximately 80% of C and N is recycled within the mud flat ecosystem (Figures 3-3A and 3-3B). Mud shrimp and fish account for most of this recycling. DIN is imported to the mud flats in equal proportion from the river, ground water, and N fixation, while much of the N export is in the form of the fish trophic species (Figure 3-3B). Little of the C is exported as fish because of the low C:N ratio of these organisms. Instead, the excess C (the fraction not exported as detritus or fish) is converted to sediment organic matter (SOM) and becomes an important resource for the *Upogebia* and *Neotrypaea*.

The analysis suggests that the riverine, groundwater, and atmospheric inputs to the mud flat are small relative to the size of the C and N recycled flows within the mud flat. While it is difficult to determine how variations in the external C and N flow would ultimately change the mud flat food web, we can surmise that the large flux of recycled material prevents the mud flat from rapidly responding to changes in external C and N inputs.

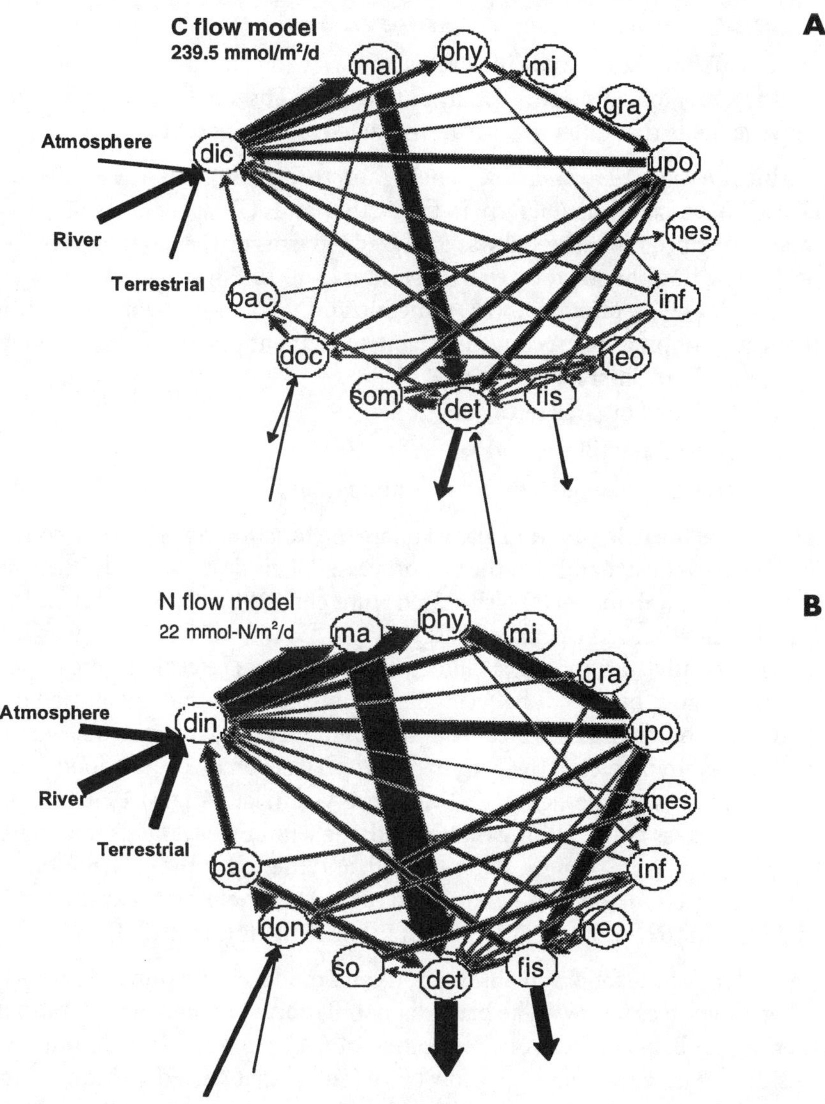

Figure 3-3 Model calculations on contribution of river, groundwater, and atmospheric DIN to an Oregon tidal mud flat, and role of functional groups in trapping DIN. Arrows between compartments indicate consumption of 1 compartment by another. Thickness of arrow is scaled to size of largest flow. A) Food web developed using data from Idaho flat on the Yaquina River, Oregon, USA. B) Condition of (A) except *Neotrypaea* and *Upogebia* mud shrimp are reduced to 5% of original biomass, similar to effect of spraying pesticide specific to the 2 shrimp.

bac = bacteria
det = detritus
din = dissolved inorganic N
don = dissolved organic N
fis = fish
gra = seagrass
inf = infauna
mal = macroalgae
mes = zooplankton
mic = benthic microalgae
neo = *Neotrypaea*
phy = phytoplankton
som = sediment organic matter
upo = *Upogebia*

Case Study III: The Lovinkhoeve Soil Ecosystem

The Lovinkhoeve case study illustrates how a steady-state food web model can be used as a tool for cause–effect analysis to explain observed differences in N dynamics between fields that differ in agricultural management practice.

During the years 1990 and 1991, a field-monitoring program was carried out at the Lovinkhoeve experimental farm in the Netherlands (Zwart et al. 1994). In this monitoring program, the soil was analyzed in terms of the organisms present, their position within the soil food web (De Ruiter et al. 1993), and the key soil process rates such as C-respiration and N-mineralization. Soil was taken from 2 different management practices: a conventional management practice (CF) and an integrated practice (IF). IF differs from CF in

1) the use of organic fertilizers,
2) reduced soil tillage, and
3) reduced pesticides and no soil fumigation.

The aim of the field program was to relate the functioning of the soil community food web to the dynamics in the soil processes. This link was made by using the detrital food web model (O'Neill 1969; Hunt et al. 1987; De Ruiter et al. 1993), which can be used to calculate energy and nutrient flows from food web interactions. A steady-state model calculated annual process rates that were close to the observed rates, but the fit between the within-year dynamics quantified by a sort of dynamical model was less clear (De Ruiter et al. 1994). Nevertheless, when the within-year dynamics in the N mineralization were analyzed, the food web model could explain a phenomenon that was observed in late August 1990 following a multiple stress event. During that period, the winter wheat fields were harvested, the crop residues were ploughed into the soil, and the soil at the CF field was fumigated. After that period, the N mineralization turned into net immobilization in the CF fields (at both depth layers), while not in the IF fields (Figure 3-4).

By looking at the food web observations and model calculations, we can explain this phenomenon as follows. The bacterial populations had high growth rates during this period, probably because of the addition of fresh, in part easily decomposable, crop residues. These residues have a low N content so that decomposition led to N immobilization by the bacteria to produce biomass. Because of the soil fumigation, most other groups of organisms were extinct or at very low abundances. In the IF fields, the bacterial growth rates are increased, presumably also leading to a net N immobilization, but this immobilization was in these fields obscured by the N mineralization by the other groups (there was no fumigation in the IF fields), especially the bacterial grazers such as protozoa and nematodes. Hence, the population dynamics within the soil food web provided an explanation for the observed effects of disturbances such as soil fumigation on ecosystem processes.

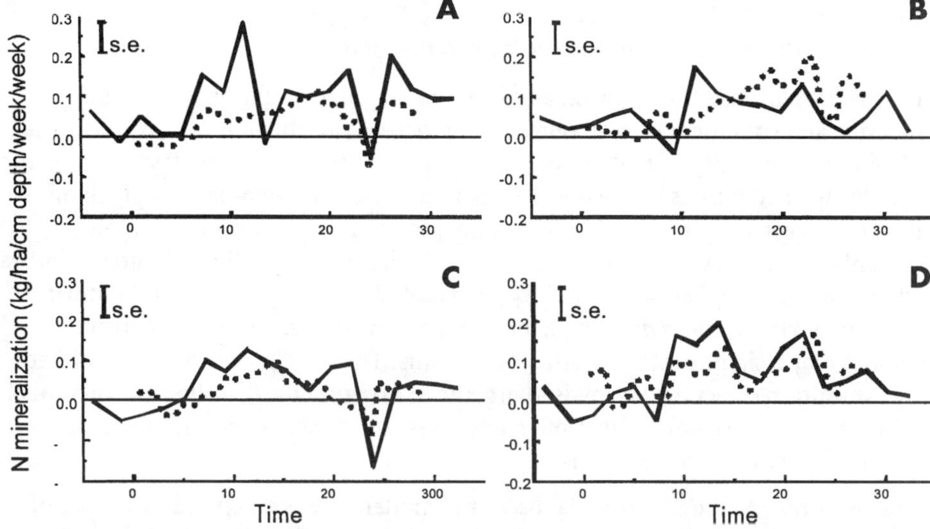

Figure 3-4 Observed (solid line) and simulated (dotted line) dynamics in N mineralization rates based on population dynamics in soil food webs as observed in fields. A) Conventional farming, 0- to 10-cm depth; B) integrated farming, 0- to 10-cm depth; C) conventional farming, 10- to 25-cm depth; D) integrated farming, 10- to 25-cm depth. Vertical bars denote experimental standard errors (SEs) of differences of means. (Reprinted from *Trends in Ecology and Evolution*, Vol. 9, De Ruiter et al., Modeling food webs and nutrient cycling in agro-ecosystems, p 378–383, copyright 1994, with permission from Elsevier Science.)

Case Study IV: An Aquatic Microcosm Food-web Model

The following case study, dealing with the potential impact of the insecticide cyfluthrin on the dynamics of plankton populations, is based on data from published sources (Hommen et al. 1993; Hommen 1998) and illustrates how food web models can be used as hypothesis-generating tools.

A dynamic food web model was constructed to describe the dynamics of a plankton community in an outdoor microcosm. Species were aggregated into functional groups, and necessary parameters were obtained from the literature. However, it was necessary to adapt literature-derived parameters to the specific environments in the microcosms. This was done by sensitivity analysis, followed by calibration of the most sensitive parameters against a dataset of undisturbed model ecosystems. In the next step, results from single-species toxicity tests provided input for the model to predict the impacts of insecticide on key functional groups within the plankton. Model predictions can be compared with population dynamics observed in the treated systems and used to determine dose–response curves for the plankton system (see Chapter 2). Once this phase is complete, the model can then be used to design other microcosm experiments, the results of which can further improve the model. Ultimately, when confidence in the model is established, it can finally be used to extrapolate to conditions beyond those of the original microcosm system.

For example, it could predict the consequences for system dynamics under stress after the addition of another trophic level to the food web.

For the present case study, the data of Heimbach et al. (1992) were used. They describe an experiment in 5-m^2 outdoor microcosms in which the dynamics of the plankton community were studied in 3 untreated systems over 230 days to check the variability of replicates. Since the phytoplankton community was strongly dominated by *Cryptomonas* sp., only 1 functional group of primary producers (phytoplankton) was defined in the model. The zooplankton were divided into Cladocera (dominated by *Daphnia magna*), Copepoda (e.g., *Megacyclops* sp.), and Rotatoria (dominated by *Keratella quadrata*), all primary consumers. Secondary consumers did not play a significant role in the test systems. The nutrient cycle was simplified by assuming a direct flow from dead organic matter into the dissolved nutrient pool. Water temperature, solar radiation, and the test substance concentration in the water served as forcing functions.

The dynamics of the 5 state variables of the model were represented as a series of differential equations containing parameters such as maximum production rate, maximum consumption rate, respiration rate, temperature optimum, and half saturation constants for the uptake of nutrients and for prey. Temperature and solar radiation data were taken from measurements. As initial conditions, estimates of the other model parameters were taken from another microcosm experiment (Hommen et al. 1993) and from the literature. A sensitivity analysis was conducted to identify those parameters that were most important for the model output and that were fitted to the mean dynamics of the systems by the Downhill-Simplex algorithm (Nelder and Mead 1964; implementation taken from Press et al. 1989). In the next step, the uncertainty of the parameter values was assigned on the basis of observed variability of the 3 replicates. The coefficients of variance (CVs) for the mean densities of the 4 modeled groups in the experiment were calculated as 29, 15, 29, and 51 for phytoplankton, Cladocera, Copepoda, and Rotatoria, respectively. Assuming the same CV for all the population-related parameters, Monte Carlo simulations showed that a value of 7% gave the best overall fit for the variability of replicates (Figures 3-5A and 3-5B).

The model apparently showed a tendency to overestimate the biomass of phytoplankton (Figure 3-5A) and to underestimate the biomass of cladocerans (Figure 3-5B). One reason for this may be the simplification of the food web, neglecting for example to account for the fact that Cladocera can use food resources other than phytoplankton (e.g., periphyton and microorganisms).

The potential effects of the pyrethroid insecticide cyfluthrin were the starting point for a "virtual" experiment with the food web model. The half-life of cyfluthrin in water is ~ 4 d for water with a pH of 8 (Heimbach 1987). Toxicity data were available only for *Daphnia magna* (LC50 0.1 to 2.7 µg/L, with most values below 0.2 µg/L), fish, and green algae (EC50 > 10 mg/L). The LC50 values for copepods and rotifers were estimated according to expert judgment, based on assumptions that Copepoda

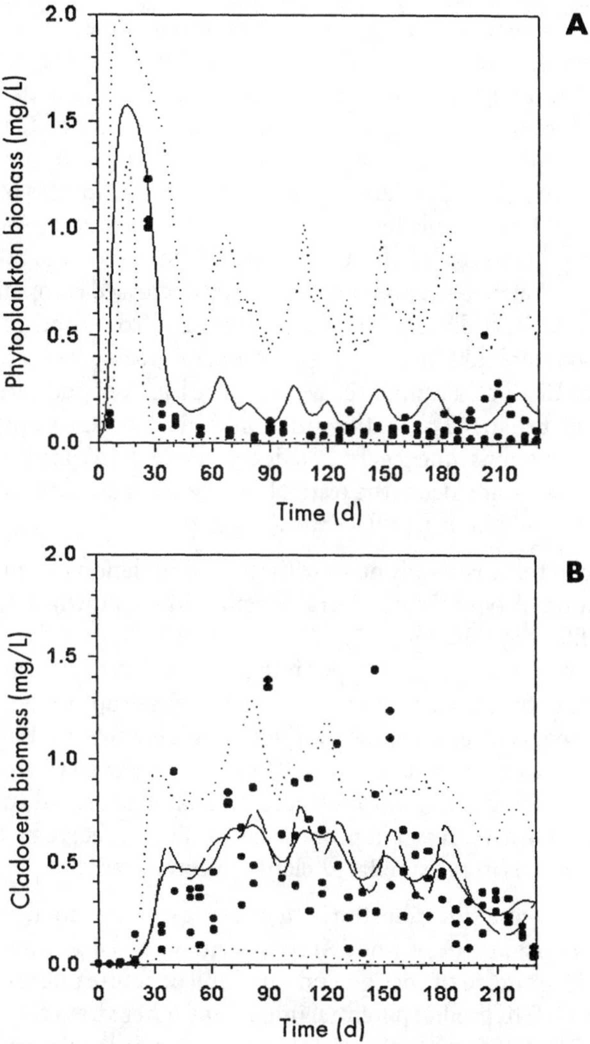

Figure 3-5 Comparison between microcosm observations (points) and food web model simulations for A) phytoplankton and B) Cladocera. (Solid line = mean of 500 Monte Carlo simulations; dotted lines = range of the simulations).

are less sensitive than Cladocera (an LC50 of 0.5 µg/L was used) and that Rotifera are insensitive to pyrethroids (48-h no-observed-effect concentration [NOEC] >> than highest simulated test concentration simulated).

These toxicity data were incorporated in the model as additional terms describing biomass loss due to the toxicant. Hence, effects were modeled as state changes. For simplicity (and because no information was available about the shape of the dose–

response curve), a linear dose–response function was assumed, with a single application of the test substance on day 50. Simulations covering application doses in the range of 0 to 10 µg/L resulted in shifts in mean population densities. Furthermore, the model results also suggested that the mean biomass of Cladocera would start to decrease around 0.07 µg/L and be most affected by an application dose of 0.3 µg/L (Figure 3-6). At higher doses, however, the mean biomass of Cladocera increases again, which can only be explained when we consider the food web links in the model: Copepoda were also affected by the test substance, but at low concentrations, this effect is balanced at population level by decreased competition for food with the Cladocera. In addition, Copepoda can benefit from another indirect effect, the increase in phytoplankton due to suppression of grazing by Cladocera. Up to a certain level, the Cladocera cannot recover because the Copepoda are too abundant, but at doses higher than 0.3 µg/L, the situation shifts. Now the Copepoda begin to be affected, with the consequence that Cladocera again dominate the system after cyfluthrin has disappeared from the water. If application dose increases further, the insensitive rotifers will in turn replace the Cladocera.

The next phase in this case study deals with the extrapolation to a more realistic exposure situation. A typical agricultural practice for cyfluthrin is 9 applications of 56 g/ha every fifth day (Heimbach 1987). Assuming a 5% spray drift input and a mean water depth around 1 m, an input into small surfacewater ecosystems of 0.028 µg/L can be estimated for each application. Due to the exposure regime, the simulated cyfluthrin concentration in the overlying water is above 0.01 µg/L for a period of 50 days, and peak concentrations of 0.04 µg/L are reached (Figure 3-7). The model indicates that effects on the crustacean plankton are small and that the exposure regime does not result in pronounced indirect effects caused by shifts in food web interactions in the simulated plankton community.

In conclusion, this case study illustrates that once a dynamic food web model is developed for a certain type of (model) ecosystem, virtual experiments can be performed to 1) optimize the design and sampling of future micro- or mesocosm experiments and 2) to predict potential impacts of other stress regimes. Only recently have dynamic food web models been constructed to analyze toxicant impacts in experimental systems (e.g., Swartzman and Rose 1984; Bartell et al. 1992; Hanratty and Stay 1994; Traas et al. 1998; Park 1999a). These models, however, have yet to be used as decision tools within a regulatory framework. Their validation status needs to be improved before they can be used in higher-tier risk assessment of new chemicals, and this should be the focus of much future research in aquatic and terrestrial systems.

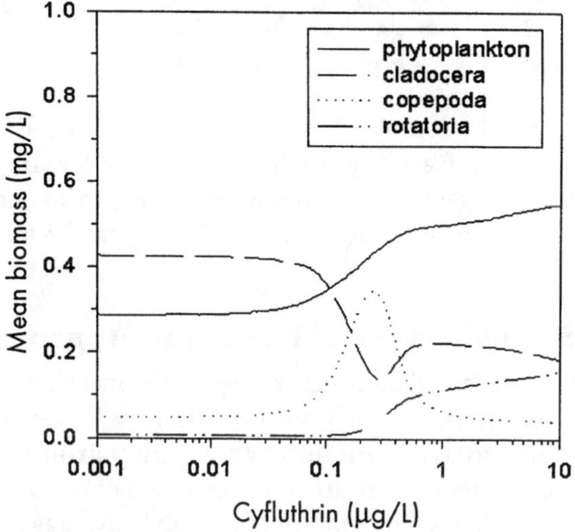

Figure 3-6 Relationship between mean biomass of several food web functional groups (phytoplankton, cladocerans, copepods, rotifers) after application of pyrethroid insecticide cyfluthrin (single application on day 50) as predicted from a food web model

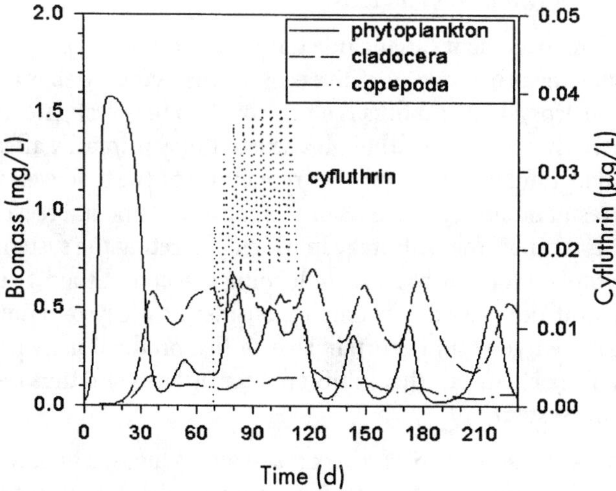

Figure 3-7 Deterministic simulation of 9 repeated applications of pyrethroid insecticide cyfluthrin (0.028 μg/L) on cyfluthrin concentrations in water column and on resulting biomass dynamics of phytoplankton, cladocerans, and copepods

Applying the Food Web Approach in Environmental Management

The approaches and examples contained in this chapter indicate the rapid progress of research in applied food web ecology. In this final section, the aim is to highlight the future role of food web science in environmental management and to give some indication of the practicability of these approaches in the practice of environmental management.

Advantages and Limitations of Food Web Approach

The food web approach presented in this chapter provides mechanistic relationships between stressors, populations, and their role in ecosystems. In general, the effect of a stressor or group of stressors is determined by comparing an impacted ecosystem to a similar reference ecosystem. There are numerous ways of calculating food web relationships, each of which has different limitations and advantages (see "Assessing single and multiple stress effects within food webs," p 101). Each of these food web analyses can be used to directly compare stressed and reference ecosystems, or alternately the food web models can be coupled to density-dependent models that describe the relationship between a stressor and individual populations within the food webs (see also Chapter 2). The difference between the stressed and reference food webs provides a quantitative description of the stressor's effect, while the density-dependent model shows the level of bioaccumulation and biomagnification of the stressor in individual populations.

The goals in risk management can include the protection of target species, yet more often they include the protection of entire ecosystems. A food web is a functional description of an ecosystem and thus is a valuable tool to investigate the impacts of stressors at the ecosystem level. Although a stressor may primarily affect 1 population or functional group in a food web, the remainder of the food web will inevitably be altered as a result of altered patterns of feeding and elemental recycling. One example may be a management strategy in which protecting the system from an increase in phytoplankton due to nutrient loading is accomplished by enhancing zooplankton production. Another example, considering ecosystem function, might be managing a system to ensure a certain level of fish production by protecting the phytoplankton from chemicals that inhibit photosynthesis and thus reduce primary production in the system.

Food webs can also be useful tools for risk assessments focused on a single species (or population). For example, a top carnivore may be exposed to a contaminant through bioaccumulation, which makes it necessary to assess the flow of the contaminant through the entire food chain. Finally, a species may not be directly affected by stressors but only indirectly through links within the food web, for example, if a predator's key prey resource is sensitive to the stressors.

Of course, when species are directly affected by a stressor, a food web approach might not be necessary for risk estimation. Moreover, even when a food web approach may theoretically be useful, available data may not be sufficient to conduct an analysis. The HCN (1997) stated that a lack of relevant data is the "greatest obstacle to use food web models in ecotoxicological risk assessment." Usually toxicity data are available for only a few species, making extrapolation between species necessary. In addition, food web models require a great deal of ecological knowledge of the species being considered (e.g., consumption and/or production rates, mortality rates, prey preferences), which is unnecessary in standard risk assessment methods such as species-sensitivity distributions or the risk quotient approach. In addition to the need for more data on individual species biology, food web approaches also require increased knowledge of species interactions.

Moreover, theoretical food web studies have indicated that perturbations on species in food webs often have unforeseen consequences, which may seem counterintuitive. For example, Yodzis (1996) found that the consequences of perturbations were not always predictable terms of scale or direction. From his analyses, it appeared that approximately 10% of the effects were counterintuitive, for example, that a reduction in the size of a predator population leads to a reduction in the size of the prey population.

Recent workshops on ecological risk assessment (Higher-tier Aquatic Risk Assessment for Pesticides [HARAP], Community Level Aquatic System Studies Interpretation Criteria [CLASSIC]) have introduced the concept of the ecologically acceptable concentration (EAC) for use in interpretation of model ecosystem experiments. At these workshops, the process of population and community recovery, observed during the study itself or predicted to be likely under natural conditions, has been identified as the most important criterion for establishment of an EAC. If adopted by regulators, the use of recovery indicators would mark a major shift in the emphasis of risk assessment, from ensuring no effects of a toxic substance (e.g., based on the NOEC of the most sensitive species) toward allowing small, negligible direct effects. If this approach is adopted, then it becomes even more imperative that a food web approach is incorporated into the regulatory framework, to ensure that possible indirect effects can be evaluated simultaneously with direct effects.

Finally, it should be emphasized that our ability to verify food web models is limited. Many existing models describing the dynamics of food webs have never been challenged with experimental data (HCN 1997). On a more positive note, an important recent finding is that a long-held simplifying assumption of classical food web analyses—that the ecosystem is at a steady state—has now been confirmed using constrained optimization methods. Similar to many ecological analyses, the steady-state assumption results from using linear analysis in a nonlinear world. The steady-state assumption is restrictive: We cannot use this analysis in rapidly fluctuating ecosystems. However, when changes in the numerically dominant population in the analysis are known, the models are able to predict changes in food web structure under nominal conditions of "trophic species" growth (or decline) (Jackson and Eldridge 1992). A trophic species in this context is an aggregation of organisms with similar function properties (see "Case Study II: Oregon mud flat ecosystem," p 109). Greater impediments to the development of food web models in oceans and other large impoundments are the physical processes of dispersion and advection. These transport processes are difficult to formulate without extensive physical oceanographic information and generally make food web analyses impractical in highly dispersive environments, but they constitute an active area of current research in oceanography.

Practical Guidance for Implementing Food Web Approach

The risk assessment process can be divided into 4 different phases:

1) problem formulation,
2) exposure assessment,
3) effects assessment, and
4) risk characterization (Norton et al. 1992; Suter 1993b).

During the development of a food web study, it often becomes apparent that we lack basic information on feeding relationships and life-history information or that the current dogma concerning these relationships does not hold up in light of new food web information. Under these conditions, prospective simulations can often lead to new hypotheses and experiments to fill model data requirements.

The development of a research program that includes successive iterations of food web analyses from program conception to its end could be an effective way to manage and focus research on the effects of toxic substances on ecosystems. Through the judicious use of field observation and laboratory studies suggested by the food web analysis, it would be possible to develop detailed ecosystem descriptions with relatively limited resources. This could replace the current "random-walk" approach to ecosystem monitoring, where many redundant parameters are measured at substantial cost and with little analytical value. With existing methods, it would be feasible to use food web analyses through successive stages of multiple stressor risk assessment. How this approach might be integrated within the tiered approach outlined in Chapter 8 is not yet clear, although it is certain that food web classification (see "Spinning food webs using functional classification," p 86) has a role to play here, not least in terms of the choice of bioassay species and influence of trophic habits and trophic position on species susceptibility (see "Trophic transfer within food webs," p 91).

During problem formulation, sometimes also called "hazard definition" (Suter 1993b) or "hazard identification" (European Commission [EC] 1996), food web models are used to screen for contaminant effects using available datasets and information from the literature. Through this process, the risk assessor can identify the most sensitive endpoints that can be measured, for example, in a mesocosm experiment or a whole lake ecosystem.

If transfer of a contaminant through the trophic levels is important, food web models are used to predict bioaccumulation or biomagnification (see "Trophic transfer within food webs," p 91). On the other hand, in effects assessment, food web models can serve as powerful tools to support experimental investigations (we mentioned their role as planning and screening tools for experiments). In subsequent stages of the risk assessment process, they can also be used to analyze the results of the experiments. Hence food web models are used to develop hypotheses, to design experiments, and then to interpret the experimental results, and by comparing observations and model predictions, to verify model prediction. In addition, food web analyses can serve as extrapolation tools to answer questions such as "What is the ecological acceptable concentration for specific endpoints?" and "On what time scale could we expect an ecosystem to recover from an encounter with a stressor?"

One further role for food web models in risk assessment may be extrapolation between ecosystems. Suter and Bartell (1993) have suggested the possibility that a calibrated and verified model for an experimental system (e.g., a mesocosm) can be re-parameterized to describe another ecosystem (see "Case Study IV: An aquatic microcosm food-web model," p 113) .

Finally, in the risk characterization phase, integrated fate and effects models can be used to estimate the risk of a specific hazard and to consider uncertainties about exposure and effects.

Conclusions and Recommendations for Future Research

The arguments presented here demonstrate that an understanding of food webs and their structural and functional components can reveal the "wiring behind the board" of ecotoxicology. Far from being simply the concern of a small group of ecologists, food webs are ready to begin take their place as the central framework in ecological risk assessment. However, great effort is needed in the following areas:

- Combining food web models and chemical fate models will ultimately allow the prediction of effects from an understanding of how fate processes expose and affect organisms within food webs.
- Developing a more sophisticated methodology for functional classification of species in ecosystems must be accomplished. As part of current international programs on the assessment of biodiversity, there is a need to begin the synthesis of "ecosystem users' manuals," containing information about species' roles in local food webs and overall ecosystem function. This approach would be a functional taxonomy, building on the wealth of ecological information in the published and gray literature, with the goal of identifying and protecting keystone species that have key functional importance within communities.
- Advocating for food web models as diagnostic tools in environmental assessment is urgently required. The use of functional feeding classification in

monitoring programs is already being developed for the marine environment; similar efforts are required for terrestrial systems.

- Our understanding of food webs has a temperate, northern hemisphere bias. The increasing pollution problems in tropical systems, notably from metal mining and pesticide use, mean that the research required to develop similar approaches for tropical systems will require even greater effort and increased training and support for scientists from emerging nations.

While these research areas explicitly focus on food web ecology, they also reflect the central themes of ecotoxicology: chemical fate, leading to toxicological effects in an ecological context. The next 10 years of ecotoxicological research pose a significant challenge for researchers. Perhaps the greatest prize will be the development of a truly ecologically focused science, with the food web approach as its central paradigm.

CHAPTER 4

Food Webs: Interactions and Redundancy in Ecosystems

Peter C. de Ruiter, John C. Moore, Bryan Griffiths

In the field of environmental sciences, ecology is considered to play a key role, being the discipline in which organisms, populations, and communities are studied in interaction with their biotic and abiotic environment. Of special interest to ecologists is how organisms, populations, and communities respond to environmental disturbances. The primary focus is not to understand how environmental stress factors may cause adverse effects, but rather to understand the mechanisms of adaptation and selection processes that have led to the structure and functioning of these organisms, populations, and communities.

Communities, Ecosystems, and Disturbances in Environmental Ecology

The increasing awareness of environmental problems has directed much of ecological research toward the question of how environmental stress factors, related to human activities, alter the structure and functioning of ecosystems. Examples of environmental issues that inspire ecologists are global change, environmental pollution, acidification, and enrichment. This field of ecology is referred to as "environmental ecology" (Pahl-Wostl 1995) or "stress ecology" (Barret and Rosenberg 1981). Many international biological programs aim to encourage this environmental orientation of ecology, for example, the International Biosphere Geosphere Program (IBGP) and the IBGP project Global Change in Terrestrial Ecosystems (GCTE).

Although environmental ecology started several decades ago, some ecologists still conclude that ecology has not been able to fulfill its assigned "task" to an extent that might have been expected. Pahl-Wostl (1995) puts forward the principal argument that it is the progress in ecology that hampered its application to environmental problems because "ecology still lacks a common definition and/or perception in

Ecological Variability: Separating Natural from Anthropogenic Causes of Ecosystem Impairment.
D.J. Baird and G.A. Burton, Jr., editors.
ISBN 1-880611-43-0

how ecosystems function, and hence, a common approach to define and establish disturbance in terms of changed ecosystem functioning." On the other hand, it can be argued that "the glass is half full," pointing to the impressive amount of knowledge and understanding ecology has generated about how ecosystems function under disturbance regimes and how communities and populations are able to survive stress events. A central concept in these ecological studies is "stability," which is described in terms of resistance, the ability to withstand a perturbation or stress, and resilience, the ability to recover rapidly (McNaughton 1994). A summary of all these studies should start with the pioneering contribution by Odum (1969), who argued that an ecosystem can always be considered to be in a particular succession stage and that a disturbance will set the ecosystem back in stage. This setback can be seen from the frequency and abundance of pioneering species. This concept has further been developed in terms of trophic complexity and energy use efficiency (Ulanowicz 1996). A second contribution is that by May (1972, 1973), who caused ecologists to rethink relationships between community complexity and stability, as his theoretical models indicate a negative relationship between complexity and stability rather than the positive relationship assumed by many ecologists. The models by May (1972, 1973) encouraged many theoretical and empirical studies on complexity, patterns of interaction strength among populations and stability (e.g., Menge and Sutherland 1976; Pimm and Lawton 1977; Cohen 1978; Paine 1980, 1992; Yodzis 1981, 1988; Moore and Hunt 1988; Lawton, 1989; Polis 1991; DeAngelis 1992; De Ruiter et al. 1995; Ulanowicz 1996). All these studies have contributed to our fundamental understanding of the behavior and functioning of ecosystems by revealing ecosystem properties that play a key role in the evolution of natural communities, so that we can understand how and why ecosystems and communities are organized and how they function under disturbance regimes (see also Chapter 3). These studies should also provide a scientific basis for understanding adverse effects and risks of human activities on natural ecosystems, which in turn provide the basis for political and social decisions about how best to treat our natural environment (e.g., Polis 1999).

Ecotoxicology as Part of Environmental Ecology

In the 1970s, ecotoxicology started as a field of environmental ecology, with a primary aim being to contribute directly to the Organization for Economic Cooperation and Development (OECD) and U.S. Environmental Protection Agency (USEPA) chemicals testing programs. In this framework, ecotoxicologists were merely assessing effects of chemicals on single individuals and populations under laboratory conditions, for example, in terms of predicted effect concentration (PEC). This approach has come under criticism because the outcome of the single-species laboratory tests do not always relate to field conditions. For example, the tests do not take into account temporal and spatial heterogeneity and other kinds of environmental attributes that determine the chemical speciation of contaminants (see also Chapters 5, 6, and 7). The tests do not take into account the biological properties of a system, such as bioaccumulation and magnification of contaminants in food chains, or the indirect effects of contaminants through interactions among populations. As a consequence, both policymakers and scientists became increasingly interested in evaluating risks of contaminants in terms of altered community structure and ecosystem functioning.

As a first step to bridge the gap between laboratory single-species tests and complex ecosystems, several extrapolation methods have been developed, including the formulation of the HC5, which is the hazardous concentration for 5% of the species (Van Straalen and Denneman 1989), based on the theoretical work by Kooijman (1987). These extrapolation methods have been criticized for their inability to serve as real measures of community and ecosystem effects, in part because of underlying assumptions such as the nonrandomness of species (Forbes and Forbes 1993) but also because extrapolation methods do not take into account the structure and organization of real communities and ecosystems, in terms of interactions among populations and between populations and ecosystem properties. These interactions are considered to be central to understanding how the dynamics of populations and communities are altered through the direct and indirect effects of pollutants, and hence to be central to understanding and predicting how many species will reveal observable effects (Chapters 2 and 3; Van Straalen and Løkke 1997).

This raises the obvious question of why ecotoxicology did not build on the many theoretical and empirical findings in the field of environmental ecology. One reason might have been that policymakers wanted rapid answers to ecotoxicological studies, risk assessments, and criteria development, and there was no time for lengthy ecological experiments. Another reason is that ecology itself lacks a common definition or perception of how ecosystems function and hence lacks a common approach to define and establish disturbance in terms of changed ecosystems (Pahl-Wostl 1995). This is in line with the argument by Kareiva et al. (1996), who stated that ecotoxicology has only weakly addressed fundamental ecological issues because basic ecology itself has only recently begun to provide the necessary tools for such practical work: "Toxicology symposia and texts make frequent reference to principles of ecology, citing everything from food webs, to disturbance theory, to ecosystem homeostasis. In practice, however, ecotoxicology is largely toxicology with ecology added as a 'seasoning' as opposed to a main ingredient — an important applied science dominated by dose–response curves, vulnerable species assays and reductionist laboratory experiments, and bereft of ecological theory." Van Straalen and Løkke (1997) added to this statement that ecotoxicologists have not received much attention from fundamental ecologists for their ecological questions: "This situation hides the potential danger that ecotoxicologists are going to use theories and concepts that are considered to be out-of-date." For this reason, Van Straalen and Løkke (1997) organized a workshop on ecological risk assessment of contaminants in soils, focusing on recent developments in environmental ecology that can be used for putting more "eco" into ecotoxicology. One of these recent developments was the concept of a food web approach, which was given attention in the contributions by Moore and De Ruiter (1997b) and Van Wensem (1997). Based on the outcome of this workshop, the Health Council of the Netherlands (HCN 1997) initiated a study on the feasibility of taking a food web approach in ecotoxicological risk assessment.

Food Webs: Why and How?

As special descriptions of biological communities, food webs focus on trophic interactions between consumers and resources. Food web interactions are important to the dynamics of populations because these dynamics are largely a function of benefits derived from the acquisition of energy (and nutrients) and losses due to predation. Food webs, therefore, provide a way to analyze the dynamics of various populations in the context of the stability of the community as a whole. Moreover, the analysis of disturbance effects in food webs embraces the overall assemblage of direct effects on sensitive groups and all kinds of indirect effects through trophic interactions. Such indirect effects include not only those that occur through changes in the dynamics of interacting populations but also those that occur through the transfer and magnification of contaminants through food chains or through changes in the abiotic environment.

Food webs open the door to translating ecotoxicological effects on particular populations to changes in a particular pathway within the cycling of matter, energy, and nutrients (Chapter 3; Verhoef and Brussaard 1990; De Ruiter et al. 1994). The cycling of energy and nutrients and the structure of food webs are deeply interrelated, in that the trophic interactions represent transfer rates of energy and matter (Hunt et al. 1987; De Ruiter et al. 1993, 1994; Hairston and Hairston 1993). Hence, by means of the population dynamic descriptions of the trophic interactions in food webs, we can analyze the stability of communities and the disturbance effects on communities as well as on ecosystem processes (DeAngelis 1992; Moore et al. 1993; De Ruiter et al. 1995). Moreover, the connection between effects on communities and on ecosystem processes makes it possible to evaluate redundancy in ecosystems. "Redundancy" means that ecosystem processes do not have to change following species extinction because the surviving species take over the role of the extinct species.

To fully benefit from the food web approach, we should be aware that ecotoxicology makes some specific demands on food web ecology. The first demand has to do with the ambition to translate effects on populations and communities (see also Chapter 2) to effects on ecosystem processes. Such translation requires a direct connection between community and ecosystem functioning. A connection can be made when, in the food web descriptions, taxonomic species are aggregated into functional groups (sensu Moore et al. 1988). Functional groups in food web analyses embrace all species that share the same prey and predators. Functional group descriptions of food webs imply that community and ecosystem are described in the same "currencies," namely in terms of material pools (population sizes) and material flows (feeding rates). Each functional group then represents a specific component in the cycling of matter, energy, and nutrients, and effects on groups can be translated into effects in terms of altered contributions to an ecosystem.

The second demand has to do with the way we should define stability and equilibrium in ecotoxicology. Generally, "stability" is defined as the extent to which the system can withstand disturbance (resistance) and the rate with which it recovers from disturbance (resilience) (McNaughton 1994). A "disturbance" is described as a departure from the original equilibrium state, and a system is stable when it is able to return to the equilibrium state. The equilibrium state of the system is then the collection (vector) of the original population sizes. In ecotoxicology, disturbances are the consequences of stressors. These stressors may directly "kill" some organisms, hence they may change the state of the system. But in most cases, the principal effect of stressors will be changes in the functioning of organisms in terms, for example, of altered growth and death rates. Then disturbances mean changed rates, rather than changed states. A change in rate implies that the original state is unchanged but that it is no longer in the equilibrium state because the changed rates determine a new equilibrium state. Stability now means the ability of the system to reach the new equilibrium without losing species. This difference between disturbance of state and rate can also be seen from the way we use models that evaluate the stability of a system, such as the community matrix model proposed by May (1972):

$$dx/dt = Ax \qquad \text{(Equation 4-1)},$$

where x is the vector describing the state of the system in terms of the departures of the population sizes (designated as $X - X^*$) from the equilibrium state (designated as X^*), and A is the community matrix in which the elements describe the per capita effects of the populations upon one another under equilibrium conditions. A state disturbance can be formalized in terms of particular values in vector x. Stability then means that x will return to 0, that is, to the original equilibrium (X^*). A rate disturbance can be formalized in terms of changed values in matrix A. Stability then means that the original equilibrium state X^* is no longer in equilibrium and that under the guidance of the changed A, a new equilibrium state will be reached. In ecotoxicology, we most often deal with rate disturbances.

Food Web Theory in Ecotoxicological Practice

This section consists of 2 parts. First we present the outcome of a series of food web studies in which empirically based equilibrium descriptions of food web structure are constructed, combined with an a priori modeling exercise of how disturbance might affect food web structure and ecosystem processes. The second part gives the outcome of an ecotoxicological experiment in which the effects of environmental contamination are evaluated in terms of altered structure and stability of food web and ecosystem processes.

Equilibria, Interaction Strengths, and Stability in Food Webs

A first step in the analysis of disturbance effects on food webs and ecosystem processes was to obtain equilibrium descriptions in which community and processes are connected to each other. Examples of such equilibrium descriptions are available, such as for a series of belowground, detritus-based, food webs from agricultural and native soils (Figure 4-1A; Figure 4-3B; De Ruiter et al. 1993). The food webs used were as follows: a shortgrass prairie web from the Central Plains Experimental Range (Nunn, Colorado, USA from Hunt et al. 1987); two webs from arable farm systems on the Lovinkhoeve Experimental Farm, Marknesse, The Netherlands, one farmed by conventional methods, the other by integrated farming, substituting inorganic fertilizer with organic manure and employing less tillage and reduced pesticide application (Brussaard et al. 1988; De Ruiter et al. 1993; Zwart et al. 1994); two webs from Horseshoe Bend Research Site, Athens, Georgia, USA from arable farming systems with and without conventional tillage (Hendrix et al. 1986, 1987); and two webs from Kjettslinge Experimental Field, Uppsala, Sweden from barley fields with and without N fertilizer (Andrén et al. 1990). The average annual population sizes of the functional groups were taken as the equilibrium population sizes and the annual average feeding rates among the groups as the equilibrium energy flow rates. The population sizes were recorded through field monitoring (Hendrix et al. 1986, 1987; Andrén et al. 1990; De Ruiter et al. 1993; Zwart et al. 1994). The energy flows were derived from observed population sizes and from data on death rates and energy conversion efficiencies through modeling. The basic assumption underlying the calculation of feeding rates was that the annual (equilibrium) feeding rate should balance the annual death rate through natural death and predation (O'Neill 1969; Hunt et al. 1987):

$$F_j = \frac{d_j \; B_j + M_j}{a_j \; p_j} \qquad \text{(Equation 4-2)},$$

where F_j is the feeding rate (kg C ha^{-1}y^{-1}) , d_j the specific death rate (y^{-1}), B_j the average annual (equilibrium) population size (kg C ha^{-1}y^{-1}), M_j the death rate due to predation (kg C ha^{-1}y^{-1}), a_j the assimilation efficiency, and p_j the production efficiency. For polyphagous predators, the feeding rate per prey type (F_{ij}) was based on the relative abundances of the prey types and on prey preference:

$$F_{ij} = \frac{w_{ij} \; B_i}{\sum_{k=1}^{n} w_{kj} \; B_k} F_j \qquad \text{(Equation 4-3)},$$

where F_{ij} is the feeding rate by predator j on prey i, and w_{ij} is the preference of predator j for prey i over its other prey types. These calculations of feeding rates started with the top predators, which suffer only from natural death, and proceeded working backwards to the lowest trophic levels. By this methodology, estimates of energy flow rates for all trophic interactions were obtained. The equilibrium descriptions obtained in this way show that population sizes (De Ruiter et al. 1993) and energy flow rates (De Ruiter et al. 1998) decrease with trophic level, indicating trophic pyramids in all webs (Figure 4-1B). Furthermore, the modeled energy flow rates were verified at the level of overall C mineralization (De Ruiter et al. 1993) and were found to be close to the observations, indicating that the equilibrium descriptions were close to reality (De Ruiter et al. 1993).

These equilibrium descriptions enabled the construction of community matrix models (sensu May 1972, 1973) with which community stability and disturbance effects can be analyzed. The elements in the matrices refer to the per capita interaction strengths among the trophic groups. Values for these interaction strengths were directly obtained from the equilibrium population sizes and feeding rates (De Ruiter et al. 1995) by assuming Lotka-Volterra equations for the dynamics of the functional groups:

$$X_i = X_i\left[b_i + \sum_{j=1}^{n} c_{ij} X_j\right] \qquad \text{(Equation 4-4)},$$

where X_i and X_j represent the population sizes of group i and j, respectively; b_i is specific rate of increase or decrease of group I; and c_{ij} is the coefficient of interaction between group i and group j. The matrix elements (α_{ij}) are defined as the partial derivatives near equilibrium: $\alpha_{ij} = (\partial X_i \ / \ \partial X_j)$. Values for the interaction strengths are derived from the equilibrium descriptions by equating the death rate of group i due to predation by group j in equilibrium, $c_{ij} X_i{}^* X_j{}^*$, to the average annual feeding rate, F_{ij} (Equation 4-1), and the production rate of group j due to feeding on group i, $c_{ji} X_j{}^* X_i{}^*$, to $a_j p_j F_{ij}$. With equilibrium population sizes, $X_i{}^*$, $X_j{}^*$, assumed to be equal to the observed annual average population sizes, B_i, B_j, the per capita effect of predator j on prey i is

$$\alpha_{ij} = c_{ij} X_i^* = -\frac{F_{ij}}{B_j} \qquad \text{(Equation 4-5)},$$

and the per capita effect of prey i on predator j is

$$\alpha_{ji} = c_{ji} X_j^* = \frac{a_j \ p_j \ F_{ij}}{B_i} \qquad \text{(Equation 4-6)}.$$

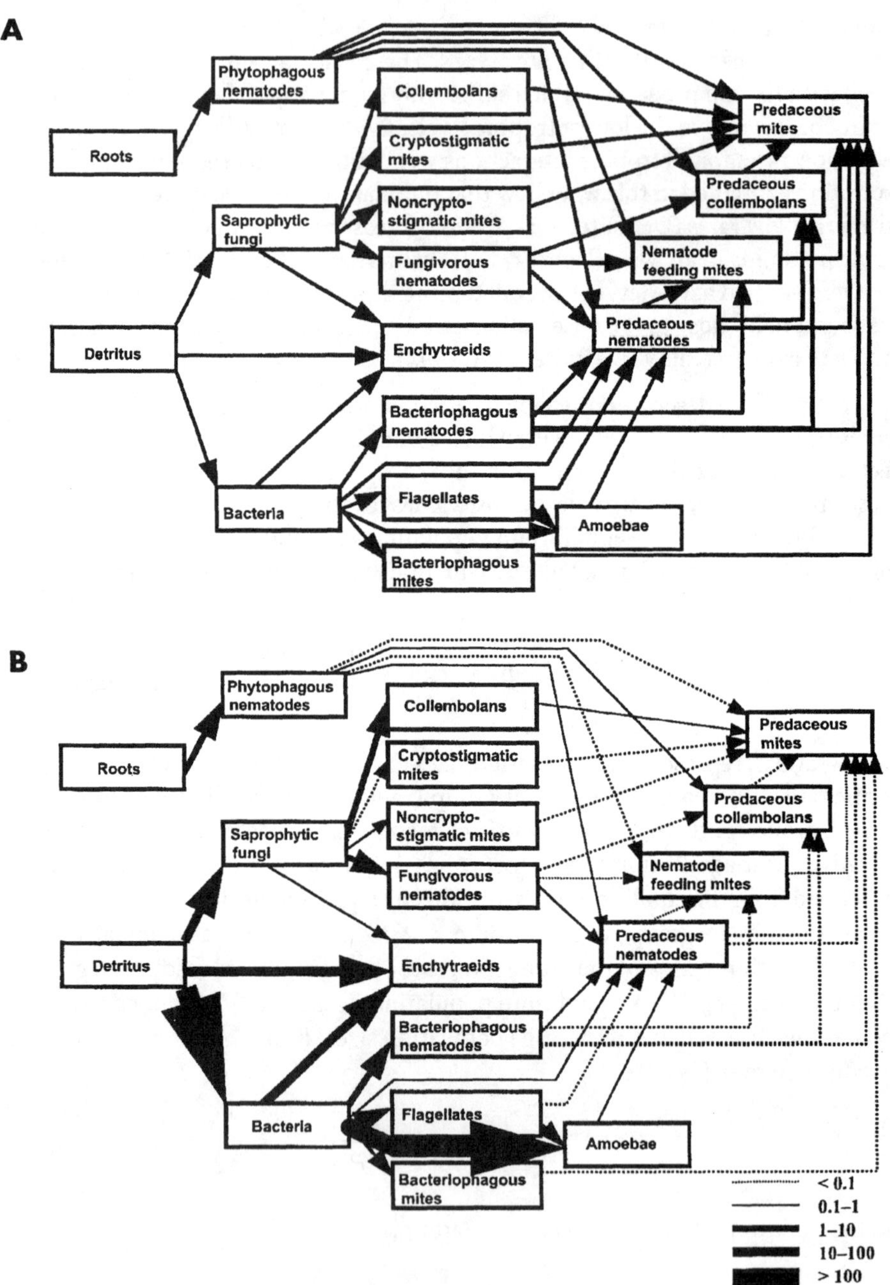

Figure 4-1 Belowground food web from Lovinkhoeve experimental farm: A) Food web architecture. B) Equilibrium energy flow rates (kg ha^{-1} y^{-1}). Species are aggregated into functional groups (sensu Moore et al. 1988) based on food choice and life-history parameters. Detritus referred to all dead organic material. Material flows to detrital pool, e.g., through death rates and excretion of waste products, are not represented but were taken into account in material flow calculations and stability analyses.

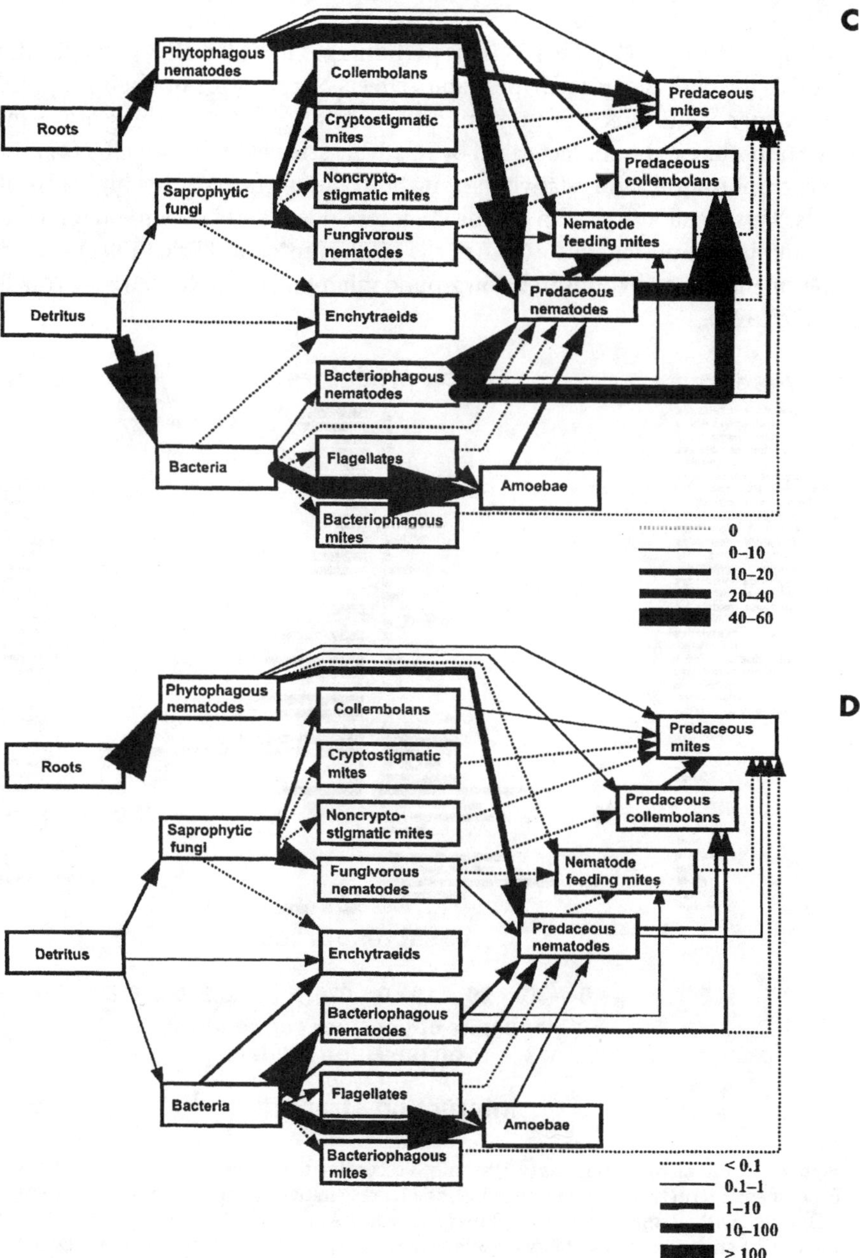

Figure 4-1 *(continued)* C) Impact of disturbing an individual interaction on community stability, expressed in terms of % matrices that became unstable. Sensitivity analysis done with matrices in which mean of diagonal terms was close to (1% below) critical value for stability. D) Impact of disturbing an individual interaction on overall C mineralization rate (kg ha^{-1} y^{-1}). (Reprinted from *Plant and Soil*, Vol. 157, 1993, p 263–273, Calculation of nitrogen mineralization in soil food webs, De Ruiter PC, Van Veen JA, Moore JC, Brussaard L, Hunt HW with kind permission from Kluwer Academic Publishers.)

When interaction strengths are derived in this way, they appear to be ordered along the trophic position (Figure 4-2). This patterning is found in all 7 webs (De Ruiter et al. 1998) for both the negative per capita effects of predators on their prey and the positive per capita effects of prey on their predators. This patterning of the interaction strengths can be characterized by relatively strong top–down effects at the lower trophic levels and relatively strong bottom–up effects at the higher trophic levels. This result agrees with the idea that top–down and bottom–up forces act on communities simultaneously (DeAngelis 1992; Hunter and Price 1992) and with experimental observations in belowground microbial-based food webs (Wardle and Lavelle 1997).

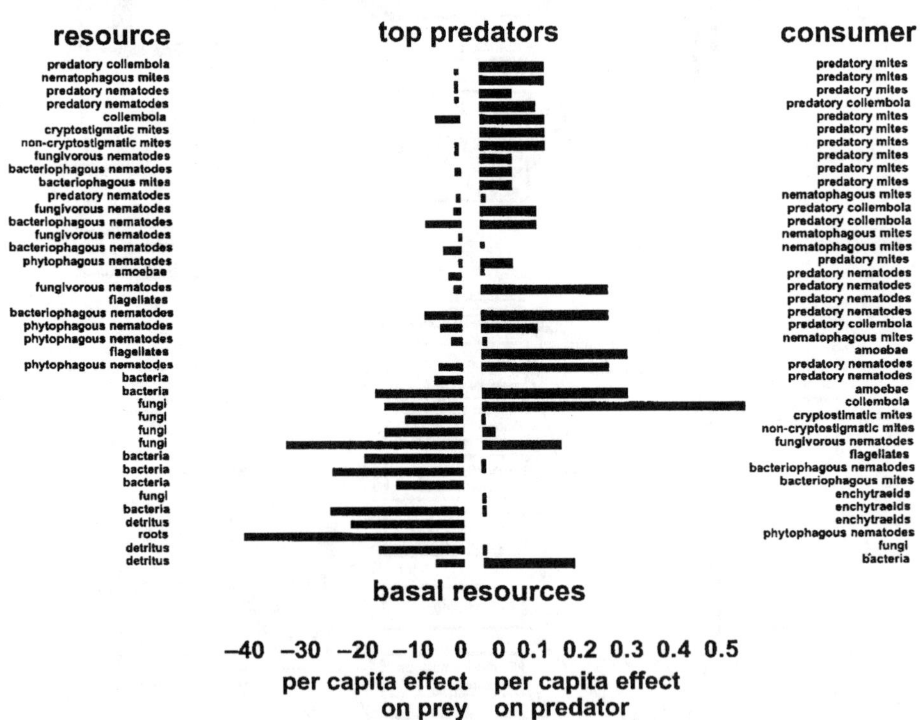

Figure 4-2 Interaction strengths (y^{-1}) arranged according to trophic position in web of Lovinkhoeve experimental farm (conventional farming), as a representative example for other food webs (Reprinted with permission from De Ruiter PC, Neutel AM, Moore JC. Energetics, patterns of interaction strengths, and stability in real ecosystems. *Science* 269:1257–1260. Copyright 1995 American Association for the Advancement of Science.)

Such patterning is important for the stability of the food webs. This was shown by establishing the effect of disturbing the matrix elements on the stability of the matrix (Figure 4-3). This disturbance was modeled by randomizing the values of the matrix elements over the matrix (Figure 4-3A) according to a method described by

Yodzis (1981), which leaves the food web structure (placing of the non-zero terms) and the logical pairing of the elements unchanged. The stability of the matrices was established by evaluating the eigen values of the community matrices; when all real parts are negative, the matrix is stable and the food web can be considered to be locally stable. The comparison between the stability of the disturbed matrices and the undisturbed ones (Figure 4-3B) showed that the undisturbed matrices are far more likely to be stable than their disturbed counterparts, indicating that the pattern in interaction strengths (Figure 4-2) governs the stability of the food webs. This result is important for our understanding of disturbance effects on communities, especially when we are able to identify which general properties of food webs generate this patterning. As the interaction strengths are derived from the energetic organization of the webs in terms of pools (population sizes) and flows (feeding rates) of energy (Equations 4-5 and 4-6), aspects of the energetic organization are indicated as being the basis of the patterning, such as energy channels (Moore and Hunt 1988; Moore and De Ruiter 1997a) or population size distributions (De Ruiter et al. 1995; Neutel 2001). The energetic basis of the patterning of the interaction strengths opens the door to evaluating and predicting risks of disturbances from the ways in which these disturbances affect the energetic organization of food webs. Our analysis is comparable to the analyses of Ulanowicz (1996), who, following Odum (1969), quantified disturbance effects along the axis of successional development specified in increasing trophic complexity and energy use efficiency.

Another interesting notion that arises from this result concerns the relationship between energetics and community stability. This connection was primarily meant to translate effects on communities to effects on ecosystem processes. But because the stabilizing interaction strengths were the direct result of patterns in energy pools and flows, the connection between energetics and community structure also revealed properties that govern the stability of the food webs.

The second step in disturbance–effect analysis was performed by varying the values of only 1 particular pair of matrix elements, referring to 1 interaction within the interval. This kind of disturbance simulates single press perturbations, which have been studied experimentally (e.g., Paine 1980, 1992; Wootton 1994) as well as theoretically (Yodzis 1988; DeAngelis 1996). These studies evaluated the importance of indirect effects and identified keystone modules, such as food web modules that are important to stability, and indicator modules, such as groups that become extinct after destabilization and whose presence or absence may indicate stress. The effects of the single-interaction perturbations were expressed in terms of probability that the matrix becomes unstable (Figure 4-1C). The results of this analysis show that the effects of the perturbations on stability were not equally large for all interactions and that they correlated neither with feeding rates nor with interaction strength (compare Figures 4-1B, 4-1C, and 4-2). The results are in agreement with the experimental observations (Paine 1980, 1992) that changes in energetically unimportant links could lead to a substantial decrease in stability, while changes in the strength of some energetically important links had no effect at all. The results

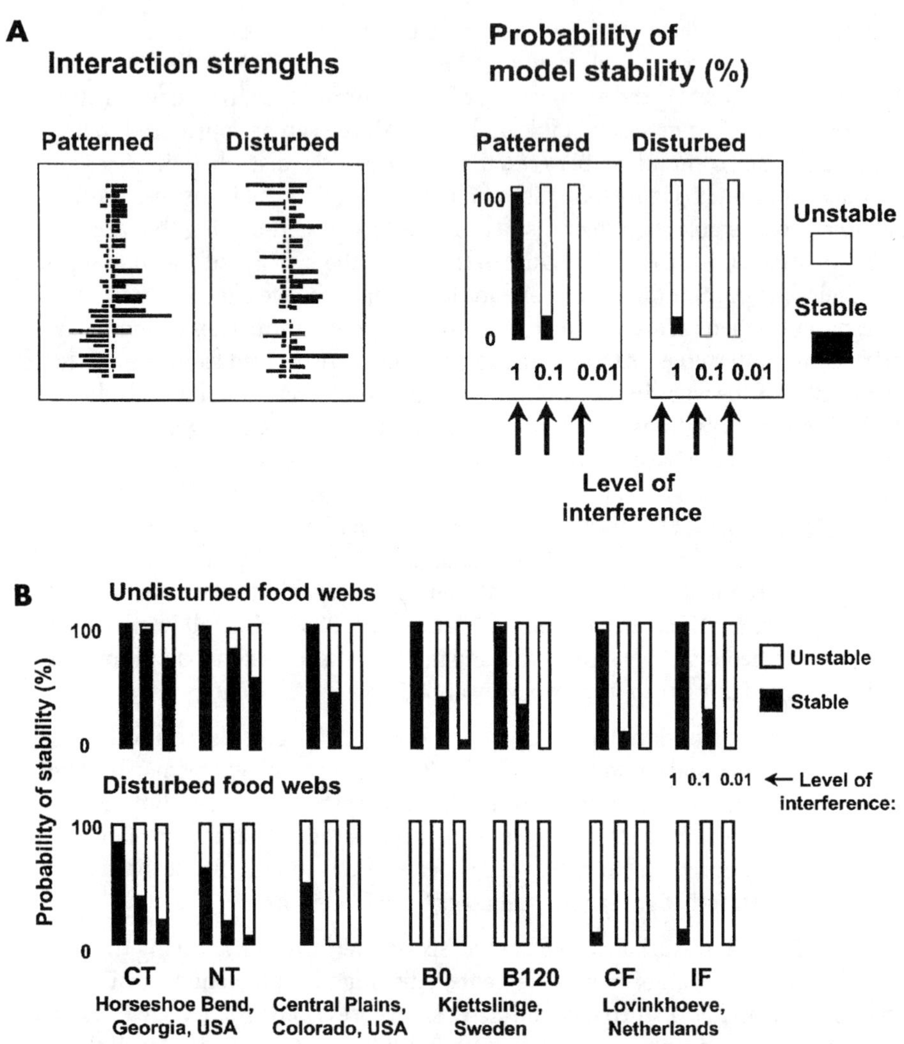

Figure 4-3 Effect of disturbing pattern of interaction strength on stability of 7 food webs. A) Scheme of disturbance and stability (black fraction in bars denotes % stable matrices based on 1000 runs. B) Comparison between undisturbed and disturbed webs. In undisturbed webs, element values were sampled randomly from uniform distribution with intervals $[0, 2\alpha_{ij}]$ in which α_{ij} is value derived from observations. Diagonal matrix elements referring to intragroup interference were set at 3 levels of magnitudes (s_i) proportional to specific death rates (d_i) with $s_i = 1.0$, 0.1, and 0.01 for all groups equally. Elements referring to feedbacks to detritus were derived in same way as trophic interactions. In disturbed matrices, values of matrix elements were randomly permuting non-zero pairs of elements; comparison was based on 1000 runs.

also agree with the theoretical observations (DeAngelis 1996; Yodzis 1988) that press perturbations are highly indeterminate with respect to size and direction. These results could not, however, be interpreted in terms of keystone modules in general, as the results from the Lovinkhoeve food web (Figure 4-1C) were not representative for the other webs in which other modules were indicated to be important to stability (De Ruiter et al. 1998).

Next, the models were used to simulate dynamically which groups are likely to become extinct after destabilization. It appeared that certain groups have a much higher probability to disappear than other groups (Figure 4-4). Moreover, approximately 50% of the disappearances were the consequence of indirect effects. This can be seen from the extinction of groups that were not involved in the disturbed interaction, which is not directly affected by the perturbation. The groups that had a high probability to become extinct might serve as indicator groups; however, the results of the Lovinkhoeve analysis (Figure 4-3) were not representative for the other webs.

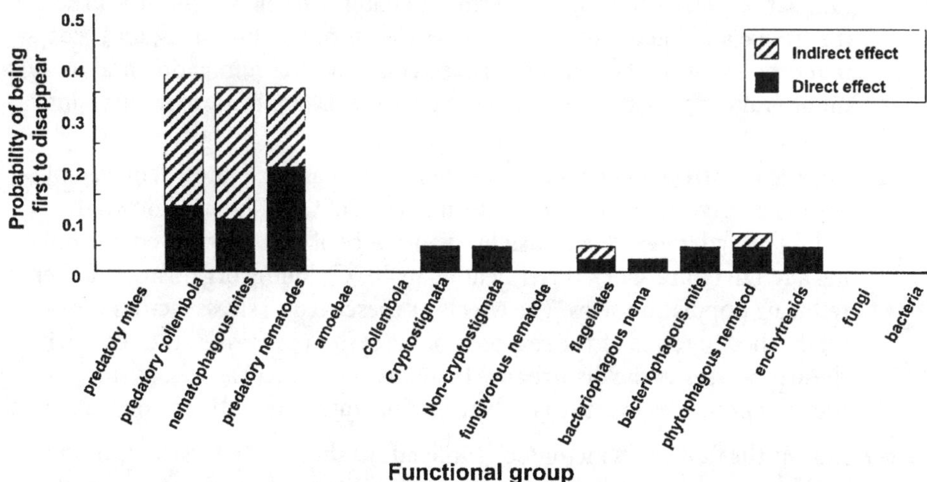

Figure 4-4 Extinction probabilities of functional groups after destabilization (probability of being first to become extinct). Direct effects refer to extinction of groups directly affected by the perturbation; indirect effects refer to extinction of groups not directly affected by the perturbation.

A final single-perturbation analysis was carried out to establish the effects of perturbations on ecosystem processes (such as the overall C respiration rate [Figure 4-1D]). Again, it appeared that disturbing an interaction that contributes importantly to energy flow and C mineralization (Figure 4-1B) does not necessarily lead to a large effect on these processes (compare Figures 4-1B and 4-1D). The analysis of single-interaction perturbations shows that indirect effects occur and that they might be important even at the level of species extinction. However, the results were scattered and different webs generated different results (Moore et al. 1996), hence generalities in terms of keystone and indicator components were hard to identify.

Experimental Assessment of Ecotoxicological Effects on Community Structure and Stability of Ecosystem Processes

Experimental investigation of the effects of environmental stress factors on the stability of communities and ecosystems requires a specific kind of experimentation, that is, stress-on-stress experiments. In such experiments, the question is whether a first disturbance event affects the way in which a community can withstand a second disturbance event. In this section, the outcome of such an experiment is described to show how effects of disturbances on community and ecosystem stability can be analyzed experimentally (Griffiths et al. 2000). The experiment was carried out using soil ecosystems that were submitted to the following stress-on-stress regime (Figure 4-5A):

1) The first stress was a soil fumigation that exposed the soil to chloroform vapor for 0 h (unfumigated control), 0.5 h, 2 h, and 24 h. The effects of fumigation were established in terms of changes in soil community structure (microbial and microfaunal diversity as well as population sizes) and ecosystem process rates. The soil was analyzed directly after fumigation and 5 months after fumigation, to allow for possible recolonization by surviving organisms.
2) The second stress was either a persistent disturbance of rates, achieved by adding a heavy metal (copper in the form of $CuSO_4$), which is known to reduce growth rates, or a transient disturbance of states, achieved by applying a temperature shock (brief heating to 40 °C), killing organisms and hence reducing population sizes. The effects of these second stress factors were established in terms of the response of ecosystem processes: the sensitivity of these processes to the disturbance (stability–resistance) as well as the time it took the processes to return to their original rates (stability–resilience).

The results of the first stress factor (soil fumigation) showed that the soil community was affected progressively as fumigation time increased, leading to the extinction of many trophic groups, species, and varieties, hence reducing strongly the soil biodiversity at fumigation times > 2 h (Griffiths et al. 2000). In fact, there was a steep decline in biodiversity between the 0.5-h and the 2-h treatment, and there were only small differences between the control and the 0.5-h treatment and between the 2-h and the 24-h treatment. The diversity of bacteria (measured through deoxyribonucleic acid [DNA] fingerprinting and ability to decompose various organic matter compounds) decreased, as did the biodiversity of the microfauna (protozoa and nematodes) at the level of number of trophic groups, phyla within trophic groups, and taxa within phyla. There was no recolonization during the 5-month period. The fumigation also affected the soil processes but not as it affected community structure. Some process rates increased (e.g., microbial growth), and the decomposition rate of plant residues and other processes decreased (e.g., nitrification, denitrifica-

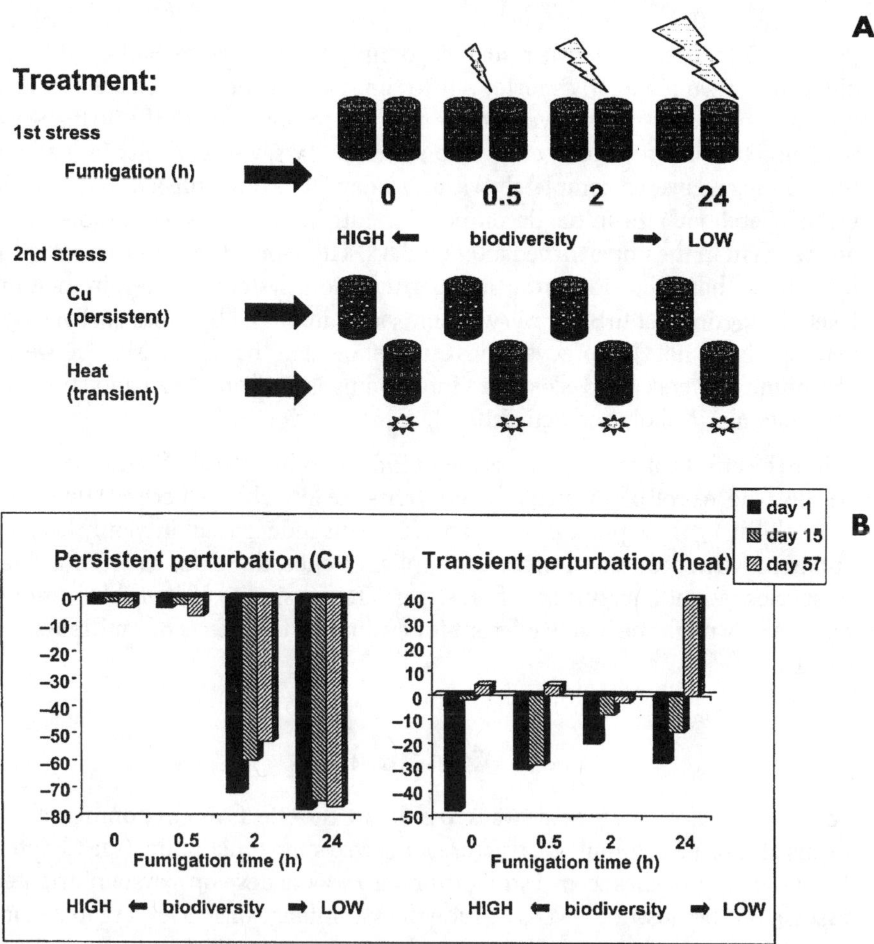

Figure 4-5 A) Design and B) results of stress-on-stress experiment. The first stress was soil fumigation of varying duration: persistent (Cu-addition) or transient (heat-shock). Response to first stress was measured in community structure and ecosystem processes. Response to second stress was measured in altered % C respiration from decomposition of rye grass material after 1, 15, and 57 d.

tion, and methane oxidation). Moreover, the process rates recovered relatively rapidly to their original values. The results of the first stress might, therefore, indicate a sort of redundancy: Although many groups disappeared, ecosystem processes were hardly affected (Griffiths et al. 2000).

The effects of the second stress, however, showed that the processes in the disturbed soils were much more sensitive to stress than were the undisturbed control soils. These effects were measured in terms of changed soil respiration rate (from the decomposition of grass residues) on 3 occasions after the second stress factor (1, 15, and 57 d). The undisturbed soils were hardly affected by the Cu addition, while respiration in the disturbed soils decreased strongly to < 70% of the original value (Figure 4-5B). There was little resilience to this persistent perturbation, although a hint can be seen of slightly shorter return times in the undisturbed soils. Soils given the transient heat stress showed a clear trend of resilience, with the disturbed soils regaining the pre-stress level of function after 57 days, while activity in the undisturbed soils recovered completely within 15 days. The undisturbed soils also showed some "overshoot": the initial decomposition rate in the perturbed sample was greater than in the unperturbed after 57 days. These results are consistent with the hypothesis that, following a transient perturbation, systems can regain their original level of function but that recovery (resilience) is impaired by a loss of diversity (McNaughton 1994). Moreover, the results show that the relationship between the community diversity and ecosystem functioning is far from direct and straightforward (see also Mikola and Setälä 1998).

While the effects of the first stress might indicate a high level of redundancy because the loss of biodiversity did not translate into changed ecosystem processes, the stability of these processes was affected. This reduced stability only became apparent when the second stress was applied. Hence, there was a sort of hidden loss of species redundancy after the first stress. The possibility of invisible loss of redundancy might be highly relevant for evaluating the effects of multiple stress events.

Conclusions

Food web approaches provide ways to evaluate the effects of environmental stress in terms of altered community structure and ecosystem functioning (see Chapter 3). Food web approaches seem a straightforward way to develop a system-oriented risk assessment because food webs include the assemblage of species, organized in networks of trophic interactions, and take into account the direct effects of pollutants and the indirect effects of ecological interactions. Keystone or indicator modules are still hard to identify because studies have generated different results.

Food web approaches can be used to analyze the effects of disturbances on the stability of communities and ecosystems and to identify properties of food webs that govern stability, such as the nonrandom patterning of interaction strengths. These patterns are related to general ecosystem properties such as energy channels (Moore and Hunt 1988; Moore and De Ruiter 1997b) or trophic pyramid population size distributions (Neutel 2001). Such relations open the door to understanding and predicting the effects of disturbance, based on alterations in the energetic organization of communities. Moreover, because the energetics and dynamics of populations are inextricably related, food webs provide a way to translate effects on communities to effects on ecosystem processes and reveal energetic properties that are important to a community's ability to withstand disturbances.

The only proof that food web approaches make sense in ecological risk assessment can be obtained by experimentation that generates results that would have been missed otherwise, for example, the possibility of a hidden loss in species redundancy. Such results also seem relevant to the field of environmental ecology because they show that apparently unaffected ecosystems lose their quality and health during the course of disturbance events (Chapin et al. 1997).

Chapter 5

Role of Environmental Variability in Evaluating Stressor Effects

Samuel N. Luoma, William H. Clements, Ted DeWitt, Jeroen Gerritsen, Audrey Hatch, Paul Jepson, Trefor Reynoldson, Ronald M. Thom

In this chapter, we discuss how environmental variability affects the exposure of organisms and ecological systems to stressors, and we give guidance on how to understand the influences of stressors. We consider the characteristics of environmental variability and the issues relating to the measurement of environmental variation. We discuss how to select the optimal indicators of ecological response in a variable natural environment. Finally, we suggest some approaches to incorporating environmental variability into resource management. In all cases, we employ examples and case studies to illustrate principles.

Variability in Exposure to Stressors

Exposure to a stressor can often determine ecological response, so proper characterization of exposure is essential to understanding the effects of a stressor. Exposure to stressors depends on both abiotic processes inherent to the ecosystem and on proximate inputs of the stressor. Taken alone, knowledge of pollutant loads to a system, or determinations of pollutant concentrations in water, sediments, or biotic tissues, is not sufficient to characterize pollutant exposures at the ecosystem scale. Spatial and temporal complexity in abiotic processes must also be considered.

Many properties of ecosystem variability are difficult to predict. But "dynamic stabilities" also exist, when a factor varies in a repeated, somewhat predictable fashion (i.e., variability is nondirectional). Success in identifying the occurrence and cause of a stress begins by identifying the most important driving processes in an ecosystem and by capturing the dynamic stabilities and events driven by those processes. It should also be recognized that variation of natural processes (including extreme episodes of natural disturbance) can be stressful themselves, if ecosystem properties range outside the limits to which the organism, population, or community is adapted.

Ecological Variability: Separating Natural from Anthropogenic Causes of Ecosystem Impairment.
D.J. Baird and G.A. Burton, Jr., editors.
ISBN 1-880611-43-0

Physical, geochemical, and biological processes can all contribute to variability in exposure to stressors:

- Variability in physical processes—Ultimately, variability in an aquatic ecosystem is driven by climate, hydrology, water movement, water–sediment interactions, and/or other physical, geochemical, or biogeochemical processes. All such processes vary in time and space and drive spatial and temporal variation (on several scales) in exposure to anthropogenic stressors. Ambient physical conditions (i.e., temperature, current, wind, rainfall, sediment deposition) also affect the spatial and temporal distribution of the constituents of habitats (i.e., substrate texture and relief), food, and the chemical environment (i.e., nutrients, contaminants, salinity). Spatial and temporal variation in physical conditions result from long-term or regional-scale processes (e.g., climate, weather, geological history), local-scale processes (i.e., land forms, air–water and water–substrate boundary conditions, land use, resource extraction), and small-scale processes (e.g., air or water boundaries with biotic structures, such as trees, seagrasses, beaver dams, or burrows).
- Variability in geochemical processes—Ambient geochemical conditions can be sources of stress (e.g., salinity change) or can influence exposures to pollutant stress (e.g., regulating contaminant bioavailability). Spatial and temporal variations in geochemical processes often are linked dynamically to variation in physical processes (e.g., temperature, water flow, deposition of sedimentary material) and biotic processes (e.g., microbial and plant uptake and transformation of chemicals, bioirrigation by burrowing animals). Effects of stressors on biota can feed back on geochemical processes, thereby enhancing or diminishing the rates at which biota are exposed to stressors.
- Variability in biological processes—Biological processes that occur within and among species can complicate routes of exposure to stressors for individuals. The scale of impact will vary as a function of the abundance of the organism, its spatial range and use of ecological resources, and its abilities to modify the local chemical and physical environment. For example, species often referred to as "ecological engineers" (e.g., trees in forests, burrowing shrimp in mud flats, corals in tropical reefs) may have a general effect on the exposure of co-occurring species to stress. At the microscale, physical and geochemical conditions may be influenced by neighboring organisms. How an organism feeds can greatly influence its exposure to pollutants (see Chapter 3).

Impacts of Physical Processes on Pollutant Exposure

Physical processes can mitigate pollutant exposures. At a broad scale, sedimentation buries contaminants; hydrodynamics and physical dilution can redistribute contaminants from input sources in predictable patterns (Axtmann et al. 1991; Diamond 1995). Small-scale physical characteristics of rivers, estuaries, or coastal zones are also important because they add heterogeneity to the resulting gradients. For example, local dilution around tributary inputs to mine-impacted rivers can create refugia from contamination in zones where mixing of the 2 water bodies is not complete (Axtmann et al. 1997). Changes in physical processes can uncover buried contamination, or less predictable, major events such as floods or spills can have long-lasting influences on the ultimate distribution of pollutants. For example, extreme flood events are critical to the ultimate distribution of mine tailings in flood plains and riverbeds.

Understanding the influences of physical processes on pollutant exposures and effects at the ecosystem scale is a difficult challenge. Physical processes are usually modeled over large scales, but the response of the potential stressor (e.g., a chemical contaminant) is typically studied at much smaller scales. For example, models of hydrodynamic processes in estuaries often have spatial resolution at the kilometer scale and temporal resolution at the tidal scale, and these are integrated regionally. Contaminant data from the same system will have local resolution (meters) and are rarely collected on more than a monthly basis, and thus it is rare that data are sufficient to coherently integrate at a regional scale. The differences in scale complicate the resolution of hydrodynamic influences on temporal and spatial variability of contaminants within the region.

Integration of scales can be aided by identifying the dynamic stabilities of transport and depositional processes (Diamond 1995). Mass balance models, hydrodynamic models, or multimedia models (Mackay and Diamond 1989) link physical processes to biological or geochemical processes of interest (Lucasz et al. 1999a, 1999b). For example, in multimedia models, air, water, soil, and sediment are treated as bulk compartments. Expressions are included for influences of emissions, advective flows, and intercompartment exchange. Diamond (1995) showed that concentrations of polychlorinated biphenyl (PCB) and As in the Bay of Quinte in Lake Ontario were controlled by rapid advection in the bay. As a result, the bay would clean itself relatively quickly if source inputs were reduced. Sediment–water exchange, primarily through particle movement, maintained elevated concentrations beyond those predicted by water residence times alone. The importance of sediments as a source depended upon the rate of permanent removal by burial or chemical transformation. The Bay of Quinte study identified specific sources of pollutant input that needed control and sources for which control would have little effect. Diamond also found that an unidentified source of substantial magnitude existed in the system.

Impacts of Geochemical Processes on Pollutant Exposure

Bioavailability is the relationship between pollutant concentration in the environment and the dose of a pollutant that each organism receives. Internal exposure of an organism to a pollutant stressor is not related to total environmental pollutant concentrations in a simple manner (Sunda and Guillard 1976; Mackay 1982; Di Toro et al. 1990). Biogeochemical processes influence chemical form and thus the bioavailability of pollutants. These biogeochemical processes are variable in time and space, just as total concentrations of the pollutant are variable.

The factors that influence pollutant bioavailability are chemical specific, environment specific, and species specific, complicating generalizations about bioavailability at a given site. For example, the characteristics of the pollutant determine its interaction with environmental conditions. Organic chemicals that are more hydrophobic and more lipophilic are concentrated more efficiently by biota and interact strongly with natural organic materials. Metals bioaccumulate differently and differ widely in speciation responses to a single set of geochemical conditions. In addition, geochemical characteristics of the water column and the sediments vary spatially and even temporally among systems. Such variations influence chemical speciation, partitioning between solid and solute forms, and associations on or within particulate material, and those influences are specific to each pollutant. Biological characteristics of the species may influence chemical and environmental characteristics in different ways or in different places. For example, in aquatic systems, animals that feed at the sediment surface may be particularly at risk to contaminants recently accumulated in detritus; those that feed at depth may be more exposed to sulfide-rich sediments or aged sediments. Both of the latter affect bioavailability. Alternatively, animals that feed in intertidal zones or near the top of the water column may be exposed to higher levels of ultraviolet (UV) radiation that may exacerbate the toxicity of some contaminants (e.g., polycyclic aromatic hydrocarbons [PAHs]).

Incorporating influences of environmental variability into assessments of bioavailability is just beginning. Metal availability from sediments is an especially difficult problem. Processes are difficult to observe, or experimentally simulate, at the scale relevant to many biota because redox gradients are spatially complex in 3 dimensions and temporally variable (Meyer et al. 1998). Samples for geochemical analysis of sediments and pore waters are typically collected at a more coarse scale than the microhabitat scales experienced by many benthos. The geochemical and ecological influences of the living component of sediments (diatoms, meiofauna, bacteria) are also difficult to separate. Microflora, microfauna, and meiofauna are the major food sources in sediments and have many important influences on sediments and the exposures of macrobiota to stressors associated with sediments (Decho and Luoma 1996).

Impacts of Biological Processes on Pollutant Exposure

Exposure to a stress is not just a function of localized exposure to a single medium such as contaminant concentrations in sediments or water (Clark et al. 1990; Luoma et al. 1992; Luoma and Fisher 1997). A population's exposure to a chemical in its environment is

- integrated among media,
- influenced by ecosystem-level biological or ecological processes,
- a function of the range and population biology of the species, and
- influenced by the habitat preferences of the species.

Pollutants are an example of a class of stressor that is distributed among solution, suspended particles, sediments, pore waters, and specific (living and nonliving) food sources. Food can be as important an exposure route for pollutants as water (Thomann et al. 1995; Wang et al. 1996). Thus each species' exposure to pollutants is determined by how that species "samples" the complicated water, suspension, and sediment milieu. In biological terms, consideration of singular environmental media (e.g., dissolved chemicals, sediment quality) is artificial and can inhibit understanding of exposure to stress. The combined variability of the components of the milieu will affect the variability of exposure to the stressor, and this is a powerful argument for the use of in situ exposure and test systems (see Chapter 1).

Ecosystem-level biological or ecological processes can affect the exposure of the food web to stressors (see also Chapters 3 and 4). For example, south San Francisco Bay, California, USA has a recurrent spring phytoplankton bloom that occurs once per year. Metal concentrations and partitioning between dissolved and particulate phases changed rapidly during the bloom in 1994 (Luoma et al. 1998). Concentrations of dissolved Cd and Zn were rapidly depleted (Cu was not), and particulate Cd concentrations (per unit mass) increased as the plants took up these metals. A mass equivalent to between 60% and 90% of the total annual input of Cd, Ni, and Zn from local waste treatment plants was removed from solution by the phytoplankton during the bloom. Bioaccumulation modeling and associated experiments (Lee and Luoma 1998) predicted that Cd exposures in bivalves would increase during the bloom if the animals selectively switched to a diatom diet enriched in Cd.

Ecological range is important in determining population responses to stress. Mechanisms (e.g., migration) exist to compensate for stress in a small portion of the range of a population. For example, where the spatial distribution of a stressor is patchy, a common occurrence, the geographic scale of pollutant impact may be smaller than the geographic range of motile species (e.g., fish; Sindermann 1996) or the geographic range of the larvae of some sessile species. One result is that populations of adults can persist in stressed habitats through immigration of tolerant individuals from less stressed habitats, even if the local population is incapable of sustaining itself over time. Similarly, if populations are absent from a few habitats where exposures to stress are high, the overall effects on population abundance might not be measurable if the species have a large spatial range (Sindermann 1996 cites some commercial fish populations).

Variability in Ecological Responses to Stress

Variation occurs at all levels of biological organization and is an attribute of any biological endpoint measured in a study. In biology, variation is essential for environmental adaptation, selection, and evolution. Understanding the natural variability in a biological variable is essential to separating changes in the response of that variable to stress from fluctuations that occur in the absence of stress.

Unfortunately, natural fluctuations in fundamental biological properties are often poorly known. For example, understanding whether a population is declining in abundance requires understanding the natural fluctuation in its numbers over time (Blaustein, Wake, and Sousa 1994). Considerable effort and much debate have also been devoted to the question of identifying attributes of a "stable" population or community (see Chapter 2 for some examples; Connell and Sousa 1983; Chesson 1986; Chesson and Case 1986).

Biological properties are stochastically variable, but a repeatable "dynamic stability" is also common, and characterizing both can be critical. For example, "condition index" (CI) is defined in estuarine bivalves as the weight of tissue within an animal of standard shell length. Condition varies naturally in bivalves, increasing as glycogen and lipid are added to the gonadal tissue mass during the early stages of seasonal maturation (Buekema and DeBruin 1977). These energy reserves are lost as gametes are released and the animal reproduces. For some bivalves, tissue mass of animals with the same shell length can change as much as 50% with the seasonal reproductive cycle. CI is also responsive to any environmental change that affects the energetics of the animal. In experiments, reduced availability of food results in reduced CI (Bayne and Widdows 1978). Exposure to pollutants or natural stresses can also affect condition (Couillard et al. 1995). But detecting the response of CI to stressors in nature requires separating the stress response from the natural cycle.

Figure 5-1 shows CI in the bivalve *Macoma balthica* at 2 sites in south San Francisco Bay, designated as PA and SJ/SV. Data were monthly sample collections between 1988 and 1998. Averaging the proportionate change by month (called "detrending the interannual data") (Figure 5-2) illustrates the typical seasonality of condition. The annual addition of glycogen and lipid occurs from March through July, as reproductive tissues ripen or mature. Then energy reserves are consumed through the rest of the year, partly in association with the release of gametes and recovery from reproduction. The seasonal cycle is similar in period and timing at PA and SJ/SV (Figure 5-1). An important feature, which becomes evident after several years of intensive sampling, is that the amplitude of the seasonal cycle in condition varies from year to year and between sites. Animals at PA tended to reach a better peak condition (Figure 5-1) than animals at SJ/SV. Variation in amplitude, variation in period, and changes in the characteristics of the seasonal cycle represent responses to environmental variability and/or stress. The baseline seasonal cycle must be well known to detect these responses.

Temporal and Spatial Scaling Issues

How can natural variability be partitioned from stressor effects, and how should this best be achieved? The answers depend on the questions of interest in a particular study and on the spatial and temporal scale at which those questions operate.

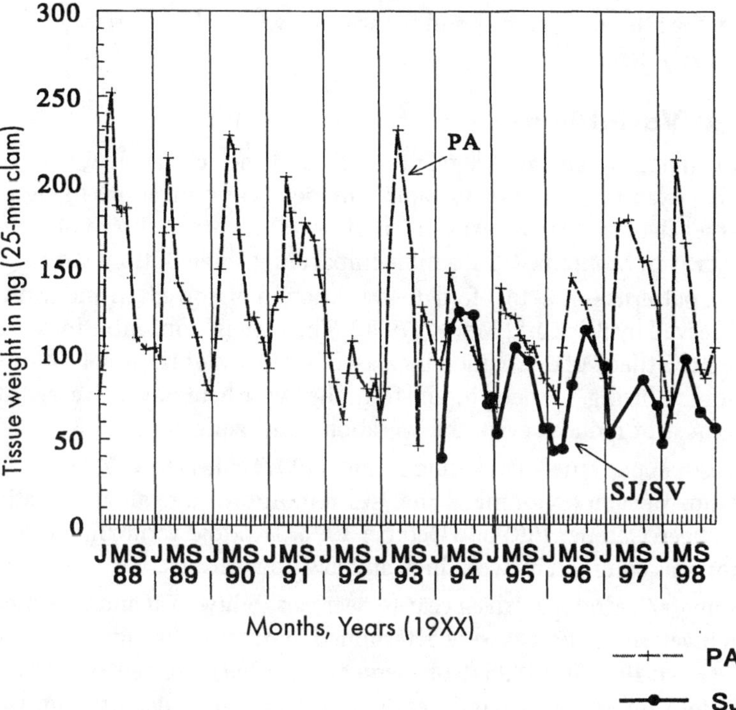

Figure 5-1 CI in clam *Macoma balthica* from 2 sites in San Francisco Bay, California, USA, 1988 to 1998

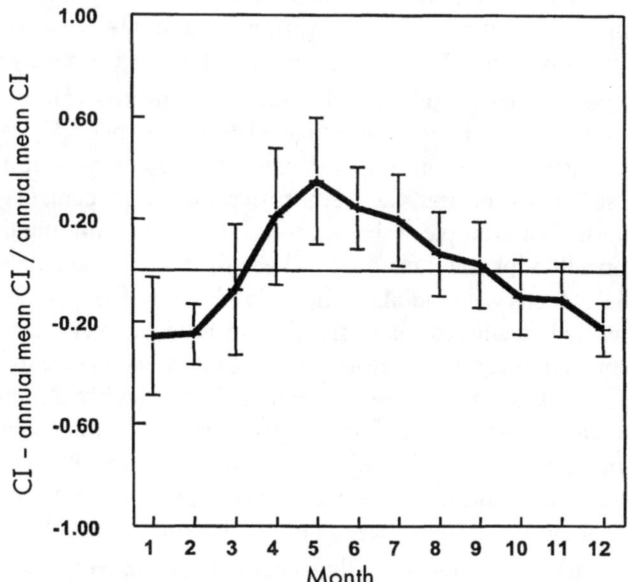

Figure 5-2 Monthly average CI of *Macoma balthica* from San Francisco Bay, based on data in Figure 5-1

Therefore, the relevant variables and the scale on which these variables are examined must be clearly defined.

Scales of Variability

Variability in nature can have fractal properties: At each closer view, a new scale appears. For example, temporal variability in metal concentrations in sediment or biotic tissues (which are good exposure indicators) occurred at several scales in the Clark Fork River, Montana, USA, a mine-impacted, cobble-bottom stream:

- Diel patterns—A 2- to 3-fold diel variation in dissolved Zn and Mn was observed by Brick and Moore (1996). The highest concentrations occurred at night. Either Mn dissolution increased because nighttime conditions were more reducing, or metal inputs from the hyporheic zone increased at night because of reduced evapotranspiration in the zone.
- Seasonal patterns—Drake and Moore (1997) collected < 64 μm sieved sediment samples for metal analyses at monthly intervals at 4 locations between October 1991 and October 1996. Two-fold variability in concentration was common between high and low flow periods.
- Climate/hydrology-driven year-to-year variability—Cd and Cu concentrations in invertebrate tissues were determined in August, during low river discharge, between 1986 and 1995 (Hornberger et al. 1999). Concentrations varied 4- to 10-fold among years in mid-reach stations. Tissue concentrations were lowest during years of the lowest annual river discharge and highest during years of the highest annual river discharge ($r^2 = 0.85$ at 1 site). Presumably, this reflected differences in particulate metal input from the contaminated floodplains of the river. Interannual variability in tissue concentrations (among 10) was 3- to > 10-fold higher than within-site variability (in 1 y).
- Event-driven changes—Spills, accidents (e.g., tailing pond failures; van Geen and Chase 1998), and floods (Macklin et al. 1997) are periodic events that can result in disastrous episodic inputs of contaminants to mine-impacted rivers. On time scales of centuries, these events may ultimately control metal distributions. For example, a single flood can deposit contaminants that remain on a floodplain for centuries. The sum of smaller events may also be influential. Metals were mobilized from floodplains of the Clark Fork after a large river ice jam and associated flooding in 1996. Such events occur on a decadal basis in this ecosystem and may periodically enhance metal contamination in deposits in the system. Episodic pulses of highly contaminated runoff once entered the Clark Fork as surface runoff from contaminated floodplains every summer during strong thunderstorms (Phillips and Spoon 1990). Many such episodes were accompanied by fish kills in the river, probably of sufficient frequency to limit maturation of locally breeding fish.
- Longer-term trends in human influences—Human activities can also add variability to stress. For example, Hornberger et al. (1997) found declining

metal concentrations in sediments adjacent to remediation projects within a decade after the projects began. Sites farther downstream than 30 km of the remediation showed no decline from earlier levels of contamination.

No single study can consider all scales of variability. But a conceptual model of the system must exist that includes recognition of changes in the exposure to stress, in the response of interest, and in the natural system. Then the scales of variability critical to the question must be identified and included in the sampling design. It is crucial, in experimental design, to constrain characterization of environmental factors to aspects relevant to the study question. On the other hand, failing to incorporate flexibility in study designs can result in overlooking aspects of variability that are critical for management interpretation. The components of variability should be examined for relevance and scale, and those that are known to be important, or known to be potentially confounding, identified. If potentially important factors are not well known, it is probably best that they are overmeasured in the initial phases of the study until their influence is better understood. Then a decision can be made whether or not to continue measuring them, and at what scale. Environmental factors that are unlikely to bear on the question may be eliminated from the routine sampling protocol, but this decision probably should be revisited on occasion. Figure 5-3 shows a series of practical questions that might be applied in sample design. Most important is the principle that inclusion of natural variability is most manageable when the scale of the question is well matched with the scale of sampling.

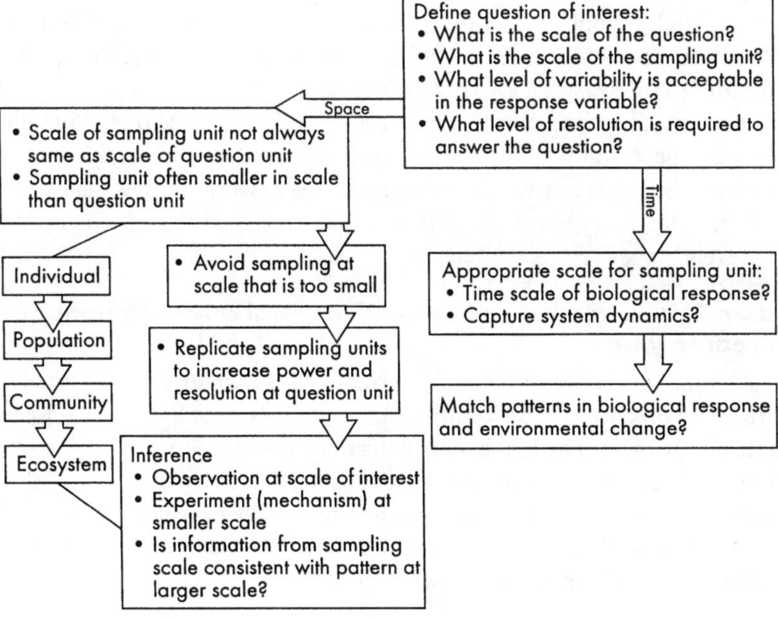

Figure 5-3 Conceptual relationship in space and time between study questions ("question units") and response variables ("sampling units")

Identifying Critical Temporal Scales: San Francisco Bay Case Study

A 25-year study at mud flats in south San Francisco Bay illustrates several patterns of temporal variability associated with patterns in driving forces. The management-related questions each required different temporal sampling scales.

Question 1: How does hydrological variability influence metal concentrations in bay sediments?

Hydrology varies daily, seasonally, and year to year. Capturing any one of these scales of variability requires frequent sampling at the temporal scale below the scale of interest. Figure 5-4 shows seasonal variability in concentrations of Cr, V, and Ni at a south San Francisco Bay site (SJ/SV), as depicted by near-monthly sampling. The primary source of these 3 metals to the sediments is the terrigenous soil in the watershed (Hornberger et al. 1999). The time series followed similar patterns for the 3 metals, suggesting their variability was linked to the same driving force. In 3 wet years (1995 to 1997), peak concentrations of metal occurred in the months of highest freshwater inflows to the estuary, between January and June. In a dry year, 1994, no January to June peak occurred, however. Concentrations typically were at a minimum during the dry season of the year (July to December). Mechanistically, winds and runoff inputs combine to drive the trend. January to March runoff brings in particles with higher metal concentrations, and a greater dominance of finer-grained particles occurs in sediments because diurnal winds are weakest in this season. Fine-grained sediments are progressively winnowed from the bed in the summer and fall seasons when strong daily winds resuspend sediments and mix the bay (Thomson-Becker and Luoma 1985; Schoellhamer 1996). Because of the irregularities in the seasonal cycle, sampling only seasonally would not have identified the pattern. Sampling monthly, at a scale smaller than the question unit (Figure 5-3), was the successful strategy. Diel variability may also occur in this mudflat environment, but the relative predictability of the seasonal cycle suggests it was probably not necessary to sample with a frequency greater than monthly to identify seasonal trends.

Question 2: Does a progressive trend in metal concentrations occur from year to year?

It was essential to sample with sufficient frequency to identify the seasonal cycle, in order to avoid misinterpreting trends, especially since only 5 years of data were available (Figure 5-4). The higher concentrations observed in 1995 compared to the previous year were most likely a return to seasonality rather than a unidirectional increase in concentration. Thus, intensive sampling over several climate regimes helped clarify seasonality, which was necessary to help separate variability in seasonality from unidirectional trends.

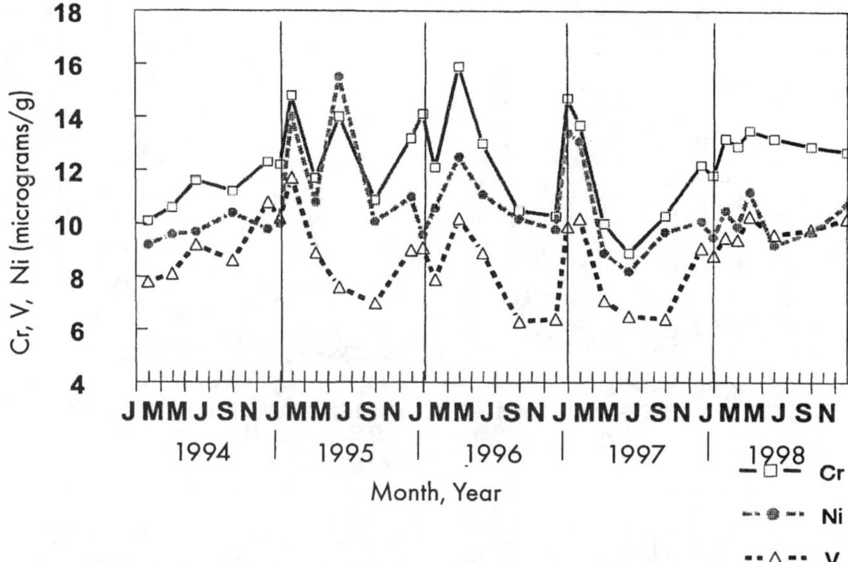

Figure 5-4 Cr, V, and Ni in sediments of San Francisco Bay, 1994 to 1998

Question 3: Can we detect the influence of episodic events against a varying background?

Episodic events may be one of the most important features of the variability of exposure to stress. Sampling to identify a source that could be episodic must be of sufficient frequency or properly timed to identify episodes. For example, Hg concentrations in sediments at SJ/SV varied quite widely between 1994 and 1998, from 0.3 to 3.0 µg/g dry weight (dw; Figure 5-5). The highest concentrations occurred only following some large storms and high freshwater inflow events (January and June 1995 and 1998; January 1996 and 1997). Because there is an abandoned Hg mine in the watershed of the SJ/SV site, erosion of mine tailings seems the most likely source of the Hg. It is notable that the sponsor of this study was initially interested in supporting only 3 samplings per year. It is unlikely that schedule would have uncovered the pulse inputs of Hg, the primary source of this contaminant to the estuary. A schedule of 6 to 7 samples per year did identify the episodic inputs and illustrated the need for flow-weighted load evaluations from the mine-affected stream. Sampling on near-daily time scales during the critical time of year would also be necessary to better characterize the fate of the Hg.

Question 4: What sampling frequency is optimal to determine long-term trends?

Many monitoring studies are conducted with the goal of identifying decadal-scale trends. In such studies, the length of the time series is critical to sensitivity in trend identification. Figure 5-6 shows the 25-year annual mean concentrations of Ag in

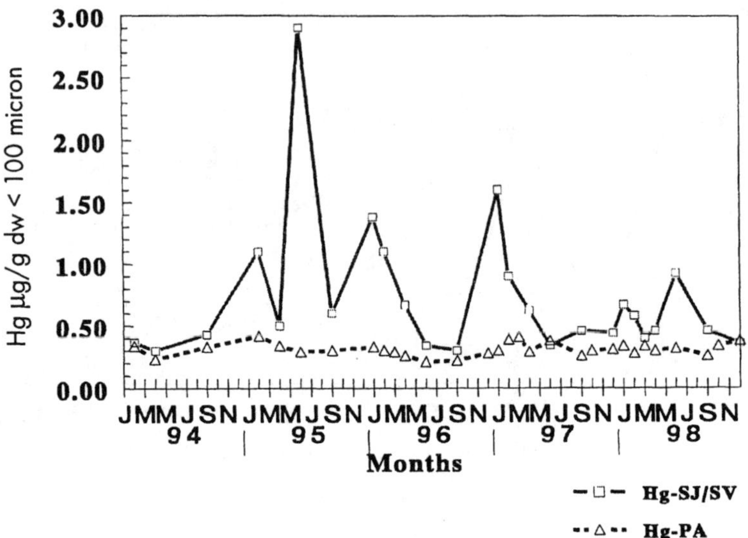

Figure 5-5 Hg in sediments of 2 sites in San Francisco Bay, 1994 to 1998

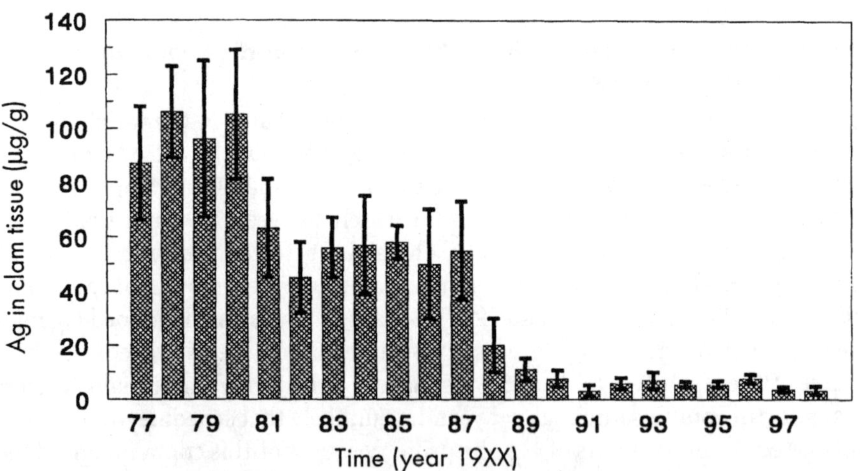

Figure 5-6 Annual mean Ag concentrations in tissues of clam *Macoma balthica* from San Francisco Bay, 1977 to 1998

sediments at PA. The long-term decline in Ag occurred because of improved waste treatment processes at a nearby waste treatment facility (Thomson et al. 1984). Again, the scale of sampling and interpretation (annual) was necessarily smaller than the scale of the question (decadal). Each bar represents the standard deviation (SD) from the running mean of 9 to 11 samplings in each year. In retrospect, sample collection on the monthly time scale was probably not necessary to detect the long-term trend. Less data would have sufficed, but only if they were collected during the same season in each year. Infrequent sample collection would have delayed trend detection, and failure to at least stratify by season could have led to spurious conclusions, especially early in the study. The dramatic nature of the long-term changes at PA also emphasize that directional change in human impacts on ecosystems can occur. Ecological studies must recognize that stressors from human activities often are changing unidirectionally as societally driven environmental management goals change (Luoma et al. 1998; van Geen and Luoma 1999).

Scaling Management Questions to Appropriate Sampling Units

Successfully linking stressor, response, and factors affecting their variability requires matching scales. The question of interest may differ from the unit at which the question can be sampled and studied ("sample unit" in Figure 5-3). In many cases, the question unit encompasses a greater scale than the sample unit (see Figure 5-3). Failure to match scales can impair or prohibit causal inference. For example, atmospheric deposition of sulfates from fossil fuel combustion, a major component of acid rain, occurs on a subcontinental scale (eastern North America, northern and eastern Europe, eastern China) over a period of many decades (Schwartz 1989; National Acid Precipitation Assessment Program [NAPAP] 1991). Because of the wide area and long time span, suitable control or reference environments were difficult to identify. The natural acidity of some low alkalinity waters added to the challenge of understanding effects of pollutant deposition (e.g., Schindler 1988). Demonstrating a causal relationship between acidic deposition and biological effects was controversial to the extent that small-scale, short-term studies were employed to understand large-scale, long-term effects. The controversies were not a shortcoming of the investigations but resulted from an inability to directly replicate the scales of the natural experiment (regional sulfur oxide emissions and their consequences), combined with the lack of sufficient environmental monitoring data collected before widespread acidification was suspected.

Often a stressor varies over a large scale, yet operates on biological responses at the local scale. In these situations, the question unit should encompass a scale over which the stressor varies from high to low values or is both present and absent. For example, in examining the effects of toxic metals in streams, reaches should be selected with a range of toxic loads from little or no loading to extreme loading (e.g., Clements 1994). Processes that vary on the same scale as the stressor, and that may affect the strength of the stressor, should be considered in the study design. One way to address the question is with sampling units selected in space or time such that the stressor varies but the other controlling variables are held constant (stratification). Variability in biological response within sample units is generally not of interest, nor of use, in addressing the primary question. Variability at a local scale may of course confound or reduce the power of the primary analysis.

In some cases, the biological response varies at a large scale while the stressor varies at a much smaller scale, another type of mismatch of scale. One example is the study of effects of pesticide use in agroecosystems by applying different pesticide treatments to small (0.5-ha) fields. The indicator of biological response is survival of insect species. Many such species have geographic ranges of hundreds of hectares. If the insects can disperse into and recolonize the plots, then the ability to detect local effects is confounded by the mismatch of small sample units with insect populations (the question unit in this case) that exist at much larger spatial scales than the sample units.

Matching Management Scale to Study Scale: Fraser River Case Study

For many management scenarios, the level of decision-making is often at 1 or 2 scales above that at which much work has been done. This mismatch can result if data were collected to address questions other than those of interest to decision-makers. For example, a study of the Fraser River was specifically designed to provide a prospective dataset for assessing biological response to multiple stressors across a 250,000-km^2 catchment. An objective of the study was to describe and explain natural variation across the catchment so that this could be accounted for when assessing sites exposed to stresses. The variation for 119 families of aquatic benthic invertebrate from 260 sites in the basin was described in ordination space.

The range in variability at the catchment scale is shown in Figure 5-7A, partitioned based upon the similarity within faunal assemblages. Differentiation of a site exposed to stressors was determined by comparison with the variability observed among the sites, as measured by faunal assemblage composition. Therefore, the scale of interest for describing the variability in assemblages is the catchment (Figure 5-7A). As the spatial scale becomes smaller at the subcatchment level (Figure 5-7B) and within a reach (Figure 5-7C), the variability is reduced. In this case, the replication unit is the site (reach). Committing resources to describing variability within the reach, a common strategy in stream sampling, is unnecessary for the objective of the study. The appropriate scale at which to sample repeatedly is the site (reach) level, in order to increase power in understanding the high variability within a catchment.

Temporal variability was examined at 2 scales, relative to that observed across the catchment (Figure 5-8). Seasonal variation is presented for 5 sites and can be seen to be much greater than annual variation (over 3) at a single site. Therefore, in explaining variability in community assemblages, effort should be spent in either capturing this variability or discounting it when conducting site assessments. Eliminating the consequences of seasonal variation can be assured by sampling potentially stressed sites at the same time the reference site data are collected; this is the approach employed by the River Invertebrate Prediction and Classification System (RIVPACS) study design (Wright et al. 1984). Alternatively, reference sites should be sampled

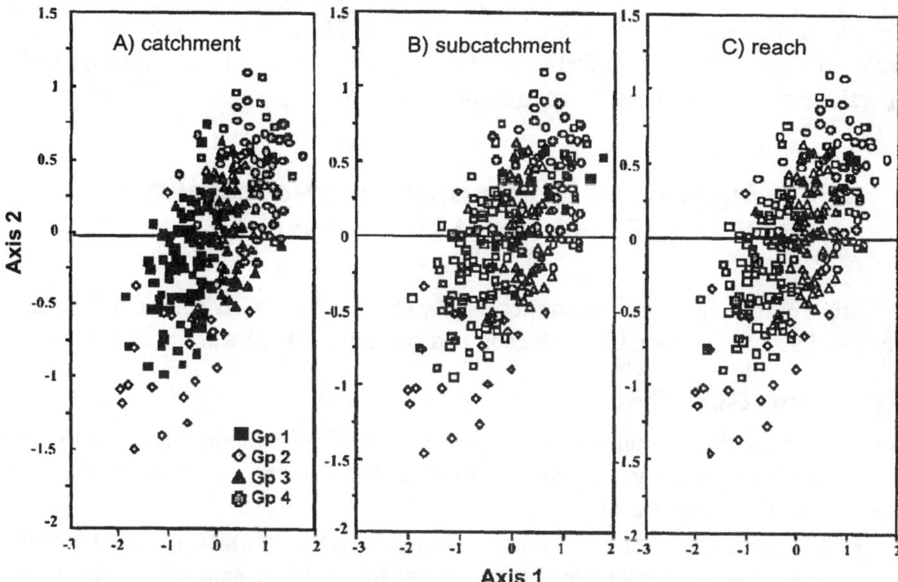

Figure 5-7 Range in variability of benthic community assemblages from the Fraser River, Canada, as measured by site ordination at 3 levels of spatial scale: A) catchment, B) subcatchment, and C) reach

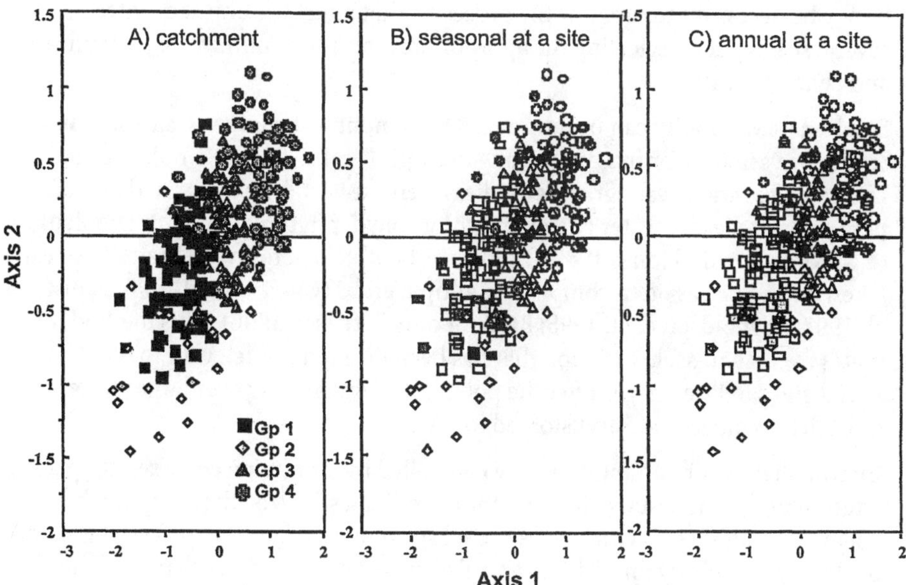

Figure 5-8 Range in variability of benthic community assemblages from Fraser River, Canada by site ordination at catchment level A and at 2 temporal scales: B) seasonal at a site and C) annual at a site

across multiple seasons (see example in Chapter 6). The latter is more expensive but makes the application of predictive models more flexible. The former is more cost-effective but reduces the utility (power) of the dataset.

Measuring Environmental Variability: Experimental Design Issues

Experimental design is critical in successfully characterizing environmental variability. The following section illustrates some important principles about that design.

Sampling Guidelines

Variability at different scales may be either characterized by sampling or minimized (and made irrelevant) by appropriate stratification or sampling methodology. There are 3 general rules of thumb:

- Minimize (by sampling methodology) the variability that operates below the spatial or temporal scale of the object of interest (i.e., sampling unit).
- Stratify or select components that operate at the scale of the sampling unit.
- Characterize components that operate at a higher scale than does the sampling unit.

It may be necessary to monitor the processes that affect the object of interest on several scales before selecting the appropriate scale for minimization, stratification, and characterization.

Small-scale variability can be minimized by sampling at a larger scale than the source of variation, or by compositing multiple "grabs" of the sampling gear. Limnological and oceanographic studies often use vertically integrated grabs or pumped samples for water chemistry, chlorophyll, phytoplankton, or zooplankton (e.g., Knowlton and Jones 1989). Studies of benthic macroinvertebrates in streams, lakes, and estuaries often composite multiple grabs (e.g., Holland 1990; Paulsen et al. 1991; Reynoldson et al. 1995; Barbour and Gerritsen 1996). This method integrates over small-scale variation that has been deemed not relevant to the study. Aerial and satellite images offer the potential of integrating over larger scales than most field ecologists are accustomed to.

Intermediate-scale variability can be controlled by stratifying or by selecting a single stratum under the assumption that the central questions of the study can be adequately addressed within the selected stratum. In estuarine studies, it is common to stratify on salinity for all biota and also on substrate and depth for benthic invertebrates (e.g., Weisberg et al. 1993, 1997). Stream sampling protocols for benthic macroinvertebrates may consist of a single habitat type, such as riffles (e.g., Kerans and Karr 1994) or may consist of composited sample sweeps taken in consistently defined multiple habitats (e.g., Barbour et al. 1999; Maxted et al. in press). Lake benthic sampling may be stratified by depth zone (Johnson 1998). Biological variables subject to seasonal temporal variability are sampled frequently enough to obtain an estimate for each relevant season, or they are sampled during an index period when observed values are relatively stable. For example, the Environmental Monitoring and Assessment Program (EMAP) sampling program selected a midsummer index period (1 observation for each routine site) for benthic macroinvertebrates and fish (Holland 1990; Paulsen et al. 1991). Sampling for chlorophyll and other constituents in lakes and estuaries typically requires at least monthly sampling during the growing season to obtain a growing season average (e.g., Knowlton and Jones 1989). An exception to this is nutrient (total N, total P) concentrations in temperate lakes: A relatively stable estimate can be obtained from a single sample in early spring, during spring water mixing, and before early diatom blooms have depleted nutrients (e.g., Linthurst et al. 1986). Vegetation can often be sampled on annual or longer intervals.

Some characterization of sources of variability is unavoidable in studies of nature. Not all influential variables can be stratified or minimized by adjusting sampling methodologies. Variability in natural or anthropogenic factors also can be exploited to generate experiments, but those experiments cannot be interpreted if the influential variability is not itself characterized. For example, in a long-term trend study, it would be necessary to characterize seasonal variability over several years in San Francisco Bay sediment–metal concentrations before we could safely stratify by season (Figure 5-4). The characterization shows that stratified fall sampling of Cr, Ni, or V would give values less subject to year-to-year variability than would spring sampling. If the question is what is the source of Cr or Ni, estuarine variability created an experiment that allowed development of a hypothesis. The seasonal peak in concentration suggested freshwater inflows as a source. The lack of a spring peak in concentration in the drought year further verified that source. Characterization of seasonality over multiple years was the appropriate sampling strategy.

Statistical Power

In general, most studies aim for the maximum statistical power possible in order to maximize confidence in conclusions. Prior information on the magnitudes of spatial and temporal variability in an object of interest (e.g., an exposure or response variable) can be used to determine the number of samples that must be collected to detect a true difference in the variable across treatments, with a pre-defined level of statistical significance (i.e., confidence). Power analysis is used to estimate the number of samples that will be needed to detect an effect of a certain size (say, 20% loss of taxa richness) with a given confidence (say, 90%) and a given power (say, 90% probability of detection). The tradition of 95% confidence (or $p < 0.05$) in statistical tests of significance is arbitrary; there is no scientific requirement that confidence be set at 95%. Indeed, for environmental protection, where we wish to prevent adverse effects, power may be more important than confidence (Peterman 1990; Fairweather 1991).

Assessment of effects and power analysis require careful consideration of effect size. Demonstration of statistical significance is increasingly possible with a very large sample, but the effect size, the actual magnitude of the change, is usually of more interest. Effect size is the estimation of the biological and ecological significance of an effect. There are few simple rules to guide conclusions about the important size of an effect, however. For example, we can ask, "How many species must be extirpated from a habitat before we consider it a significant loss?" There may not be an objective answer to the question, unless some other system function or property (e.g., ability to recover from disturbance) can be shown to change with the loss. In general, protection from loss or degradation is justified by the societal value placed on intact, natural ecosystems, as well as by the goods, services, and natural capital provided by those systems (e.g., Goodland and Daly 1996). But quantitatively relating these considerations to the amount of species loss is uncertain (see also Chapters 3 and 4).

In any field study, there are sampling effort tradeoffs between sampling thoroughly at a few sites (intensiveness) and sampling sparsely at many sites (extensiveness). An increased sampling effort to characterize small-scale variability in time can compromise our ability to characterize larger-scale variability in space or to extrapolate responses to sites outside the study area. Clear statement of the question or sampling objective is necessary to identify the acceptable tradeoff. A priori identification of the appropriate spatial, temporal, and biological scales at which the objects of interest operate may also help evaluate where to expend the available effort, but choices between tradeoffs are inevitable (see "Identifying Critical Temporal Scales: San Francisco Bay Case Study," p 150). Power analysis is invaluable for helping to determine the necessary sampling effort (and costs) for both intensive and extensive study designs.

Study Design Issues: Reference Condition Example

The fundamental concept behind the reference condition approach is to identify the range of features that characterize unimpaired communities from a database of biological and environmental attributes measured at unimpaired sites. The database is then used to develop predictive models that match environmental variables to biological features. A set of environmental measurements are then made at a new site and used in the model to predict the expected biological condition. A comparison of the actual biological condition to the predicted condition allows an assessment of impairment at the new site.

Reference sites must be selected carefully because they form the benchmark against which test sites will be compared. The reference sites should represent the full range (variability) among minimally impaired conditions that can be achieved at sites anticipated to be ecologically similar. The determination of the reference condition from reference sites is based on the premise that sites least affected by human activity will exhibit biological conditions most similar to those at natural, pristine locations. The reference condition is described using biological attributes. It is important to recognize that there is no single reference condition in nature. Thus the reference condition is defined by a set of possible reference states and is quantified for the comparisons using a probabilistic model based on environmental site attributes.

Three basic characteristics exist for describing a suitable reference condition (Hughes 1995):

1) It must be politically acceptable and reasonable.
2) It should represent a sufficiently large number of sites or areas of reference within waterbodies.
3) It must represent important aspects of natural conditions. Detailed examples of the reference condition approach are presented in Chapter 6.

How to cope with environmental variability in field studies involving multiple stress impacts

- Capture natural variability:
  - Minimize (by sampling methodology) components that operate below relevant spatial or temporal scale.
  - Stratify some components (especially at the scale of the sampling unit).
  - Characterize the most important components, especially if they operate at a higher scale than does the sampling unit.
- Carefully state questions and identify critical sampling scales.
- In design, maximize power and identify effect size.
- Sample at units smaller than the question unit.
- Oversample unknown sources of variability in initial phases of the study.
- Constrain long-term characterization of environmental factors to those most likely to have important influences.

Environmental Variability and Selection of Ecological Response Variables

Ecological response variables are selected for a variety of reasons. Investigators may be interested in one or more particular species, biological assemblages or communities, habitats, or ecosystem functions. In other cases, investigators may be interested in evaluating risk from a particular stressor in a particular habitat. In many cases, an indicator species, assemblage, habitat, or function is used as a surrogate for a wider diversity of ecological variables. Ecological response variables can also be useful with regard to understanding influences of environmental variability.

The selection of a response variable or an indicator should be guided by the question of interest, by the scales relevant to that question, and by the nature of environmental variability that affects the stressor and the response variables. We discuss 4 critical issues related to the selection of appropriate ecological response variables:

1) identifying appropriate taxonomic scales for indicators,
2) using single species as ecological indicators,

3) using assemblages of species as indicators, and
4) using ecological functional groups as indicators.

We also provide examples of how knowledge of natural environmental variability affects the utility of each type of indicator.

Identifying Appropriate Taxonomic Scale

A main purpose of monitoring studies is to provide data to support scientifically based management of habitats for the restoration or maintenance of ecological function or integrity (Noss 1990). Aside from monitoring the relevant variables at the relevant scales, a program should describe a baseline or reference condition and characterize natural fluctuation over time. It should also include an experimental component that evaluates the mechanistic effects of anthropogenic and other stressors (Kremen 1992). The details of monitoring will be strongly associated with the size and shape and the biological and physical heterogeneity of the habitat in question and also with the relative impacts of anthropogenic and natural disturbance regimes and local and global climatic changes. The choice of the taxonomic scale for an indicator of ecological response must be suitable for all these monitoring criteria. Species or species assemblages are often the best choice as indicators of community composition, structure, and ecological function, although there are many other biotic and abiotic indicators available at all biological levels (Noss 1990).

What is the unique value of studying species or species assemblages? First, such studies permit direct assessment of management goals. Single-species monitoring allows tracking the status of a species at risk. Monitoring species assemblages allows evaluation of biodiversity. Second, such studies may provide direct or indirect insights into ecological function, although this is narrow (i.e., investigation of trophic interactions) compared with direct measurements of functions such as nutrient cycling. The purpose and value of using a species or an assemblage as an indicator is, therefore,

- to observe biotic response to environmental stress,
- to provide early warnings of natural responses to environmental impacts, and
- to measure the success of management or restoration options (Noss 1990).

What general properties should indicators exhibit? Once the questions of interest are clearly identified, then one general consideration is important: The indicator should facilitate continuous assessment over the range of stresses encountered (Noss 1990). To enable this, it is vital to consider aspects of habitat and environmental heterogeneity explicitly. For example, assemblages that encompass a range of dispersion (low to high vagility, local to widespread distribution) and that differentially occupy a wide range of available subhabitats (specialists and generalists) might be expected to display a greater range of sensitivities than do more narrowly selected sets of taxa (Terborgh 1974). Explicit rules for the selection of an appropriate indicator can be useful. The rules should permit matching of biological or ecological attributes to the nature of the stressors and the associated variabilities in habitat and environment. No single selection criterion (e.g., sensitivity, size, specialization, turnover rate, or range) can be used for all questions (Landres et al. 1988). However, question-specific guidance can be established to increase the probability of making an appropriate choice.

Organism-at-risk as Indicator: Amphibian Population Declines

In evaluating environmental stressors, we often assume that the response of hardy species will indicate the response of more sensitive species at risk. However, in some situations, we are unaware of an environmental problem until a change in the status of a sensitive indicator species alerts us to a problem. Amphibian population declines present an example of this phenomenon. In recent decades, extinction rates, population declines, and range reductions of amphibian species appear to have increased (e.g., see Crump et al. 1992; Blaustein, Wake, and Sousa 1994; Pounds et al. 1997).

The unique biological attributes of amphibians contribute to their sensitivity to environmental pollution. For example, amphibians typically have a biphasic life cycle, laying eggs in water where the larvae develop into a mature terrestrial form. Thus amphibians are sensitive to environmental impacts in both the aquatic and terrestrial habitats (Blaustein, Wake, and Sousa 1994). Adult amphibians typically have thin, moist skin that is relatively permeable to pollutants; thus they may be especially sensitive to some types of pollutants. In addition, adult amphibians may have difficulty colonizing new sites because of their high fidelity to a particular site and because they must remain near water as they migrate (Blaustein, Wake, and Sousa 1994). No surrogate species can replicate this unique combination of sensitivities. Amphibians may also be indicators of indirect effects in biological communities. Amphibians are critical links in the food chain as both predator and prey in many meadow, forest, stream, and pond ecosystems. Therefore, amphibian population declines could have marked consequences for local ecology.

Sensitive organisms such as amphibians may be early indicators of environmental damage that would otherwise be detected only after the damage is so severe as to be irreversible (Lubchenco et al. 1991). The global nature of amphibian population declines may be an indicator of some previously undetected effect of complex environmental disturbance. Because of the multiple sensitivities of amphibians, they could be more sensitive than typical surrogates to combinations of multiple interactions among stresses induced by a common general disturbance. In general, preserving natural habitat is vital to preserving a declining species. With larger habitat area, demographic stochasticity of a population may be reduced, making the population more resistant to environmental stochasticity (Gilpin and Soule 1986; Shaffer 1987).

Directly studying an organism at risk presents many challenges, including determining the natural population variability, the variable etiologies that can be associated with population declines, and the apparent global scale of problems such as amphibian declines. It is essential to separate natural variability from declining populations in order to understand whether populations are exhibiting trends over time or are simply fluctuating within a normal range (Connell and Sousa 1983; Blaustein, Wake, and Sousa 1994). Long-term studies of amphibian populations have reached different conclusions in this regard. The contradictions occur partly because baseline information is so rare. Studies have suggested that no declines were occurring (e.g., Meyer et al. 1998), others concluded that declines were occurring and could be linked with stressors (e.g., Pounds et al. 1999), and still others suggested that declines were occurring but that no stressor could be implicated directly. For example, a study of the common frog (*Rana temporaria*) in Switzerland over 2 decades revealed that populations were not declining except in specific locations in which fish were introduced (Meyer et al. 1998). Pounds et al. (1997) developed statistical models for an amphibian assemblage in Costa Rica, concluding that amphibian population declines in this region are not within the range of natural variability. Studies of a diverse amphibian assemblage in the southeastern U.S. over 16 years highlighted the difficulties of even determining whether populations are fluctuating within the range of natural variability (Pechmann et al. 1991).

The global scale of amphibian population declines makes this issue particularly challenging. However, the global scale of this issue also creates substantial difficulties in identifying cause and effect. It is impractical to conduct manipulative experiments at the global scale. Identifying and understanding population declines requires both observational and experimental approaches at the local scale.

Populations that are in decline appear to be influenced by different factors in different locations. For example, the young of some amphibian species in the Pacific Northwest are sensitive to ambient levels of UV-B radiation, and some of these species have comparatively low levels of the UV repair enzyme photolyase (Blaustein, Hoffman et al. 1994). Some species that are sensitive to UV radiation also appear to exhibit some range reductions or population declines in parts of their range (Blaustein, Hoffman et al. 1994). Some populations of the Cascades frog (*Rana cascadae*) appear to be declining in California in a manner that may be linked with introduced species and habitat loss (Fellers and Drost 1993). In Australia and Panama, incidence of chytridiomycosis fungal infection may be related to mortality associated with population declines of several amphibian species. Several studies have found interactive effects of UV radiation with other environmental stressors, including PAHs (Hatch and Burton 1998), carbamate insecticide (Zaga et al. 1998), fungus (Kiesecker and Blaustein 1995), and pH (Long et al. 1995). Other experimental studies have investigated interactive effects of pH and metal toxicity (Horne and Dunson 1994); pH, metals, and water hardness (Horne and Dunson 1995a); and pH, metals, and dissolved organic C (DOC; Horne and Dunson 1995b).

Overall, the global scope of the problem of amphibian population declines is challenging to understand from the local scale of most studies. The mix of etiologies requires going beyond simple, single-cause generalizations to explain the global-scale problem. It is unlikely that any single factor will explain all amphibian population declines. Multiple environmental factors probably interact to affect populations at different locations in different ways.

Indicator Taxon Approach: When Is Study of a Single Taxon Advantageous?

It is commonly asserted that evaluations of stressor effects should focus on the significant effects. "Significance" is usually defined as changes in populations or communities. The variance in biological responses increases with the level of ecological organization because of accumulating complexity in the number and types of intrinsic biological interactions (Luoma and Carter 1991). Responses at higher organizational levels may also be manifested over longer time scales than those at lower levels. Because of such challenges, it may be difficult to adequately consider some important influences of environmental variability (e.g., variability of exposure) if studies are completely focused at higher levels of organization. Studies that consider more than 1 indicator and both higher and lower levels of organizations can help overcome such deficiencies. Alternatively, stress–response relationships at the whole-organism level and/or on individual taxa can be used as indicators of broader effects. Reasons for using a single taxon as an indicator or to accompany studies of populations or communities include these:

- Such studies can be practical to carry out over the detailed time and/or space scales necessary to identify roles of multiple stressors or to separate stress from influences of environmental processes.
- If studies consider endpoints that are relevant to the population or community, conclusions about higher-level effects can be inferred.
- Improving understanding of the stress effect can set the stage for effective evaluations at more complex levels of organization or of more complex questions (i.e., as diagnostic aids).

Example 1: Effects of TBT on dogwhelk *Nucella lapillus*

Studies that defined and verified the effects of the antifouling agent, tributyltin (TBT), on gastropods provide examples of how the study of a single indicator taxon can progressively lead to understanding of impacts of a stressor. This example illustrates the synergy between laboratory and field studies and the usefulness of detailed study of one or a few indicators. The proof of TBT effects resulted from observations at different levels of biological organization, but some key findings resulted initially from studies of a single species (as described by Bryan et al. 1986; Gibbs and Bryan 1986):

- Abnormalities in the genitalia of female predatory snails (*Nucella lapillus*) were initially observed in abundance by zoologists familiar with these species.

Field studies demonstrated the co-occurrence of imposex (development of male sexual characteristics in females) and elevated bioaccumulated TBT exposures in snails over broad spatial scales.

- Laboratory studies then showed the mechanism. TBT exposures caused the specific physiological response in specific vulnerable species, at very low concentrations.
- Population-level significance of the effect was inferred from the endpoint (females with the defect were proven to be sterile).
- The onset of the effect, along with TBT bioaccumulation, was shown when animals from an unpolluted environment were transplanted in a TBT-contaminated environment.
- Specific reactions to TBT were linked to damage in other aquatic benthos.
- Observations demonstrated that the populations of the species most sensitive to TBT were in decline over wide geographic areas.

After the details of TBT effects were well understood, the science developed rapidly, expanding the geographical scale and expanding knowledge of the threat to a variety of invertebrates.

Example 2: Impact of heavy metals on *Macoma balthica*

Constraining studies to an individual taxon can also partition effort so that it is practical to develop intensive, interpretable time series over long periods. For example, near-monthly sampling of the clam *Macoma balthica* was conducted over 25 years at the Palo Alto mud flat in San Francisco Bay (Nichols and Thompson 1985; Hornberger et al. 1999). Other data were also collected simultaneously: analysis of sediments for metals and geochemical characteristics, salinity determinations, consideration of relevant hydrologic and hydrographic data available from other sources (Hornberger et al. 1999). CI (average weight in a standardized shell length) was monitored as a response variable, and animals were archived that were later employed to determine reproductive condition. The strategy sacrificed spatial resolution for temporal intensity (Luoma and Phillips 1988). The benefit was a body of work that identified and explained long-term reductions in metal exposure after the Clean Water Act of 1972 was implemented. More important, the design facilitated observations of recovery from reproductive damage as exposure to metals declined; the legislation generated an experiment in nature, and carefully designed time-series studies took advantage of the experiment.

Criteria for selecting an indicator taxon are well known (Phillips 1977). An important aspect for temporally intensive long-term sampling (e.g., *M. balthica* at Palo Alto) is sufficient abundance that frequent sampling is not onerous. It must be relatively easy to collect sufficient data to capture variability at critical time scales if the study is to be perpetuated. The details of this case study are described in Chapter 6.

A Priori Selection of Indicators: Focusing on Target Assemblages

Specific assemblages within communities can be used as effective biological probes. One challenge in community studies is to determine which component of the community to monitor. No program can be exhaustive in its taxonomic scope. Indicator taxa or assemblages offer tractability and reliability and provide data that can be extrapolated to other elements of the ecosystem. Monitoring only a component of the community can optimize the time, effort, and cost of sampling.

In theory, the selection of an indicator assemblage can also enhance the resolution of the monitoring program (Kremen 1992, 1994). Use of indicators increases the likelihood of overcoming the problem of separating the stressor "signal" from the background "noise," or separating variation observed in reference habitats from variation caused by anthropogenic stressors. For example, assemblages of terrestrial arthropods are proposed as indicators of habitat quality and intensity of anthropogenic disturbance, for conservation planning and management (Kremen et al. 1993). This exploits the fact that arthropods are the most diverse component of terrestrial systems, occupying a diversity of niches and habitats over a range of spatial and temporal scales. Arthropod assemblages also offer the potential to probe a wide and variable range of environmental factors, by virtue of their morphological and functional diversity. Their spatial and temporal distributions span the ranges occupied by plants and vertebrates, and they include finer-grained patch sizes and distributions, more complex seasonal and successional sequences, and patch dynamics with more rapid turnover (Kremen et al. 1993). Other beneficial attributes include wide ranges in body size, vagility, growth rate, and life history and the fact that certain assemblages are specifically adapted to the narrow microhabitat characteristics of threatened or fragile habitats (e.g., old growth forest floors). Careful clarification of the critical question will help identify the assemblage best suited as an indicator. Separating variation in reference habitat from variation caused by an anthropogenic stressor can be accomplished by exploiting reference sites as controls (see also Chapter 3 for a consideration of alternative approaches to species classification).

Monitoring for the purpose of determining ecological integrity requires that both biological diversity and ecological complexity be captured. It is therefore arguable that vertebrate and plant species should be incorporated within a monitoring program. Kremen (1994) argues that selection of narrower target assemblages of biogeographically informative taxa gives the best opportunity to identify environmental patterns and thus aids understanding of the distribution patterns of a variety of species. For example, assemblages that result from evolutionary radiations within a region may be biogeographically informative, by virtue of their high species richness and endemism within the region. Such groups are particularly well distributed in the tropics. They are of special significance for conservation because they offer localized resolution of communities, habitats, and ecotones, as well as areas of endemism and hotspots of biodiversity.

A series of steps for the development of target taxon analysis are recommended by Kremen et al. (1993):

- Initially select 5 to 10 higher taxa (i.e., bats, birds, frogs, some vascular plant families and arthropod functional groups) that are relatively well characterized (e.g., by a local or regional species list). Also include some lower-ranking taxa (genera or species) with high diversity and endemism. Within the candidate target taxa, use additional criteria, including wide distribution, high abundance, and high beta or gamma diversity, to refine the selection.
- Subject the information value of the target assemblages to limited testing across an obvious environmental gradient or dispersal barrier. Analyze correlations between the taxa and environmental gradients and between the taxa and different taxonomic assemblages. Use strength of correlation as a further selection criterion.
- Sample selected assemblages over all relevant major habitat types and environmental gradients. Use this information to identify areas of endemism or to select a minimum number of sites by complementarity, to represent the full range of species or habitat types.

One significant benefit of targeting an assemblage or taxa may be the increased resolution of stressor impacts because of the reduction in the natural variability associated with sampling a larger number of species. The Malagasy satyrid butterfly subgenus *Henotesia* demonstrated higher resolution of known topographic and disturbance effects than did the entire butterfly assemblage (Kremen 1994). This may be because the larger, more vagile species were removed from the analysis when the target taxon was isolated for analysis. Dispersive species could be less suited as candidate target taxa because their range may include unsuitable habitats or excursions through unsuitable habitats, which may obscure biogeographical or ecological patterns. Target taxa also may not represent subtle but perhaps important characteristics of the system that may contribute significantly to ecological stability and diversity. For example, studies that do not target insect species that pollinate or disperse the seeds of a wide variety of plants will miss "keystone mutualists" that provide critical resources for large numbers of other species (see Kremen et al. 1993). This underlines the importance of assessing indirect effects in food webs, outlined in Chapter 4.

Case Study: Macroinvertebrate Assemblages in River Monitoring

Entire assemblages can also be studied to detect biological impairment. The biological variables that are measured are the taxonomic composition and the abundance of an assemblage captured by a certain sampling protocol, for example, benthic macroinvertebrates in rivers. The indicator variable to determine impairment is an index calculated from the taxonomic composition data. Examples include comparisons of observed to expected taxa (e.g., Reynoldson et al. 1995) or an index of biotic integrity calculated from ecological metrics (e.g., Karr et al. 1986; Barbour et al. 1999). Regardless of the index used, the approach rests on the comparison of the index value at a given site to the value expected at "least stressed" or reference sites. In these studies, comparison with reference sites is not pairwise but is achieved by comparison with a reference condition, represented by samples collected from many sites within a region, or even a model developed from outside knowledge of species' individual responses to stressors. Several factors may affect the species composition of the assemblages, the value of the index, and the response of the index to stressors. These typically include such things as geographic region, catchment area, stream gradient, latitude, elevation, habitat factors, water color, and pH. These environmental factors are examined for their effect on the biological variables being measured in the reference site dataset, and if biologically significant, a predictive rule or model is developed to properly classify any given test site with the appropriate reference condition, as determined by the environmental factors. Models and rules to predict reference condition include discriminant models to predict groups (e.g., Reynoldson et al. 1995); simple associations of geographic regions, sediment type, or depth (e.g., Barbour et al. 1996); and regression models of salinity or water chemistry (e.g., Engle et al. 1994; Gerritsen et al. 1999).

The assemblage approach has strengths and limitations:

- Selection of a community assemblage is useful (essential) if the responses and stresses are unknown or multiple.
- No single species is necessarily sensitive or responsive to all stressors and exposure routes; therefore, attempting to detect response in many species (community) increases the likelihood of identifying such a response. However, measurement at the community level also makes detection of signal more difficult because of the greater degree of natural variation.
- The community level integrates across time and response, so the approach detects cumulative effects.
- The assemblage approach provides limited diagnostic capability. This limitation may reflect a lack of knowledge rather than an inherent inability. Contrast Europe, where there is lots of natural history information, with North America, where there is comparatively little.
- In a stepwise process, the community level should be used as the primary indicator of response. Follow-up studies at lower scales of organization (population, individual, suborganism) can aid diagnosis of cause and effect or provide mechanistic explanations for observations.

Functional Group Approach

Nontaxonomic categories, such as functional feeding groups and life-history traits, have been frequently employed to assess environmental conditions in aquatic ecosystems (Kerans and Karr 1994; Barbour et al. 1996; Wallace et al. 1996; Richards et al. 1997). For example, functional feeding groups are used in benthic macroinvertebrate feeding ecology to characterize natural variation in stream ecosystems (Vannote et al. 1980). The relative abundance of different functional feeding groups such as shredders, grazers, collectors, and predators is directly influenced by physical–chemical and habitat characteristics of ecosystems. The abundance of grazers is related to the quality and quantity of their food resource (attached algae and periphyton) and indirectly related to light availability.

Analysis of functional feeding groups has a long history in biological assessments and has a number of advantages in evaluating influences of stressors in variable natural environments. The approach is very useful in identifying specific stressors in certain situations (Barbour et al. 1996). Functional groups also occur across the extreme variability that results from patchy distribution of individual species within and between habitats. Reducing variability of a community measure increases the power to detect differences. Relatively coarse levels of taxonomic resolution are sufficient to detect effects of pollution on benthic communities (Warwick 1993). The most likely explanation is that closely related species often have similar sensitivity to the same contaminants, and therefore, aggregating species into higher taxonomic units reduces sampling variability. In addition, when samples are collected over relatively large geographic areas, higher taxonomic aggregates will be represented at more sites than will individual species. If this same pattern is observed for species within functional feeding groups, abundance of these groups may be less variable than abundance of individual species. In a recent study of the effects of heavy metals on streams in the Southern Rocky Mountain Ecoregion of Colorado, USA, Clements et al. (2000) found that abundance of 1 functional group, grazers, was highly sensitive to heavy metal contamination, and the response of these organisms was a better indicator of metal pollution than any of the specific taxa that were included in this group. Specifically, grazing mayflies were often eliminated from metal-polluted sites (Clements et al. 2000). Because periphyton is a major sink for heavy metals in these systems (Kiffney and Clements 1993), reduced abundance of grazing mayflies may result from dietary exposure to metals. Laboratory experiments in which mayflies grazed metal-contaminated periphyton demonstrated bioaccumulation and toxic effects on growth. Elimination of functional feeding groups such as grazers may also have important consequences for contaminant fate and effects in aquatic ecosystems. Sallenave et al. (1994) reported that downstream transport of PCBs was greater in experimental streams with grazers or shredders than in streams without these 2 functional groups.

A key challenge in the application of functional feeding groups is the proper assignment of taxa to the different feeding categories. Many macroinvertebrate species have generalized feeding habits and are relatively opportunistic consumers.

Feeding habits vary within a species over time, either seasonally or as a result of ontogenetic and developmental changes. Feeding habits may also vary among locations because of differences in food availability. Precise assignment of organisms to a specific functional feeding group adds important uncertainties to the application of functional feeding groups in biological assessments. Further discussion of functional classification within food webs is given in Chapters 3 and 4.

Ecosystem-level Responses to Disturbance

Because every possible indicator in all ecosystems cannot be measured, identifying general patterns of responses to disturbance across ecosystems and types of disturbance is essential. Rapport et al. (1998) developed a common suite of ecosystem indicators and suggested that "it is one of those refreshing simplifications that natural systems, despite their diversity, respond to stress in very similar ways." Characteristics of the ecosystem distress syndrome described in Rapport et al. (1998) include changes in nutrient cycling, primary productivity, species diversity, species composition (sensitive species replaced by tolerant species), and size composition. Many ecologists are uncomfortable with the analogy between individual and ecosystem responses to stress described in Rapport et al. (1998). But ecosystem-level processes are closely linked to the structural characteristics of communities, such as abundance, species richness, and community composition; functional ecosystem processes include rates of primary productivity, nutrient cycling, energy flow, and decomposition. Structure and function can also interact. Reduced ecosystem species diversity that is due to anthropogenic disturbance may have important consequences for ecosystem function (Tilman and Downing 1994). Results from model ecosystem experiments showed that depauperate communities had lower productivity and reduced CO_2 uptake compared to species-rich communities (Naeem et al. 1994), although this experiment has its detractors.

A key advantage to measuring ecosystem function is that it integrates responses of component populations (see also the food web approach advocated in Chapter 3). However, because of functional redundancy, lower sensitivity, and high variability, it may be difficult to measure how functional measures respond to subtle perturbations (Schindler 1987; Stay et al. 1988).

A hierarchy of biological indicators

- Indicator taxon: Intensive, small-scale time and/or space variability but loses relevance at higher orders of organization
- Targeted taxa: Can be used to resolve specific stress responses in complex stress regimes but is not indicative of some components of community response
- Species assemblages: Relevant, but variable. To be useful, baseline condition and its variability must be known.
- Nontaxonomic functional groups: Can integrate across patchily distributed species, but can lose sensitivity if not stress-specific
- Ecosystem responses: Can integrate across population variability, but can lose sensitivity

For example, in studies conducted in the Experimental Lakes Area, Schindler (1987) concluded that functional measures were relatively insensitive to acidification and that shifts in community composition were better early warning signs of ecosystem stress. Alterations in community composition often preceded changes in ecosystem function in these stressed ecosystems.

Evaluating Effects of Suites of Stressors

Stressors rarely occur in isolation, so developing protocols for evaluating suites of stressors is as important as the choice of stress indicators. A framework that incorporates scientific understanding, human judgment, and uncertainty can be helpful in evaluating effects of complex suites of stressors. For decision-making purposes, the framework also should provide breakpoints or thresholds that trigger actions. The system development matrix approach is an example of a framework that uses a conceptual model and a system development matrix to apply adaptive management principles to restoration of coastal aquatic ecosystems (Figure 5-9).

The initial step in production of a framework is development of the conceptual model for the system, population, or assemblage in question. The form of the generic model is this:

controlling factors→habitat/population/assemblage structure→function–process.

Model formulation forces people studying the system to identify on paper their best understanding of the relationship between particular parameters of concern (e.g., species diversity of the assemblage, net ecosystem productivity) and factors that likely control or influence that parameter (e.g., light, temperature, contaminant concentration). The process of constructing the conceptual model helps to identify key linkages between stressors and response variables.

An example of a conceptual model for a seagrass community (the eelgrass *Zostera marina* L.) is shown in Figure 5-10. The model is essentially biophysical and identifies the key factors that control the density (structure) of eelgrass meadows in Puget Sound, Washington, USA. The range of values for each of the controlling factors is based on a large dataset that consists of field and laboratory studies (Thom and Albright 1990; Thom 1995; Thom et al. 1998). If the controlling factors are within the ranges indicated, eelgrass is likely to flourish. The relationship between eelgrass biomass and Dungeness crab density (function) is less well developed but is documented (Figure 5-11; Thom et al. 1989). The dataset is not well developed enough to predict with precision the biomass of plants or the number of crabs. There is a strong indication, however, that in spring, late summer, and fall, crab density is related to eelgrass biomass. Hence, there is uncertainty in the relationship between controlling factors and structure or function (see also the similar example given in Chapter 3).

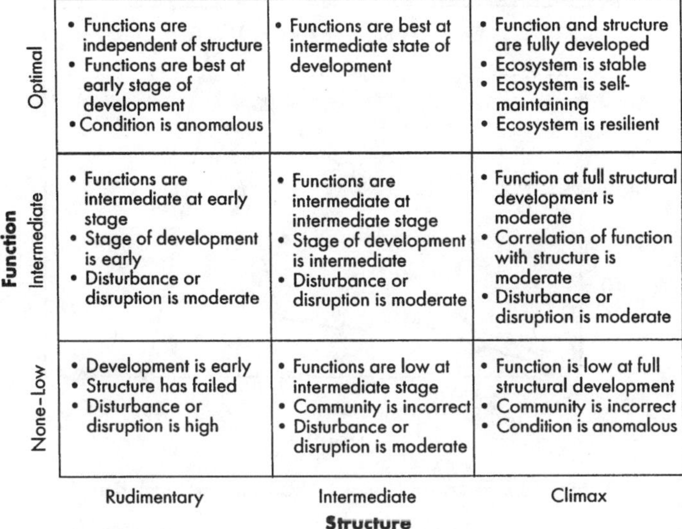

Figure 5-9 Diagram of a system development matrix

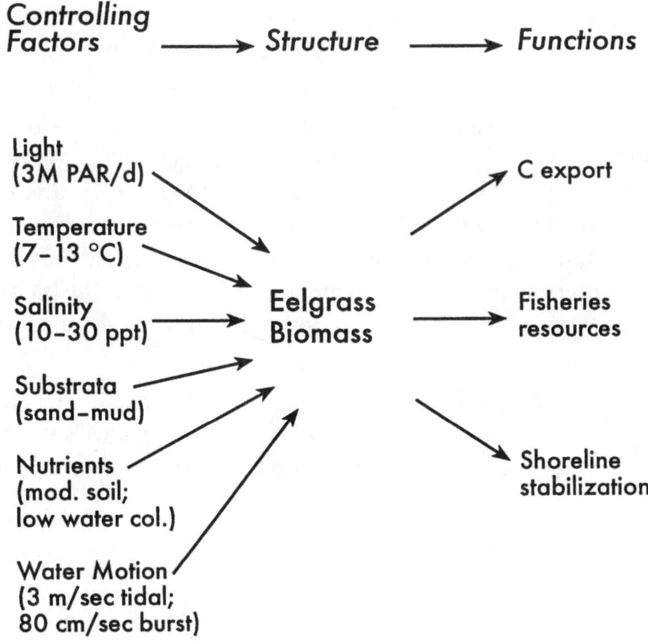

Figure 5-10 Conceptual model of seagrass ecosystem

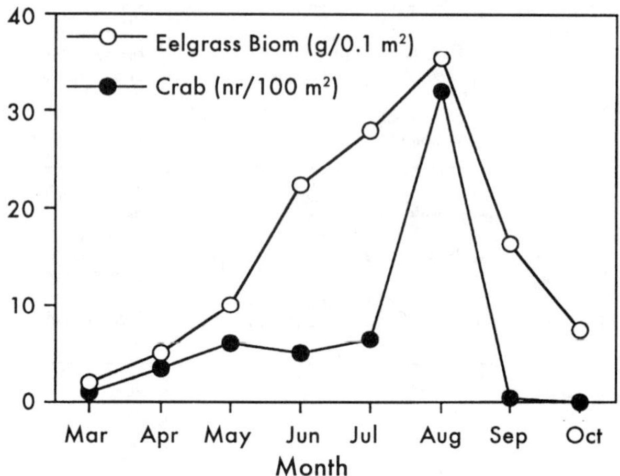

Figure 5-11 Eelgrass biomass versus Dungeness crab density

This eelgrass model is relatively simple and has a strong supporting database. Models can be developed with much less data, as long as components and connections within the model make ecological sense and components and connections that have little or no certainty are excluded from the model. Through model development, key components or connections where data are poor or lacking can be identified and recommended for additional directed research.

An example of a system development matrix is shown in Figure 5-9. Structure and function are positively correlated, as indicated by moving from the lower left box to the upper right box in the figure. Nonlinear and negative correlations are also possibilities, as is the total lack of correlation. Next, the matrix is divided into 9 boxes (system states) by dividing the structure and function axes into 3 levels. The reasoning behind the use of 3 levels is provided in Thom (1997; 2000). Although 3 levels (i.e., low, intermediate, and high) are often used to describe the states in many ecological systems, theoretically any number of divisions can be made as long as they are justified.

The axes of the matrix do not have to be structure and function. Structural parameters can be used on both axes. Time can be the *x*-axis, with a structural (e.g., species diversity index) or functional (e.g., net ecosystem productivity) parameter on the *y*-axis.

Dividing the axes into 3 levels acknowledges the fact that the fully developed assemblage, for example, is defined by a range of values for structure and function rather than by a single point, that is, it acknowledges the uncertainty. The range of values can be prescribed based on a variety of methods. In general, the range of values represents the range of normal (natural) variation or dynamic stability in the parameter.

Descriptors are developed for each of the 9 states. For example, the box at the lower left corner indicates a system that is both structurally and functionally rudimentary. In contrast, the upper right-hand box indicates a fully developed (climax) system that has the attributes of stability and resilience and is self-maintaining. A system state defined in the upper left corner of the matrix indicates some deviation from what is normal or predicted.

Examples

1) A seagrass meadow may have a low or rudimentary level of shoot density (the structural parameter of interest) but a high density of shrimp (the functional parameter of interest). The suite of reasons explaining this condition may include depletion of fish predation on the shrimp as a result of overfishing. Alternatively, the shrimp may have undergone an abnormally high reproductive effort linked to an anomalous climatic condition, a contaminant may have reduced the density of shrimp competitors, or some combination of these stressors may have occurred. These hypotheses can be used to design measurement or experimental studies to resolve the mechanism.
2) An endangered salmon population in a particular Pacific Northwest watershed is to be restored. The responsible local, state, and regional municipalities combine their efforts to restore salmon spawning, rearing and feeding, and migratory corridors (all aspects of structure) throughout the watershed at a very great cost. Following restorative actions, monitoring for a period of time reveals that the salmon population (the function) is in the medium (middle right box) as opposed to the high range. The reason the population has not recovered to predicted levels may be related to several factors, not the least of which may be overfishing in the open ocean. Overfishing is something over which local municipalities have little control.

In order to be effective and useful to decision-makers, a framework should, therefore,

- acknowledge the uncertainty in the system being studied but allow for interpretation of the condition of the system,
- apply a qualitative human interpretation (i.e., levels) to help understand a complex situation,
- force scientists to predict the various states and associated levels of parameters for variables and to erect explanations regarding why a system might enter a particular state, and
- force scientists to develop a conceptual model that incorporates the best understanding of the relationships between variables, which requires that the conceptual model become an active component of the assessment program.

Reaching Conclusions in a Variable Environment

Variability can reduce certainty in conclusions regarding the effects of multiple stressors on biological systems. Uncertainty is derived from 3 major sources:

1) natural stochastic variability,
2) uncertainty in the relationship between controlling factors and biological responses, and
3) sampling error.

It is often difficult to quantitatively partition the effect of the sources of uncertainty or to do anything about them even if they can be quantified. Stochastic variability can and should be evaluated. Uncertainty with regard to the relationship between controlling factors and biological responses can be reduced through systematic field and laboratory studies and investigations. Issues related to sampling can be minimized by appropriate and rigorous attention to study design and execution. In the end, however, conclusions or interpretations often require a choice about the kind of uncertainty that is most acceptable.

In traditional statistical hypothesis testing, types of statistical error characterize the uncertainty associated with inference of effects or associations. Type I error establishes the traditional standard of proof in reductionist science. Type I error occurs when a study incorrectly concludes that an effect occurs, when in fact there is no effect but random error created the perception of an effect. The probability of a Type I error is expressed by α or the p-value. Minimizing Type I error requires that an effect is unambiguously demonstrated before any conclusion is drawn. The burden of proof lies with those who assert that an adverse effect is occurring and that it is caused by the stressors. Statistical inference under hypothesis testing involves only the rejection of a null hypothesis. The great fallacy that confounds our thinking is that if we fail to reject the null hypothesis (cannot demonstrate effect at $p < \alpha$), we therefore have demonstrated that no effect occurs. In fact, the only allowable conclusion from a failure to demonstrate an effect is that no conclusion can be drawn.

Minimizing Type I error increases the risk of committing Type II error. Type II error occurs when the conclusion of "no effect" is prematurely drawn, when in fact an effect has occurred, but we cannot demonstrate it at the requisite α-level. An example of a conclusion prone to Type II error would be the conclusion of "no effect" from the observation that mean density of a sensitive species, measured 10 times in a contaminated habitat, is not significantly lower (at $p < 0.05$) than in an uncontaminated habitat. Mean density may in fact be lower, but because of variability in abundances, the hypothesis test failed. Environmental variability reduces the sensitivity of studies that attempt to determine such effects and their causes, resulting in a larger risk of Type II error than would characterize a controlled experiment. Over-reliance on traditional hypothesis testing has been much criticized by statisticians (e.g., Yoccoz 1991), and many propose model estimation or maximum likelihood methods (e.g., Cox 1986) or Bayesian statistics (e.g., Ellison 1996) instead.

One response to a higher probability of Type II error in determining stressor effects in nature was development of the much-debated "precautionary principle." The precautionary principle developed, at least in part, from the environmental movement's concern with the pollution levels that had developed from policies of the past (Earll 1992). There was also doubt whether society would ever be in a position to predict the consequences of human activities for large complex ecosystems. Given such doubt, it was perceived that a precautionary approach was philosophically necessary to protect the environment whenever an adverse effect from an activity was suspected (no causal link need be proven).

Gray and Bewers (1996) criticized the lack of scientific input to the original formulations of the precautionary principle and stated a scientifically based principle. In part, they stated that the principle

> shall apply to human activities for which there exists a scientific basis for believing that damage to habitats or harmful effects on species are likely to result. Measures shall be based upon pessimistic assumptions regarding uncertainties in the measurement and prediction of effects in the environment.

By the latter, they meant that the benefit of the doubt should be weighed in favor of the environment rather than the generator of the stress, when uncertainties are present (i.e., use science to evaluate risk, but minimize Type II errors), but that scientific feasibility of the effect is an important consideration.

Environmental variability is a primary factor raising the risk of Type II error. Therefore an appropriate precautionary principle could be important to reconciling influences of that variability on management of ecosystem stressors. One approach might be to more clearly define "scientific basis" or scientific feasibility of damage. The bases of proof that a stressor contributes to an effect in nature are these:

1) a body of evidence showing associative linkages, including elimination of main confounding variables;
2) the responses to events; and/or
3) the existence of a plausible mechanism to explain the association.

Existence of any of these 3 criteria might be a basis for caution in concluding the absence of an effect. The stronger the body of evidence, or the larger the number of lines of evidence supporting a realistic impact, the more caution that might be justified. Understanding influences of natural variability is an essential ingredient to developing a scientific basis for the precautionary principle, as we refine our abilities to fully understand effects of stressors.

Summary

Variability can be viewed as either an impediment to clear understanding or an inherent part of the richness of natural systems. Environmental variability is one of the principal causes of uncertainty in assessments of risk from stressors. Variability that remains uncharacterized leads to uncertainty or even to complete masking of important effects. Perceived variability may result in criticisms that call into question the conclusions of a study. Controlled experimental results and model predictions are often dismissed by field ecologists as unrealistic and inapplicable to the "real world" with its attendant variability. A final consequence of the perception of "intractable variability" is that worthwhile studies are not even attempted. Our central thesis here is that environmental variability is not intractable, but that with judicious design, analysis, and interpretation, it can be characterized and controlled, so that the uncertainty of field studies and field experiments can be reduced and interpretations optimized. Some specific consequences and characteristics of variability in natural systems are listed below.

1) Environmental variability is not just pervasive stochastic variability, error of measurement, or experimental error. It is often definable by dynamic stabilities (properties that vary in a repeated, somewhat predictable fashion). Capturing the nature and the role of environmental variability is a critical requirement for separating and understanding causes of change in ecosystems.
2) Environmental variability has several influences in studies of effects of multiple stressors.
    - Exposure to stressors is more variable than is often recognized.
    - Biological properties that respond to stress can have an inherent variability. Onset of (or recovery from) stress may result in a directional signal, while fluctuation or variability of the property is nondirectional. The baseline cycle must be understood if we are to identify the stress response.
    - Responses to stress may be modified by variation of environmental factors, including natural disturbances.
    - Determining the influences of a stressor requires accounting for abiotic environmental factors that also might influence the response.
3) Some considerations when defining influences of environmental variability include these:
    - Capture relevant components of natural variability by
        - minimizing (by sampling methodology) the components that operate below the spatial or temporal scale of the sampling unit,
        - stratifying or selecting components that operate at the scale of the sampling unit, and
        - characterizing components that operate at a higher scale than the sampling unit (the question unit).

- Carefully state questions and identify critical sampling scales a priori.
- Maximize the power of the study design and identify the appropriate effect size (actual magnitude of change) that is to be considered important.
- Sample at units smaller than the question unit, in most cases.
- Oversample unknown sources of variability in initial phases of the study.
- Ultimately constrain long-term characterization of environmental factors to those most likely to have important influences.

4) Biological indicators are used to observe biotic response to environmental stress, provide early warnings of natural responses to environmental impacts, and measure the success of management or restoration options. The appropriate levels of organization for the indicator can vary.
   - An individual indicator taxon has the disadvantage of limited relevance to community-scale change but may be the best choice as an indicator of small-scale time and/or space variability.
   - Targeting a few taxa can increase the ability to resolve stress responses in complicated environments (e.g., increases signal-to-noise ratio). For example, a narrowly targeted assemblage may be highly effective at expressing response to a specific stress.
   - Entire assemblages of a community type may also be used as indicators of biological impairment, but effective use of the assemblage approach often rests on comparing the community at a given site to the characteristics of "least stressed" or reference sites. Multiple references (a population of references) are essential in such comparisons.
   - Nontaxonomic categories such as functional feeding groups can be used as bioindicators of impairment that integrate across the patchy distribution of individual species. One example is the use of grazing mayflies as an indicator of metal effects. Coarse levels of taxonomy may be suitable for functional group analysis, again reducing the spatial and temporal variability that occurs at the species level.

CHAPTER 6

Separating Stressor Influences from Environmental Variability: Eight Case Studies from Aquatic and Terrestrial Ecosystems

Samuel N. Luoma, William H. Clements, Jeroen Gerritsen, Audrey Hatch, Paul Jepson, Trefor Reynoldson, Ronald M. Thom

It can be difficult to unambiguously establish the influences of a particular stressor or group of stressors in a complex ecosystem, except perhaps when the effects are extreme (Luoma and Carter 1991). Yet this is a critical problem we face when attempting to understand the influences of human activities on ecosystems. Single experiments or studies are rarely adequate to establish cause and effect in complex ecosystems, and many of the individual approaches to demonstrating stressor effects have important inadequacies. A multifaceted body of work is at the center of most examples in which stressor effects are explained. In this chapter, 7 case studies are presented in which effects or influences of multiple stressors were explained and separated from natural variability. The goals of this chapter are to demonstrate that identification of stressor effects is tractable, although not necessarily simple, and to illustrate some specific strategies that have worked. To illustrate the range of challenges involved as a body of work begins to be established, we also present 1 case study in which the quest for cause and effect is just beginning. The examples are from several different fields of ecology, and they cover a variety of scales and a mix of disciplines.

General Requirements for Linking Cause and Effect

In a natural system, associating a biological change to the specific influences of any single variable has some specific requirements (described as "ecoepidemiology" by Sinderman 1996), including these:

- Demonstrating that the affected process is sensitive to the stressor (strength of association)

Ecological Variability: Separating Natural from Anthropogenic Causes of Ecosystem Impairment.
D.J. Baird and G.A. Burton, Jr., editors. © 2001 Society of Environmental Toxicology and Chemistry (SETAC).
ISBN 1-880611-43-0

- Demonstrating that a feasible mechanism exists to explain the effect (coherence)
- Separating stressor-induced change from background fluctuations (specificity of association)
- Eliminating confounding variables and thus unambiguously relating the detected change to the stressor of interest rather than to some co-occurring process or factor (as in a time-order association or high statistical significance compared to other explanations)
- Demonstrating the effect repeatedly in a body of work from different times and places (consistency of replication)
- Defining the expected response to a change (predictive performance).

Absolute scientific proof of an effect might require that all of the above criteria are met before allowing us to conclude cause and effect. Sindermann (1997) suggests the more reasonable position: that the scientist directly state which inferential criteria form the basis for his or her opinion and that continued attempts to better delimit pollution-induced from natural changes in populations are almost always called for. In each case study in this chapter, we emphasize how the variability of nature, the variability of the stressor, and the variability of the responses to a stressor interact to affect the ability to interpret stressor effects.

Variability of Response

The inherent or cyclic variability of measurable biologic processes must be understood in order to separate a stressor response from natural variability. Basic studies of how biochemical, physiological, population, or community metrics or measures vary on different time scales in unstressed circumstances are rare, and responses to natural disturbances also are not well studied. If the natural variability of potential stress responses can be sufficiently characterized, evaluations of pollutant effects become more effective. Data collection on multiple aspects of a problem and at least some baseline of understanding from undisturbed circumstances can be important to resolving effects of stress.

Variability of Nature: Confounding Factors

Observation of a response in a contaminated ecosystem, or even correlation with a gradient, may not be sufficient to prove pollutants are the cause of impairment. Confounding factors and compensatory complexities must be overtly considered. For example, Martin et al. (1984) found statistically significant changes in energetics (scope for growth [SfG]), changes in condition index (CI), and reproductive impairment in mussels deployed along a pollutant gradient in San Francisco Bay, California, USA. The responses correlated with tissue burdens of polycyclic aromatic hydrocarbon (PAH) and some metals. Suspended solid concentrations were not considered, however, and a gradient in suspended particulate matter (SPM) was subsequently shown to co-occur with the observed response. Nutritional deficit could not be eliminated as at least a partial explanation of the observed response because of the very high inorganic SPM. Confounding variables clouded the certainty of conclusions about the nature of any pollutant effect. Frustration with ecosystem complexities can lead to abandonment of the field study approach. For example, follow-up studies in San Francisco Bay (controlled or field studies) could have helped clarify Martin et al.'s (1984) results; but no such studies were ever conducted. This problem of correctly attributing cause and effect is further illustrated by the example of kelp decline in southern California.

Case Study I: The Giant Kelp Forests of Southern California

This case study describes the effect of domestic sewage and grazers on kelp forests in southern California. Kelp forests dominate much of the rocky coastline from Baja, Mexico through Alaska. The dominant species in southern California, which forms the canopy of the forests, is the giant kelp *Macrocystis pyrifera* L. This species of brown algae produces several dramatically long stipes with attached leaf-like blades from a root-like structure called a "holdfast." Individual stipes with blades can reach 40 m in length as they extend to the surface of the water. Because this plant is tremendously productive and contains commercially valuable compounds such as potash and colloids, it has been harvested for almost a century. The interest in kelp as a resource was fueled by World War I when potash supplies from Europe were interrupted. Under the auspices of the U.S. Department of Agriculture, kelp resources were surveyed and mapped along the coastline from California through Alaska in the period of 1914 to 1920. Post-war tracking of kelp area was continued by the kelp industry and later by the California Department of Fish and Game. This monitoring provides a long-term dataset on fluctuations in kelp bed canopy coverage, the most readily mappable indicator of kelp distribution.

Based on monitoring data, large changes in kelp canopy coverage were documented between 1940 and 1960 (North et al. 1964). These variations are in excess of those found before about 1930. A major El Niño event in the years 1957 to 1958 resulted in the virtual extinction of kelp along the Palos Verdes Peninsula, in the Los Angeles metropolitan area. Subsequent El Niño events have been well studied and show a significant impact on kelp distribution (Dayton et al. 1992).

Concern for the kelp resource sparked a research effort by Dr. Wheeler North of the California Institute of Technology beginning in the early 1960s. North's research found vast wastelands where the sea bottom had formerly been occupied by kelp. The rocks in these wasteland areas were dominated by huge densities of sea urchin, a major grazer of kelp and other seaweed species (North et al. 1964). Furthermore, North found that, in the absence of kelp, the urchin population was sustained by ingestion of the particulate organic matter that covered the bottom. The origin of this organic matter was terrestrial; the source was sewage effluent (1970 mean solids discharge = 450,000 kg d^{-1}) from the outfall at White's Point on the peninsula (Meistrell and Montagne 1983). Further inquiry showed that a major predator of urchins, the sea otter, had been hunted to extinction by the early 1900s. Hence, free from predatory pressure, the urchin population was able to increase to a size sufficient to devastate the kelp forests surrounding Palos Verdes. Sewage particulates, which were delivered at a relatively constant rate through the discharge, essentially allowed the urchins to remain dense enough to graze any kelp plants that may have been recruited to the area. Laboratory experiments showed that the effluent itself was somewhat toxic to the plants (North et al. 1964).

To restore the forests, North killed sea urchins by using hammers and lime. His team then transplanted adult kelp plants from Baja into areas where sea urchin control had been carried out. Although early efforts showed some success, a steady increase in the kelp was not seen until Los Angeles County implemented strict effluent treatment procedures that removed organic matter in the mid 1970s. Partial recovery of the forest patches around the Palos Verdes peninsula was apparent by 1980 (Figure 6-1; Meistrell and Montagne 1983). A secondary effect has been the recovery of wintering shorebirds in the area (Bradley and Bradley 1993).

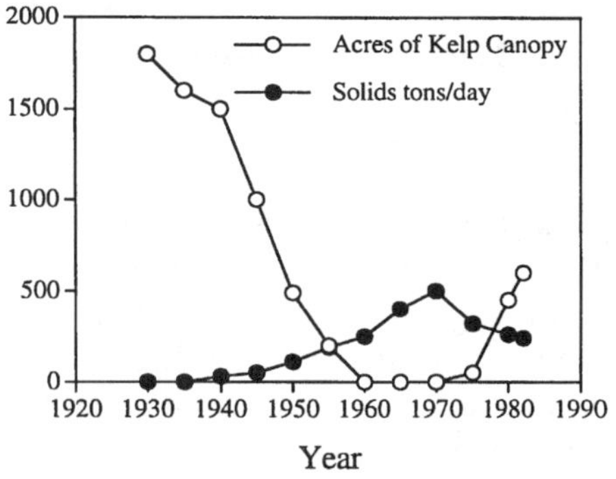

Figure 6-1 Maximal canopy of giant kelp beds off Palos Verdes peninsula, California, USA and suspended solids from water pollution control plant

Figure 6-2 illustrates a simple conceptualization of the kelp story. The stresses on the kelp were indirect (removal of otters, particulate matter discharges) and direct (effluent toxicity, harvest), as well as climatic (El Niño). The stressors largely were constant in time with regard to the scale of the impact, with the exception of the episodic climatic events. These latter events, on top of the constant stressors, appear to have driven the system into another state. There is still considerable uncertainty about factors controlling kelp abundance (North et al. 1986; Dayton et al. 1992).

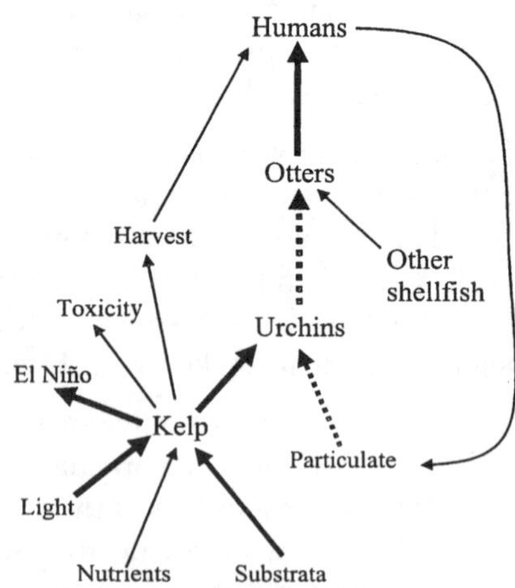

Figure 6-2 Conceptual model of kelp decline and recovery. Dashed lines indicate connections in the system that were altered or removed by human intervention.

The key impairment that was documented was the loss of kelp forest canopy area. Related changes included the loss of kelp fish assemblage and birds and the alteration of the understory assemblage (Dayton et al. 1984; Bradley and Bradley 1993).

Multi-stressor effects on kelp forests

- Human and natural stressors interacted to affect an important marine ecosystem.
- Removal of or reduction in 1 key stressor (sewage particulates) was critical to response of the system.
- Observational information and experimental studies had synergistic value in explaining cause and effect.
- Long-term monitoring established natural variation.
- Explanations could not be developed based exclusively on short-term investigations.

A key element for documenting the stress effect was the monitoring program, which spanned more than 50 years. In addition, the keen commercial interest in the kelp resources was a factor in identifying the potential problem. Partitioning out the impact of each stressor was largely based on field observations of urchin behavior. The kelp industry had instituted renewable resource harvest practices. With controlled harvest of only the upper 1 m of the stipe in a small portion of the forest patch, it was believed that harvesting was not a major cause for the decline. Harvested stipes are rapidly replaced by new stipes that grow from the base of the plant. Although toxicity was an issue, it was not believed to pose a major threat to the plants unless effluent volumes increased dramatically. Perhaps the most interesting insight came with the realization that harvest of sea otters had probably contributed to the population explosion of sea urchins.

The scale that the stressors worked ranged from the individual plant to the entire southern California region (Table 6-1). The certainty of the conclusions decreased with increasing scale.

Table 6-1 Approximate scale of different stressors on kelp communities

Level	Approximate scale	Stressor	Uncertainty
Individual plant	m^2	Urchins Toxins	Lowest
Forest patch	ha	Urchins Toxins Harvesting	Intermediate
Regional forests	> km^2	Harvesting Climatic anomalies	Higher

Case Study 2: Long-term Impacts of Agricultural Practices in Agroecosystems

There is a long history of experimentation and monitoring in agricultural ecosystems in Europe. These studies provide some valuable illustrations of how long-term implications of human disturbance can be detected or confounded by environmental or climatic variability.

Monitoring Complements Research

A number of research projects investigating the ecology and economics of integrated arable farming systems were initiated in the 1980s and 1990s (Holland et al. 1994). The impacts of pesticides, fertilizers, and agronomic and husbandry practices were analyzed in at least 13 large-scale studies in Austria, France, Germany, the Netherlands, Switzerland, and the United Kingdom. There is considerable concern about wildlife, environmental protection, and soil quality in European agriculture, and these studies effectively explored the impact of multiple stressors upon the non-pest fauna that inhabit the system, and upon the basic functioning of the soil. Agriculture may be both economically and ecologically impaired. The ultimate objective of the field investigations was to seek progress in reducing pesticide and fertilizer use by reducing impact on beneficial and wildlife species and reducing pest outbreaks, within the constraints of maintained or enhanced profitability. From an ecological perspective, impairment could be defined functionally in terms of pest outbreaks that were triggered, at least in part, by reduced predation potential. It could also be expressed in terms of the level and duration of impacts on individual beneficial or soil-dwelling species and even in terms of extinction in some cases.

Many of the studies incorporated design and measurement features that addressed questions of economics, and no studies were designed exclusively to answer ecological questions. This acknowledges the need to link research and findings to practical and economic recommendations. It also reflects that difficult tradeoffs (i.e., between the needs of agricultural economics and of ecology) may be forced by economic circumstance rather than by the preferences of scientists and economists involved in these investigations. Critical to such studies is the selection of optimum experimental scales that can resolve important phenomena with a maximum of statistical power for minimum cost.

In addition to these formal experiments, a small number of long-term vertebrate and invertebrate monitoring exercises have been undertaken in Europe since the 1970s. These provide unique opportunities to quantify underlying patterns of ecological structure and associated variability, inherent more generally within agroecosystems. The monitoring investigations are significant because they were undertaken within the same European agroecosystems as the farming system research referred to above.

Findings from Monitoring

Studies by Potts and Vickerman (1974, with subsequent additions from, e.g., Aebischer 1990), in the cereal ecosystem encompass the greatest spatial and temporal extent: sampling over an area of 62 km^2, over a 30-year period. For 1 taxon, the 365 species of Dutch Carabidae (Coleoptera), the data extend from 1890 to the present day. The more than 100 years of data on this important predatory assemblage provides a basis for analyzing the respective roles of climatic patterns (background variation) and human activity (land-use change) on the composition and distribution of the taxon (Hengeveld 1985). Potts and Vickerman (1974) demonstrated wide variation in the nature and composition of farmland invertebrate assemblages, within and between years. They also resolved clear relationships between herbivore density (cereal aphids [Homoptera: Aphididae]) and the diversity of predatory taxa, particularly Carabidae, Staphylinidae (Coleoptera), and Lyniphiidae (Araneae), suggesting limited functional redundancy within this diverse arthropod community. They predicted the occurrence of subtle but important impacts upon arthropod community structure as a result of the use of second- and third-generation pesticides. Their study influenced a decision to monitor a diverse range of polyphagous natural enemies in integrated farming studies.

Over long periods, time series analysis was used to separate summer rainfall, temperature, the proportion of the system undersown, and the effects of pesticides as explanatory variables, with a 1-year time lag, in the annual population densities of carabid sawflies (Symphyta: Hymenoptera; Aebischer 1990). Sawflies, an important food for the chicks of gamebirds, were at less than one-tenth their predicted density in an area that had received a single organophosphate pesticide spray. In the Netherlands, climatic variables were important determinants of carabid dynamics, and on a broad spatial–temporal scale, human influence appeared to be of minor importance (Hengeveld 1985). Carabidae seem to exhibit rapid spatial adaptation and possibly changes in net productivity that shift the makeup of the assemblage on a decadal time scale.

Experimental Studies: Design Tradeoffs and Scale

Holland et. al (1994) reviewed the major treatments, experimental designs, and assessment endpoints that were incorporated within 13 large-scale farming system studies. A tradeoff was apparent between the number of sites over which a study was conducted and, within sites, the area of experimental units, the degree of replication, and the degree of environmental monitoring. At one extreme was an unreplicated study with 3 treatments over individual treatment areas of up to 53 ha. At the other extreme were treatments of several contiguous fields of < 0.5 ha, with up to 5 replicates. In some studies, agronomic treatments were well replicated within split plot designs, but ecological measurements were carried out in unreplicated blocks to increase sampling area within treatments. All the studies were multiyear (2 to 6 y) and incorporated standard and reduced pesticide regimes, with variable adjustments to fertilizer regime, cover crops, crop varieties, and cultiva-

Pesticide effects on nontarget species in agricultural fields: Lessons for other nonequilibrium ecosystems

- Monitoring data defined variability; time series allowed separation of influences of multiple stressors, including climatic influences in nonequilibrial systems colonized by dispersive taxa.
- Monitoring findings influenced design of experimental studies.
- Field experiments involved tradeoffs between statistical rigor and environmental realism.
- Rare but important effects emerged from sporadic, acute toxin exposures, but intensity and extent of effect were clear only if studied in the most difficult context over large areas.
- Effects on organisms with windows of susceptibility were important.
- Reinforcement from numerous studies verified plausible mechanistic explanations for results.

tions. Some were undertaken on a single site, while others extended to as many as 6 sites to incorporate variability in soil type and climate across the agroecosystem.

Endpoints or response variables also differed, but 9 of the 13 studies investigated impacts upon nontarget, predatory arthropods, including polyphagous predators such as carabids, staphylinids, and linyphiids, in addition to impacts upon invertebrate pest populations. In all cases, beneficial nontarget invertebrate densities were higher in areas with reduced pesticide inputs compared with those that were subject to conventional pesticide spraying regimes (Holland et al. 1994). When large areas were subjected to relatively inflexible high pesticide use regimes (Greig-Smith et al. 1992), some beneficial invertebrates were rendered locally extinct for the full post-treatment phase of the investigation, and subsequent recovery, following release from spraying, was low. The level of impact on a given species was a function of life-history attributes, which affected pesticide exposure and capacity to re-invade (Burn 1992; Vickerman 1992). Predatory capacity was inhibited in the highest pesticide regimes (Burn 1992), and there was evidence that this contributed to higher pest densities in some years.

Scientific and philosophical problems arise when findings are unreplicated and cannot be supported by statistical analysis. The tradeoff, however, can be with realism; studies over larger areas may be the most realistic, even if they cannot be replicated. The Boxworth study carried out spray applications over several seasons in small groups of contiguous fields with some sacrifice to treatment replication. This contrasted to a number of investigations in which treatments were subdivided within fields, to provide a control treatment and replication within the experimental design. Investigations of the within-field dispersal patterns of beneficial invertebrates demonstrated that populations equilibrate between control and treatment plots, affecting estimates of recovery time (Jepson and Thacker 1990). Modeling of beneficial invertebrate movement patterns and population processes in sprayed farm systems demonstrates that depletions may be of longer duration when spray application is on the larger scale commensurate with commercial agriculture (Sherratt and Jepson 1993; Halley et al. 1996).

A subsequent project (see Holland et al. 1994), was designed to confirm the results of the Boxworth study in a more extensive and flexible program. They demonstrated long-term depletions (> 4 y) in Collembola, following a single organophosphate insecticide application. This study used a split field design with paired trapping grids for invertebrates in close proximity to each other. Again, treatment of adjacent fields may have damped the duration of impacts on more mobile species, which can recolonize treatment plots from control plots.

This cluster of studies has been the basis for much discussion. Evidence showed that impacts (observed and statistically verified) on smaller scales, may evolve into longer-duration, more intense effects on larger scales, where statistical power, or even the resources to repeat the investigation, is lacking. This outcome is consistent with ecological theory, but it was counter to the initial expectations of the scientists, especially since Carabidae are among the least susceptible of beneficial species to pesticides (Wiles and Jepson 1993). But a series of subsequent investigations reinforced the mechanistic plausibility of the findings of the Boxworth study.

Summary of Case Study 2

Monitoring, in these studies, revealed the influence of climate on the long-term dynamics of arthropods and the variable nature of stressors in space and time within the agroecosystem. Both experimentation and monitoring were necessary to demonstrate impacts of agricultural practices because of the compounded nature of perturbations imposed upon agroecosystems and the highly varying nature of agriculture between regions. The methodological challenges in proving impacts are immense and often require tradeoffs between statistical power, precision, and biological realism, expressed as appropriateness of scale and degree of similarity to commercial practice. In every study, some sacrifices must be made. It may only be through reinforcement between studies, each with its own particular constraints on design, and support from investigation of mechanisms through laboratory toxicology, exposure analysis, field experimentation, and modeling that uncertainty concerning the likely adverse impacts of pesticide use regimes can be minimized.

Case Study 3: Metals in a San Francisco Bay Mud Flat

The bioavailability and effects of metals were studied at a mud flat in south San Francisco Bay, located 1 km from the inshore discharge of a local suburban sewage treatment works. The study was carried on from 1977 to 1998. The sampling design was strongly influenced by recognition that environmental variability resulted in "experiments in nature" that could be exploited to better understand processes that influence metal bioavailability and effects. A study of benthic ecology and the population biology of 2 bivalves (*Macoma balthica* and *Gemma gemma*) was underway when the contaminant study was started, both of which were long-term studies (more than a decade) organized to take advantage of this ongoing study of the consequences of environmental variability.

An important aspect of the contaminant study was to characterize general environmental variability. Distinct seasonal cycles and large year-to-year variability occur as driving forces in an estuary, so sediment characteristics (e.g., grain size, Fe, Mn, total organic C) were sampled nearly monthly to allow characterization of seasonality. Other sources of variability sampled included salinity, rainfall and associated freshwater inflows, phytoplankton blooms, and discharge from the local treatment works. The choice of which variables to sample or follow was also guided by practical considerations. Measures were chosen that

- could be sustained by the particular expertise in the laboratory;
- were being determined by other ongoing programs such as weather service, stream gauge, and effluent monitoring; or
- were relevant to both exposure to metals and effects and could be routinely collected without extensively increasing sampling effort.

For example, CI, a measure of effect as well as an ancillary measure for evaluating factors that influence bioaccumulation, was determined in every sampling by adding measurement of shell length to the monthly analyses (tissue weight was determined for the metal analysis). In retrospect, dissolved metal concentrations would have been a very valuable addition. But this measure was not included in the time series because the very labor intensive effort necessary to obtain high quality data over a long period of time was unlikely to be sustained without excluding other important measures. Periodic topical studies also provided opportunities to test questions raised by the monitoring and at least partially synthesize and interpret the data at frequent intervals.

Seasonal cycles and year-to-year variability were persistent feature of all environmental factors (Table 6-2). Seasonality was linked to driving forces such as annual precipitation patterns, tidal cycles, and seasonal winds, but it was dynamically stable for most factors. After the first decades of sampling, the period of the seasonal cycles was relatively predictable (Thomson-Becker and Luoma 1985; Cloern 1996). Year-to-year variability was linked primarily to climate. Year-to-year differences in the amplitude of responses differed among factors. Over the decades, climate varied widely (from drought to 100-y floods) but without directional trend.

In contrast, directional changes in human activities occurred within the watershed over the decades. These included increases in the density of urbanization, presumably affecting urban runoff, and massive investments in improved waste treatment that reduced metal loads in effluents (e.g., from ~ 6,000 kg/y Cu in 1977 to ~ 150 kg/y in 1998). So an advantage of the multidecade sampling was that direct but nondirectional responses to climate would not co-vary with directional changes in human activities. For example, concentrations of Cu, Ag, and Zn in sediments and tissues of the bivalve *Macoma balthica* were determined near monthly; sediments

Table 6-2 Extent of variability in environmental variables at different time scales from a study of metal bioavailability and effects on bivalves at a mud flat in San Francisco Bay [a]

Environmental variable	Variability		
	Seasonal variability (range in annual mean)	Year-to-year variability (in annual min or max)	Decadal trends (over 20 y)
Precipitation (linked to urban runoff)	0 to 15 cm/month	4-fold	No trend
Salinity	12 to 28	4-fold	No trend
Phytoplankton	Seasonal bloom ~Mar to April[b]	6-fold	No trend
Sediment redox	Surface always oxic	Little visual variability	No trend
Sediment chemistry (Fe/Mn/grain size)	2-fold	~ 50%	No trend
Hg, Zn, Cr, V, Ni[b]	~ 50%	~ 50%	No trend
Cu and Ag in sediment and bivalves	~ 50%	~ 50%	5- to 10-fold unidirectional decline

[a] Data from Hornberger et al. 2000.
[b] In sediment.

were sieved to < 100 mm to eliminate grain size biases on concentration and to simulate the size of particles ingested by the deposit-feeding bivalve.

One of the biological endpoints monitored throughout the study was CI. Year-to-year variability in CI has a seasonal cycle that is tied to reproduction. Year-to-year differences in amplitude, peak frequency, and period could be indicators of disturbances or stress such as inadequate availability of food, unusual or extended episodes of low salinity, unusually warm temperatures, or effects of pollutants. Understanding the seasonal cycle of CI is essential to detecting anomalous CIs (see Figure 6-3). For example, annual mean CI can be used to characterize differences among years only if sampling is equally representative of the cycle in all years. Optimal characterization of the seasonal cycle requires a different strategy for CI (and reproduction) than for exposure variables. Reducing samples to ~ 8 per year was the lowest frequency that would capture the seasonal cycle in both.

Impairment was defined as the persistent, year-to-year occurrence of a CI below 50% of the maximum. Association with elevated contaminant exposure was a necessary condition for chemical impairment.

A subset of animals was archived from near-consecutive months for 18-month periods at 4 intervals during the study (1974 to 1976, 1979 to 1981, 1982 to 1984, 1987 to 1989). These preserved specimens were sectioned, and reproductive activity (presence of fully developed sperm or eggs) was examined in the tissue sections

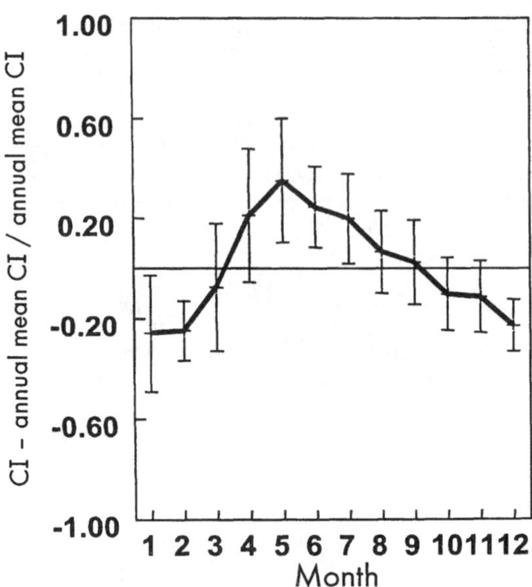

Figure 6-3 Annual mean cycle in CI expressed as proportional mean deviation from annual mean at each month, based upon near-monthly data collected between 1977 and 1999

(Hornberger et al. 2000). Seasonality in reproduction also follows a dynamically stable cycle in bivalves in San Francisco Bay (Nichols and Thompson 1985; Parchaso et al. 1997). Mature gametes are present in bivalves in most months except later summer. But frequent and seasonally representative sampling is necessary to identify whether the cycle is normal or aberrant. The best indicator of chemical impairment is probably a persistent occurrence through the years of few individuals with mature gametes. Coincidence with high metal exposure then suggests an effect from the pollutant (Hornberger et al. 1997).

Biological change was linked to metal exposure at this site. The following aspects of study design facilitated establishing that linkage:

- Change in exposure was large: Ag and Cu tissue concentrations declined from extremely high levels in 1975 to 1982 to levels about 5× background in the late 1990s.
- Persistent low reproductive activity, associated with high metal exposure, was a sign of chemical stress. Between 1974 and 1981, few animals from this mud flat ever contained mature gametes (Hornberger et al. 1999).
- Recovery further verified the likelihood of chemical stress. Reappearance of a large proportion of individuals with mature gametes accompanied decline of metal exposure.
- Persistently reduced CI was not observed. The ability to seasonally add glycogen and lipid persisted through the period of high metal exposure.

Metal effects on a San Francisco Bay mud flat: Using time series to eliminate confounding variables

- Design to exploit environmental variability.
- Recovery can be exploited to identify unidirectional trends in exposure and effect.
- Environmental data are as important as stressor data.
- Long time series allow separation of nondirectional environmental change from unidirectional changes in pollutant exposure.
- Data at 3 scales (seasonal, year-to-year, and decadal) were all valuable in understanding effects.
- Representation of seasonal cycle must be consistent if annualized means are to be compared.
- Topical studies of metal-specific responses aid diagnosis of contaminant effects.

- The seasonal period of CI was persistently changed during the period of high metal exposure and appeared to recover. Periodic fall peaks in condition were observed at reference sites and after contamination subsided at the polluted site. Fall peaks were not part of the annual mean cycle during the period of pollutant exposure. A fall peak in Cu in sediments was also evident during the period of high contamination, and this disappeared as the contamination subsided. The significance of this to the population is not yet clear.
- Topical studies of biomarkers aided diagnosis, indicating occurrence of metal-specific stress. During the period of greatest exposure to metals (1979 to 1980), spillover of Ag and Cu from metallothionein-like protein associations to fractionally greater association with low molecular weight was reported (Johansson et al. 1985). Shift of intracellular Ag into lower molecular compounds is a sign of metal movement toward toxic target sites and has been related to physiological stress in some studies. This does not prove that the population of bivalves at Palo Alto was threatened, but it does show that metal-specific stress was present. Thus, higher levels signs of stress could, feasibly, be related to this lower level sign.
- Topical study of population tolerance aided diagnosis (Luoma 1977). During the period of greatest Ag and Cu exposure, the surviving clam population was about 5-fold more tolerant to both metals than were neighboring populations that lived on mud flats and were exposed to lower concentrations of the elements. Again, the occurrence of tolerance is a specific indication that metal stress was present in the animals surviving at the PA mud flat (Luoma 1977).
- Studies with other species demonstrated mechanistic plausibility. Inhibition of reproduction in invertebrates by Cu and Ag has been demonstrated experimentally (Fisher and Hooke 1997).

Case Study 4: Heavy Metals and Benthic Macroinvertebrate Communities in Colorado Streams

Experimental designs employed in stream surveys often involve tradeoffs between spatially extensive and temporally intensive sampling (Resh et al. 1995; Wiley et al. 1997). Long-term sampling within a single watershed can provide important insights into seasonal and annual variation. In addition, long-term monitoring of a single stream can provide an opportunity to evaluate changes in ecological conditions resulting from changes in physical–chemical or habitat characteristics (e.g., improvements in water quality). However, because of the small spatial scale of most single-stream assessments, this approach may overrepresent the significance of temporal variation relative to spatial variation (Wiley et al. 1997). In addition, because single-stream surveys generally lack true replication (Hurlbert 1984), our ability to generalize beyond the specific stream under investigation is greatly limited. Finally, because the typical approach used in most single-stream assessments involves a comparison of upstream reference sites to downstream polluted and recovery sites, effects of stressors are confounded by natural, longitudinal variation (Clements and Kiffney 1995). Sampling designs that include a large number of streams at a larger spatial scale (e.g., spatially extensive designs) can be used to partition variation and identify the relative importance of different sources of variation. Because of limited funding and other logistical problems, long-term studies conducted at a large spatial scale are uncommon.

This case study will report results of temporally intensive and spatially extensive investigations that assess the effects of heavy metals on benthic macroinvertebrate communities (Figure 6-4). The single watershed study was conducted in the upper Arkansas River basin, a metal-polluted stream located in central Colorado, USA between the Sawatch and Mosquito mountain ranges. Mining operations near the town of Leadville have had a major impact on the Arkansas River since the late 1800s, when gold was discovered in California Gulch (CG). Since 1989, we have measured heavy metal concentrations in the stream, conducted acute and chronic toxicity tests, measured metal bioaccumulation in dominant taxa, and sampled benthic macroinvertebrate communities seasonally (spring and fall) from 6 to 10 stations along a 90-km reach of the Arkansas River. Stations are located upstream and downstream from Leadville Mine Drainage Tunnel (LMDT) and CG, a U.S. Environmental Protection Agency (USEPA) Superfund site (Clements 1994). In 1992, the USEPA, the Bureau of Reclamation, and the State of Colorado initiated a restoration project to reduce instream metal concentrations in the Arkansas River. Because we have collected data from reference and polluted sites before (1989 to 1992) and after (1993 to present) remediation, this research provides an opportunity to examine responses of population and community indicators to changes in water quality.

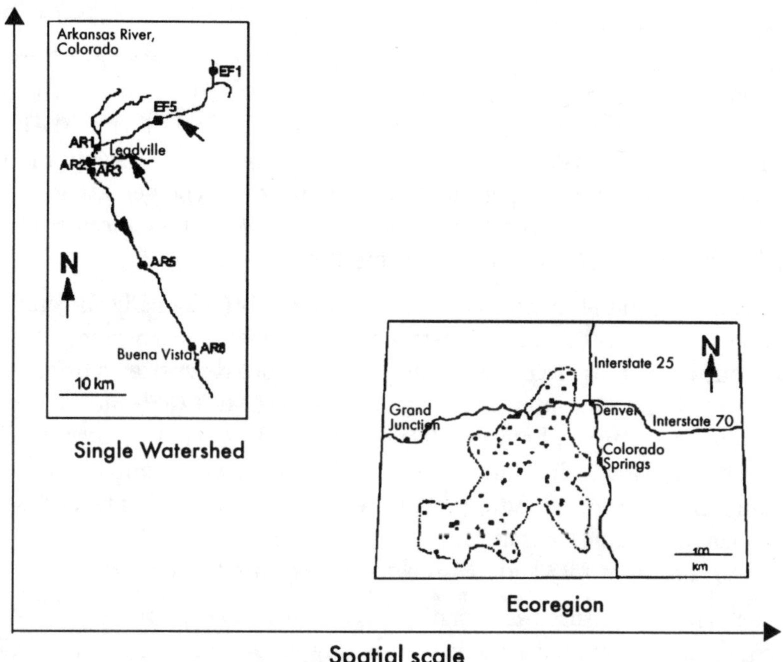

Figure 6-4 Maps of sampling locations in temporally intensive sampling of Arkansas River and spatially extensive sampling of Southern Rocky Mountain Ecoregion of Colorado, USA

A spatially extensive survey of reference and metal-polluted streams was conducted in the Southern Rocky Mountain Ecoregion in Colorado (Clements 1999). Physical, chemical, and benthic macroinvertebrate data were collected from 78 randomly selected sites in 1994 and 1995 as part of the USEPA's Regional Environmental Monitoring and Assessment Program (R-EMAP). Sampling was restricted to small streams (third- and fourth-order) and all samples were collected in late summer (August to September) during low flow conditions. Because the study sites were randomly selected, conditions in these streams are considered representative of similar streams found in the ecoregion.

Metal-polluted streams in the study area were impacted by a mixture of metals, primarily Cd, Cu, and Zn. Therefore, we used a cumulative measure of total metal concentration to examine the relationship between benthic community structure and heavy metals. We assumed that interactions among metals were additive and defined the cumulative criterion unit (CCU) as the ratio of the measured metal concentration to the hardness-adjusted USEPA criterion value, summed for all metals at a station. The CCU is given as:

$$CCU = 3\ m_i/c_i \qquad \text{(Equation (6-1),}$$

where m_i is the total recoverable metal concentration, and c_i is the criterion value for the ith metal. We placed all stations into 1 of 4 categories based on the measured

CCU. Background (unpolluted) sites were defined as stations where the CCU was < 1.0. The low metal category consisted of sites with CCU values between 1.0 and 2.0. The medium and high metal categories consisted of sites with CCU values between 2.0 and 10.0 and > 100, respectively. Results reported here are limited to the abundance of heptageniid mayflies (Ephemeroptera: Heptageniidae). Previous monitoring research and microcosm experiments have shown that heptageniids are especially sensitive to heavy metal contamination and are useful indicators of water quality (Clements 1994; Kiffney and Clements 1996).

Heavy metal levels in the Arkansas River vary temporally (seasonally and annually) and spatially (from upstream to downstream). The highest concentrations are observed during periods of spring runoff and at stations downstream from CG (Clements 1994). Zn concentrations at stations below LMDT (EF5 and AR1) prior to remediation were generally between 200 and 600 mg/L. After 1992, these levels decreased to < 100 mg/L. In contrast, remediation of CG has resulted in relatively little change in metal concentrations at downstream stations and only modest improvement in benthic communities. Zn levels have remained elevated (500 to 1000 mg/L) at stations AR3 and AR5 downstream from CG.

Heavy metal levels measured at 78 stations in the spatially extensive survey of the Southern Rocky Mountain Ecoregion were highly variable among streams. CCU levels ranged from 0.1 to 293.5. Although the CCU was < 2.0 at most (66.3%) of stations (median = 1.4), values exceeded 10.0 at 13 stations. There was little annual variation in metals levels between 1994 and 1995.

Abundance of heptageniid mayflies in the Arkansas River from 1989 to 1998 reflected changes in water quality. The density of these metal-sensitive organisms increased significantly at stations AR1 and EF5 following remediation of LMDT (Figure 6-5). In contrast, abundance of these metal-sensitive organisms has shown relatively little change at stations downstream from CG, where metal levels have remained elevated. Increased abundance of heptageniids at stations where water quality has improved and failure of these organisms to recover at stations where metal levels remain elevated support the hypothesis that heavy metals were the primary stressor in this system.

Metal effects on stream invertebrates determined by combining temporally intensive and spatially extensive study

- Microcosm experiments identified metal-sensitive indicators that could be used in the field study.
- Temporally intensive study showed increased abundance of metal-sensitive organisms after water quality improved.
- Spatially extensive study allowed generalizations about geographic extent of effects.
- Spatial distribution of elevated metal levels coincided with degraded benthic assemblages.
- Metal specificity was indicated in the spatial study when metal-sensitive species showed the greatest response to contamination.

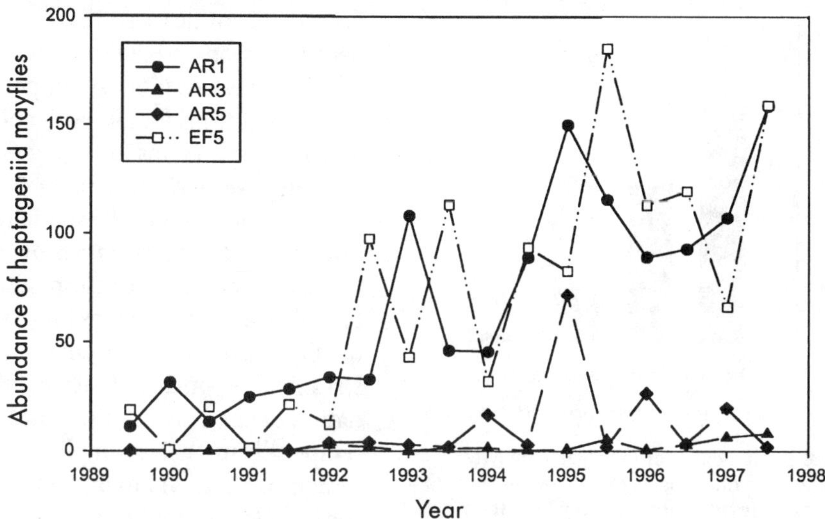

Figure 6-5 Abundance of heptageniid mayflies at Arkansas River stations before (1989–1992) and after (1993–1998) initiation of a water treatment facility in Leadville Mine Drainage Tunnel (LMDT) and California Gulch (CG), Colorado

In the spatially extensive survey, results of stepwise multiple regression for the 78 randomly selected stations indicated that heavy metal concentration was the most important predictor variable for 14 of 16 community variables that we examined. Although other physical–chemical characteristics such as water temperature, SO_4, PO_4, and conductivity were included in some regression models, they generally explained much less variation in community structure than did metal concentration. The greatest responses to heavy metals were observed for the heptageniid mayflies *Rhithrogena robusta* (Figure 6-6). Abundance of this species was significantly reduced at stations where the CCU exceeded 2.0.

The temporally intensive and spatially extensive sampling designs described in this case study provided different types of information with respect to benthic community responses to heavy metals. Microcosm experiments allowed us to identify metal-sensitive indicators that could be used in the field study. Long-term sampling of the Arkansas River provided an opportunity to evaluate the effectiveness of remediation at LMDT and CG, the 2 main inputs of metals to the system. Improvements in water quality below LMDT resulted in increased abundance of metal-sensitive organisms.

Because our long-term sampling of the Arkansas River was restricted to a single watershed, we cannot extend our findings beyond this system. However, the spatially extensive survey of streams in the Southern Rocky Mountain Ecoregion allowed us to generalize about the widespread effects of metals in this area. For example, 18 of the 78 randomly selected streams had both elevated metal levels and

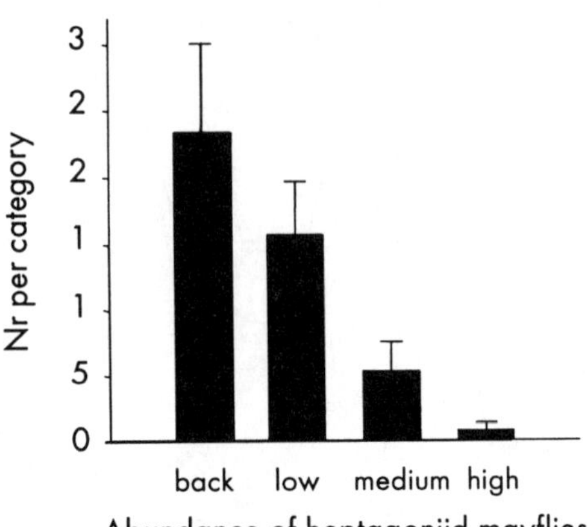

Figure 6-6 Abundance of heptageniid mayfly *Rhithrogena robusta* at background, low, medium, and high metals sites in Southern Rocky Mountain Ecoregion, Colorado

degraded benthic assemblages. Because streams were randomly selected, these results have important implications for all third- and fourth-order streams in the region. Although we lack estimates of temporal variation in the spatially extensive study, these data provide the necessary baseline information to evaluate future changes in water quality in Colorado's mountain streams. We conclude that a combination of spatially extensive and temporally intensive sampling designs is necessary to understand the influence of multiple stressors in aquatic ecosystems.

Case Study 5: Applying the RIVPACS Approach in the Fraser River, Canada

Temporal and spatial variability of biological attributes has made site-specific assessment of response to stresses difficult in aquatic ecosystems. An approach using site-specific biological targets to evaluate effects of stress in freshwater ecosystems was first developed in the UK. River Invertebrate Prediction and Classification System (RIVPACS) is a subcontinent-to-catchment scale system for assessing the condition of rivers (Furse et al. 1984; Wright et al. 1984; Armitage et al. 1987). The technique involves a multivariate statistical approach using 1) data on the structure of benthic invertebrate communities and 2) selected environmental variables. Probabilistic models are produced that resolve the issues of temporal and spatial variability and allow site-specific predictions of species assemblages to be established from reference datasets.

The methodology requires developing a large database of multiple reference sites by collecting information on the biological attributes selected (the invertebrate assemblage in RIVPACS) at each site and on the physical and chemical environment (habitat). Empirical models use this reference database to determine the probability that an assemblage of invertebrates will occur at a site with particular habitat attributes. This prediction is used as a baseline against which to evaluate sites that are exposed to either chemical or physical stress.

Reference sites are chosen to represent the diversity of conditions in the study area. The study area can vary from a single catchment to the continent. At the spatial scale of the study area, the objective is to capture as much of the natural variability as possible. To reduce sampling effort at the scale of the entire study area, it may be appropriate to use a preliminary stratification. Once the study has been bounded, then measurement at the correct scale is important. In the RIVPACS case, the scale of variation of interest is among river reaches rather than within reaches, and therefore a sampling device that integrates small-scale patchiness is preferred. The RIVPACS study integrated the small-scale patchiness by sampling with a kick net, as opposed to replicating small area samples such as Surber or Hess samplers.

The reference condition approach uses taxonomic counts (e.g., numbers of taxa present, presence or absence of taxa) or a set of derived measures (metrics) to describe the community. Multivariate methods for describing and classifying communities use either presence or absence of taxa, as in the UK (Wright et al. 1984) and Australia, or quantitative abundance counts, as in Canada (Reynoldson et al. 1995, 1997). A second approach to using reference conditions is based upon additive indexes from a subset of derived measures (e.g., number of taxa, % Ephemeroptera) and is used widely in the U.S. In the Australian scheme (AusRivAS), only family-level identifications are used. It is suggested that the family level is most appropriate in the assessment of invertebrate communities in the Fraser River catchment. However, RIVPACS uses identification to species level. In part, this is enabled by the restricted fauna in the UK and by the fact that RIVPACS is also used for conservation purposes, where species-level identifications are critical.

The Fraser River is an example of the application of the RIVPACS methodology. Critical physical, chemical, and biological variables that are believed to affect the distribution and abundance of species are measured at the same spatial and temporal scales as the community composition data. These variables should include habitat characteristics, limiting resources, and potential stressors. Examples for freshwater benthic macroinvertebrates include dimensions of the river reach, water flow rate, substrate composition, food concentrations, and water quality (i.e., pH, hardness, alkalinity, dissolved oxygen, nutrients).

The variability of environmental measures in the river differed widely. Measures of channel morphology (e.g., bankfull width, channel depth) and discharge had low coefficients of variation (CVs; ~ 5% to 10%), whereas measures of substrate composition (~ 10% to 87%) and benthic algae (~ 35%) were more variable. Characteristics of the substrate and biomass of benthic algae are naturally patchy, so the high CVs for these variables are not surprising. All water variables had CVs < 25%. The lowest CVs in the variability study were obtained for physical measurements such as pH (0.26%) and dissolved oxygen (0.91%). Nutrients and total suspended solids (TSSs) had much higher CVs (~ 22% to 24%).

We also checked the variability of the benthic macroinvertebrate collections by calculating CVs for 2 variables. Abundance (number collected per unit time) had a mean CV = 32.30%, and number of families had a mean CV = 13.38%. These values are less than typical for macroinvertebrate sampling, probably because kick net

Effects of multiple stressors in rivers: Reference condition approach to defining environmental variability

- Both habitat and biological community are temporally and spatially variable in nature.
- Complexities stemming from such variability can be resolved by developing comprehensive characterizations of benthic communities in reference areas, based on habitat characteristics.
- Reference site should be chosen to represent the full diversity of conditions in the study area.
- Habitat should be oversampled initially to capture variability; later, the number of variables can be reduced.
- Multivariate techniques are used to characterize biological communities and provide context for comparisons to communities subjected to stressors.

sampling integrates habitat patches and yields confidence in model calculation using macroinvertebrate data.

Once variability among reference sites has been captured by a sampling program, classification methods are used to describe the biological structure of invertebrate assemblages. Either presence or absence or abundance counts for each sample are used as descriptors of the benthic invertebrate community. The reference sites are clustered, and the appropriate number of groups of equivalent reference sites is selected by examining the group structure and the spatial location of the groups in ordination space. Ordination also can be used to describe and explain the variability observed among the large number of taxa that use a reduced number of new variables (ordination axes).

Once the biological variability has been described and partitioned, it is necessary to develop an understanding of the relationship between biological variability and underlying environmental variability. Because there is no a priori way of knowing which will be the best explanatory variables of biological variability, it is desirable to oversample initially. For example, it is typical to start with 12 to 25 environmental variables, within which some variables might include multiple measures (e.g., substrate and riparian vegetation). However, a reduced set of variables can eventually be used to explain the observed variation among community assemblages. In the above cases, 8 to 11 variables were reported to be sufficient.

The relationship between faunal and environmental data can be examined in a number of ways. Two useful, multivariate statistical techniques are principal axis correlation and stepwise discriminant function analysis (DFA).

Principal axis correlation determines how well a set of attributes (environmental data) can be fitted to a second set of variables (biota) in ordination space, using a multiple-linear regression. The variables can be represented as an axis on an ordination plot, and a correlation of the biota variable with the ordination of environmental variables is provided. A randomization model can be used to establish the statistical significance of the correlations. Stepwise DFA can be used to establish which environmental variables best separate sites into predefined groups formed by classifying the biological dataset. Stepwise selection of variables is usually used, and the significance level for variable entry and retention can be set.

There is often considerable similarity in the environmental variables identified as best explaining the structure of stream invertebrate communities. Of the 8 to 11 variables identified above, 7 were consistent across 5 studies from regional to catchment scales. The 7 included alkalinity, conductivity, particle size or substrate structure, water velocity, water width (or channel width), and riparian vegetation. Other variables, while probably important, either were measured at an inappropriate temporal scale (temperature) or were excluded because they confounded interpretation at exposed sites (nutrient levels).

An objective of this type of study is to develop a method for selecting an appropriate subset of the total number of reference sites to be used in assessing any new test site. The subsets of reference sites are established by grouping similar sites according biological attributes in the RIVPACS approach. DFA is used for predicting the reference group to which a test site is expected to belong. In this method, this description of the habitat is used to classify sites into the groups formed by the invertebrate data.

RIVPACS and its derivatives (AusRivAS, BEnthic Assessment of SedimenT [BEAST]) go through a selection process to identify environmental variables for the discriminant models, which use the environmental data to generate discriminant scores, and to predict the probability of group membership for a site. The final selection of the optimal predictor variable dataset is done by iteration. The accuracy of the predictions from the discriminant model can be verified by examining how well the reference sites are predicted to the correct group. The predicted groupings and actual groupings can be compared to provide a group and total error rate.

Case Study 6: Lake Acidification in Sweden

In several regions of the world, environmental agencies have established biological monitoring of lakes for the purpose of detecting biological impairment and effects of known stressors. A pilot study and a subsequent monitoring program in Sweden address approaches to those problems.

The monitoring program established in Sweden covered the entire country, including 5 ecoregions (Johnson and Goedkoop 1999). The pilot study (Johnson 1998) included 16 lakes within a single ecoregion, which were categorized a priori as "reference" or "acidified" (pH < 6). All lakes in the pilot were oligotrophic. An objective of the pilot was to assess the feasibility of the larger sampling program by answering certain design and variability questions:

1) What is the optimum sampling habitat: littoral, sublittoral, or profundal?

2) What attributes (metrics) of the benthic macroinvertebrate assemblage are responsive to acidification stress?
3) What is the power of sampling alternative habitats and of estimating alternative metrics?

The pilot study, then, attempted to "isolate change induced by anthropogenic disturbance from variation introduced by natural spatial and temporal variability" (Johnson 1998). Lakes were sampled each year for 6 years during a restricted annual index period, a 2-month window in spring (littoral), or autumn (sublittoral and profundal). The study design thus attempted to characterize interannual variability but attempted to minimize short-term and seasonal variability by always sampling in the same season. Small-scale spatial variability was characterized by obtaining 5 or 6 replicates at each site.

Results of the pilot study showed that small-scale spatial, interannual, and among-lake variability were similar within biological metrics but that metrics differed greatly in their variability. In general, abundance and biomass measures had the greatest variability, and taxa richness metrics and a benthic quality index had the lowest. Habitats also differed in the amount of variability shown by each metric; for example, the profundal habitat showed the greatest variability for all but 1 metric.

Power analysis of the metrics showed that the sample sizes (number of lakes to show a difference between groups or number of years to show a trend) required in order to attain 95% confidence in an observed trend or difference or 95% probability of detecting an effect (power) ranged from 3 to > 1000, depending on the metric and the habitat. Total abundance and biomass metrics were the most variable, and taxon richness, diversity, and tolerance or intolerance indexes were the most robust, in agreement with observations from streams. The sublittoral habitat had the greatest number of metrics with greater power. Nevertheless, even the best metrics required 8 to 15 years of observation to observe a trend of the specified effect size.

Information from the pilot study was used to help design Sweden's nationwide lake monitoring program. Samples were taken from littoral habitats, and 5 kick samples were composited into a single sample for each lake. Benthic macroinvertebrates of > 500 lakes throughout Sweden were sampled during spring and summer index periods. Lakes selected for sampling were identified as reference lakes according to the following criteria:

1) representativeness of the ecoregion;
2) uniform distribution of reference lakes;
3) no direct modifications (point source discharges, liming, etc.);
4) accessibility; and
5) existing monitoring data.

Sampling was stratified among 5 ecoregions of Sweden, ensuring that a sufficient sample of lakes would be obtained in each ecoregion and that large-scale spatial variability could be characterized. Multivariate analysis of species composition in the reference lakes revealed that the assemblages differed among ecoregions (Johnson and Goedkoop 1999).

Case Study 7: Multiple Stressor Impacts in the Great Lakes of North America

Assessing the ecological risk to biota in the Great Lakes from contaminated sediments and determining the need for remedial action have typically been based on traditional methods that use numeric chemical guidelines. The International Joint Commission (IJC) identified more than 40 geographic Areas of Concern (AOCs) where beneficial uses have been impaired. Almost all these AOCs have documented sediment contamination, but the designation is often based on relatively few chemical measurements. Few systematic data are available on the concentrations of contaminants in many AOCs, and far less information is available on the direct impacts of sediment-associated contaminants on biota. The data presented in Table 6-3 represent areal summaries of sediment contamination for Canadian AOCs (Painter 1992; Persaud et al. 1992). In many cases, large areas exceed the severe category and all areas have zones of considerable size that exceed the lowest safe level. The large size of the areas of sediments designated "contaminated" by numeric chemical criteria has been a major impediment to conducting restoration.

Table 6-3 Nature and extent of sediment contamination in AOCs, based on bulk sediment chemistry[a]

	Severe effect level (SEL)[b]		Lowest effect level[b]		
Area of concern (AOC)	km^2	% AOC	km^2	% AOC	Contaminants > SEL
Hamilton Harbor	7.9	44.2	20.0	94.5	Zn, Cr, Cd, Pb, Cu, PAHs
Bay of Quinte	4.5	0.2	132.7	22.7	Cu
St Lawrence River	0.2	0.7	6.8	51.0	Hg, Zn, Cu, Pb
Spanish Harbor	66.0	12.4	262.7	35.9	Ni, Cu
Toronto Harbor	1.4	1.8	28.5	27.4	Cr, Cu, Pb, Cd
Peninsula Harbor	0.5	2.7	5.3	17.4	Hg
St Marys River	0.9	0.3	80.9	27.2	As, Cr, Cu, Pb, PAHs
Nipigon	0.0	0.0	55.2	7.2	
Jackfish Bay	0.0	0.0	0.6	54.6	
Wheatley Harbor	0.0	0.0	< 0.1		

[a] Source: Adapted from Painter 1992.
[b] As defined in Persaud et al. 1992.

Measuring the concentrations of various chemicals in the sediments does not address the ultimate concern, namely, whether the contaminants present are exerting biological stress and/or are being bioaccumulated. In several cases, sites are identified as impacted although contaminant levels are lower than historic background concentrations, as identified from sediment cores taken from the nearby open-lake depositional basins (Reynoldson et al. 1988). A series of bioassessment techniques identified the types of stress being exerted, their severity, and the bioavailability of the contaminants present. This approach assumes that the objective of sediment management is the protection of aquatic ecosystem health from deleterious substances of anthropogenic origin associated with bottom sediments. Because "health" is a description of biological condition, it follows that biological attributes are the most appropriate indicators.

Variability is an inherent property of natural systems, and biological states within an ecosystem are dependent on environmental variability. Discriminating variability associated with natural processes from that associated with anthropogenic stress is critical if change in the biological condition is to be used to guide decisions regarding contaminated sediment. Our approach to natural variability was to identify and sample reference sites across the Great Lakes, identifying variation in selected biological endpoints and matching this to the habitat. Natural variation could then be discounted when we attempted to identify response to stress.

The range of environmental conditions found at the 252 reference sites in the Great Lakes is summarized in Table 6-4. The median sampling depth was 15 m, and 80% of the sites had depths between 5 and 63 m. The natural variability of the physical substrate was high, including sites with 0% to > 90% of each of sand, silt, and clay. Total organic C ranged from 0.01% to 12.85%, with a median of 2.16%. Metal levels also varied considerably, the median levels of Cr, Ni, Cu, Cd, and Zn exceeding the Ontario lowest effect level. Three metals had values that exceeded the severe effect level (SEL; Table 6-3), Cr, Ni, and As, and 10% of the sites exceeded the SEL for Ni.

Selection of the biological variables to be measured was based on 2 attributes. One was a set of variables that measured organisms exposed to contaminated sediment and that could express the range of expected response. The other was a response that would identify the sediment as the medium causing the observed effect. The community structure of benthic invertebrates was considered most appropriate in meeting the first criterion. These organisms reside in the sediment, and because of the many species comprising the community and the variety of life histories, behaviors, and feeding strategies, they have the greatest likelihood of displaying responses to stress. However, changes in community structure can also respond to stress other than that associated with sediment contamination, and therefore individual organism responses were also measured in laboratory sediment exposures. We measured the responses of 4 species of benthic invertebrates (*Hyalella azteca, Chironomus riparius, Hexagenia* spp., and *Tubifex tubifex*) exposed in the laboratory to sediment collected from the same sites. These data were collected from 252 reference sites distributed across all the Great Lakes during 1991 to 1993.

Table 6-4 Summary statistics for environmental variables at 252 Great Lakes reference sites

	Statistics					
Environmental variable	Average	Median	Maximum	Minimum	10th percentile	90th percentile
Water depth (m)	24.8	15	102	1	5	63
Gravel (%)	0.6	0	36.2	0	0	0.7
Sand (%)	37.0	18.8	99.8	0	1.2	95.4
Silt (%)	31.6	32.3	95.7	0	0	65.9
Clay (%)	30.7	29.9	91.1	0.1	2.0	63.8
SiO_2 (%)	60.10	59.07	93.30	19.84	46.06	76.79
TiO_2 (%)	0.52	0.51	1.92	0.05	0.22	0.76
Al_2O_3 (%)	10.06	10.48	15.02	2.39	5.46	13.59
Fe_2O_3 (%)	4.46	4.28	21.70	0.44	1.77	7.28
MnO (%)	0.16	0.10	3.29	0.01	0.04	0.23
MgO (%)	2.71	2.30	8.35	0.28	1.10	4.69
CaO (%)	5.92	3.38	32.15	0.49	1.56	12.64
Na_2O (%)	1.62	1.58	3.53	0.12	0.68	2.75
K_2O (%)	2.35	2.36	4.18	0.74	1.51	3.10
P_2O_5 (%)	0.17	0.16	0.56	0.02	0.07	0.29
Total N (µg/g)	2199.8	1460.5	12528.0	0.0	421.6	5202.7
Total P (µg/g)	684.5	563.5	7180.0	20.0	174.5	1176.5
Loss on ignition (%)	11.39	10.52	38.72	0.59	3.17	20.12
Total organic C (%)	2.16	1.75	12.85	0.01	0.39	4.88
V (µg/g)	47	38	159	4	14	88
Cr (µg/g)	45	40	123	4	12	83
Co (µg/g)	12	11	49	1	3	21
Ni (µg/g)	45	33	348	1	10	79
Cu (µg/g)	26	25	91	0	5	51
Zn (µg/g)	112	96	482	5	27	207

Table 6-4 *continued*

Environmental variable	Statistics					
	Average	Median	Maximum	Minimum	10th percentile	90th percentile
As (µg/g)	10	5	115	3	3	24
Cd (µg/g)	1	1	4	0	0	2
Pb (µg/g)	42	35	153	1	16	77
Water pH	8.0	8.0	11.9	6.3	7.3	8.7
Dissolved O (mg/L)	9.3	9.1	15.0	4.9	6.4	12.4
Alkalinity (mg/L)	78.3	75.7	118.0	38.1	45.0	108.9
Total P (mg/L)	0.0126	0.0096	0.1094	0.0014	0.0045	0.0216
Total Kjeldahl N (mg/L)	0.186	0.145	1.480	0.031	0.075	0.324
Nitrate-nitrite (mg/L)	0.221	0.249	0.416	0.001	0.019	0.349

Benthic Community Data

A total of 39 invertebrate families were recorded, and of these, 3 families are very common (> 80% occurrence): the Chironomidae (midge larvae), Tubificidae (worms), and Sphaeridae (fingernail clams). A further 3 families are slightly less common (> 50% occurrence): the Naididae (worms), Pontoporeiidae (shrimps), and Spongillidae (sponges). Almost half the families (18) are considered rare and occur at < 10% of the sites. The most abundant family is the freshwater sponge (Spongillidae), representing more than 80% of the organisms found. However, this group is frequently excluded from invertebrate enumerations because they are colonial animals and therefore difficult to compare with other organisms. The other most abundant families are the Tubificidae and Pontoporeiidae, representing > 20% of total animals found (excluding sponges). Two other families were abundant (> 10% of the total): the Chironomidae and the recent invaders Dreissenidae (zebra and quagga mussels). The majority of families (27 of 39) were not abundant (< 1% of total).

Variability among the reference sites was partitioned using cluster analysis and ordination, and sites were separated into groups characterized by similar taxa. At the family level, 5 groups of sites were identified with a strong spatial component to the site groups:

- Group 1 is characterized by a low number of animals (Figure 6-7), and the dominant organisms are chironomids. However, the chironomids are a widespread family and are found at most sites. This assemblage of families is characteristic of southwestern Georgian Bay and much of the North Channel.

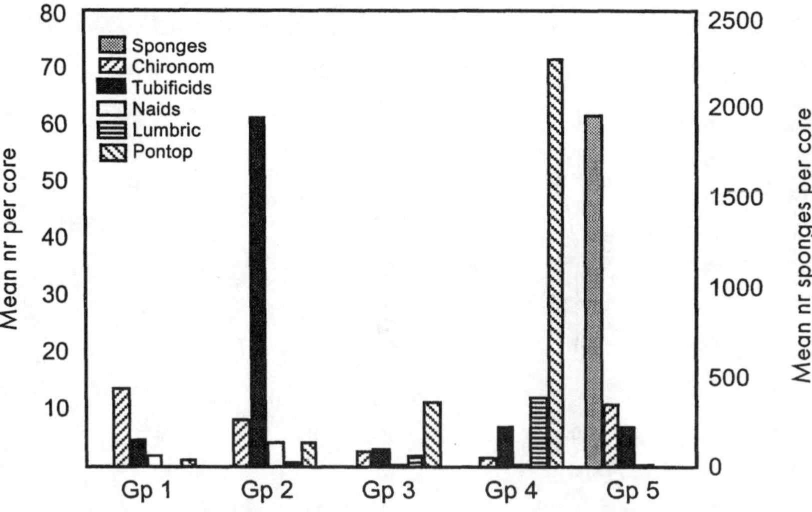

Figure 6-7 Benthic invertebrate families characterizing site groups formed by cluster analysis from 252 Great Lakes reference sites

- Group 2 is dominated by tubificid and naid oligochaetes (Figure 6-7). Both families occur in significantly higher numbers ($P < 0.05$) than in other groups. Almost 60% of these sites are located in Lake Erie, with the rest scattered through all the other lakes. The families composing this group, and the spatial distribution of the sites, suggest that it is characteristic of more mesotrophic habitats. This group also tends to be associated with water with a higher alkalinity, which is likely a surrogate for dissolved minerals, including nutrients (Figure 6-8).
- Group 3 is characterized by sites where the Pontoporeiidae are dominant and occur in greater numbers than at Groups 1 and 5 (Figure 6-7). The Chironomidae are the second most abundant family in this group of sites. This assemblage of organisms is characteristic of Lake Superior (93% of sites) and of many more-exposed Georgian Bay sites (28%), together with the remaining North Channel sites. These sites are associated with deeper water and less organic material in the sediment (Figure 6-8).
- Twenty of the 28 (71%) Group 4 sites are in Lake Michigan. This group is dominated by 2 oligotrophic families, the Pontoporeiidae and Lumbriculiidae. These families occur in significantly greater ($P < 0.05$) numbers than in other groups. These sites represent a deeper-water assemblage of organisms.
- Finally, Group 5 is unique in that it is dominated by sponges. The sites within this group are characteristically in sheltered areas such as Long Point Bay, Lake Erie, the Bay of Quinte and Presque Isle Bay in Lake Ontario; some sites in Severn Sound and Georgian Bay are also found in this group. Chironomids are also abundant in Group 5. These sites are associated with sediment with a high organic content (Figure 6-8).

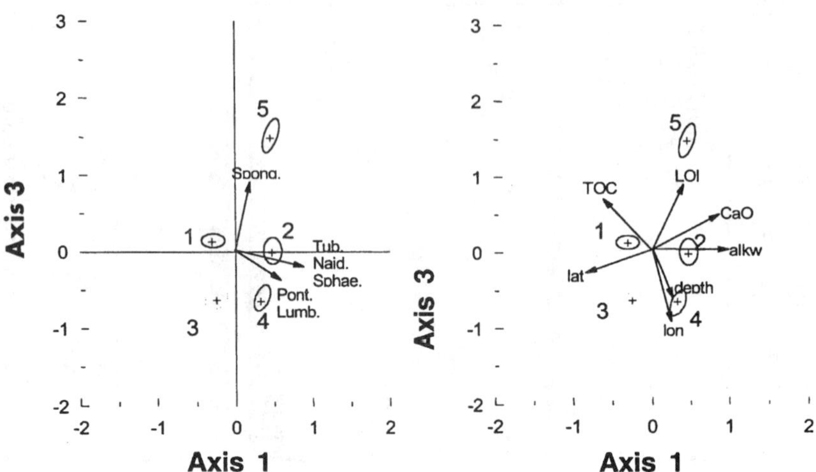

Figure 6-8 Vectors for major taxa and environmental variables shown in ordination space and site group centroids with 90% confidence ellipses (alkw = alkalinity of water, LOI = loss on ignition, TOC = total organic C)

Whole-sediment Laboratory Toxicity Test Data

Descriptive statistics in Table 6-5 show the mean, median, standard error (SE), standard deviation (SD), maximum and minimum values, range, and CVs for each measured endpoint with 4 species of benthic invertebrates exposed to 170 to 220 reference sediments collected throughout the Great Lakes over a 3-year period. Table 6-6 shows a similar set of descriptive statistics and frequency diagrams for the range in responses for each species exposed to 1 reference sediment in quality assurance studies during the 1991 to 1993 period of the study.

There was little variation in the toxicity test endpoints, which individually showed only weak relationships between 2 endpoints and sediment characteristics. In addition, when the range in response for each endpoint in a variety of sediments was compared to the range in response for the same endpoint in only 1 reference sediment (Table 6-6), few differences were noted. We therefore concluded that the range in each endpoint observed for the reference sediment database represented the natural range in responses of each organism in laboratory bioassays, and it was unnecessary to attribute different response categories in relation to environmental variation.

Using discriminant analysis, we examined the relationship between the variability in the invertebrate fauna as described by the 5 site groups formed from the faunal data. A total of 43 environmental descriptors were measured; 26 of these were examined as potential explanatory variables for the measured variation of the community. Six environmental attributes (Table 6-7) explained 70% of the variability among the 5 faunal groups.

Table 6-5 Variability in endpoints in bioassays with 4 species of benthic invertebrates exposed to reference sediments from the Great Lakes

	Chironomus riparius[a]		*Hyalella azteca*[a]		*Hexagenia* spp.[b]		*Tubifex tubifex*[b]			
Statistic	Survival (%)	Growth (mg dw/ larvae)	Survival (%)	Growth (mg dw/ juvenile)	Survival (%)	Growth (mg dw/ nymph)	Survival (%)	Hatch (%)	Cocoons/ adult (nr)	Young/ adult (nr)
Mean	85.5	0.35	86.8	0.49	95.9	2.97	98.3	58.1	9.8	28.1
Median	86.7	0.33	90.7	0.50	95.9	2.98	98.3	58.1	9.8	28.2
SD	8.9	0.07	9.9	0.13	5.2	1.02	4.7	10	1.3	9.1
CV	10.4	21.3	11.4	27.0	5.5	34.3	4.8	17.3	13.1	32.2

[a] $n = 220$ reference sites.
[b] $n = 170$ reference sites.
dw = dry weight.

Table 6-6 Variability in endpoints of 4 species of benthic invertebrates exposed to 1 reference sediment in laboratory bioassay conducted over 3 years

	Chironomus riparius[a]		*Hyalella azteca*[a]		*Hexagenia* spp.[b]		*Tubifex tubifex*[b]			
Statistic[a]	Survival (%)	Growth (mg dw/ larvae)	Survival (%)	Growth (mg dw/ juvenile)	Survival (%)	Growth (mg dw/ nymph)	Survival (%)	Hatch (%)	Cocoons/ adult (nr)	Young/ adult (nr)
Mean	87.4	0.37	91.7	0.59	97.1	5.00	98.9	56.7	11.1	36
Median	89.4	0.36	93.3	0.58	98.0	4.75	100	57.7	11.0	37
SD	8.3	0.07	7.2	0.14	4.1	0.99	0.7	5.8	0.8	8.7
CV	9.5	17.7	7.7	23.5	4.2	20.8	7.1	10.2	7.3	24.2

[a] $n = 46$.
dw = dry weight.

Table 6-7 Error rate estimates for family-level discriminant models of 252 reference sites

Error rates	Explanatory variables: depth, latitude, LOI, Al_2O_3, silt, alkalinity (water)
Total error rate (cross validation)	30.0%
Sites predicted correctly	169

LOI = loss on ignition.

Determining Impairment

Three categories of toxicity were developed for near-shore sediments in the Great Lakes, based on the results from 167 to 212 reference sediments:

- nontoxicity,
- potential toxicity,
- and toxicity.

The use of 2× and 3× the SD about the mean for each endpoint was chosen to separate these response categories. The delineations for each of the categories for each species and endpoint in whole sediment toxicity tests are presented in Table 6-8. An upper limit is provided in the nontoxic category for growth based on twice the SD of the mean. Toxicological effects on sublethal responses are usually considered negative, that is, growth or reproduction is reduced in comparison to a control. However, in areas of eutrophication or high nutrient impact, sublethality may manifest itself as an increase in biomass or production of young. This effect could have a negative impact on the structure and function of benthic invertebrate communities in aquatic ecosystems. Therefore, an upper limit for growth of *C. riparius*, *H. azteca*, and *Hexagenia* and for reproduction by *T. tubifex* has been set for the nontoxic category in this study. Although increased levels for growth and reproduction are not considered indications of toxicity, their presence should be noted.

Table 6-8 Limits derived from Great Lakes reference sites for determining toxicity of 10 test endpoints

Species	Nontoxic	Potentially toxic	Toxic
C. riparius			
Survival (%)	> 69	60 to 68.9	< 60
Growth (mg dw)	0.21 to 0.49	0.14 to 0.20	< 0.14
H. azteca			
Survival (%)	> 68	58 to 67.9	< 58
Growth (mg dw)	0.24 to 0.76	0.11 to 0.23	< 0.11
H. limbata			
Survival (%)	> 85	80 to 84.9	< 80
Growth (mg dw)	1.0 to 5.0	0 to 0.9	—
T. tubifex			
Survival (%)	> 88	84 to 87.9	< 84
Hatch (%)	40 to 78	30.8 to 39.9	< 30.8
Cocoon/adult (nr)	7.2 to 12.3	5.9 to 7.1	< 5.9
Young/adult (nr)	12.0 to 45.6	3.6 to 11.9	< 3.6

dw = dry weight

While there are a number of possible approaches to matching reference and test sites, our approach is again a multivariate method in which the entire community assemblage can be used. In this method, the reference and matching test sites are ordinated and plotted in ordination space. The distribution of the reference sites provides the range of variation in unimpaired communities. The community at the test site can then be compared to this normal variability, with a given probability of belonging to the reference group, using probability ellipses built around the reference sites. The greater the departure from reference state as measured in ordination, the greater the impact; however, determining the actual degree of impact and what departure from the reference state defines an unacceptable impact is ultimately a subjective decision.

A large river water quality survey conducted in the UK in 1990 provided the impetus for the development of methods to circumscribe the continuum of responses into a series of bands that represented grades of biological quality (Wright et al. 1995). This is a simplification of what is a continuum of responses in sites, ranging from good to poor biological quality. Despite the simplification, it was seen as an appropriate mechanism for obtaining a simple statement of biological quality, allowing broad comparisons in either space or time that would be useful for management purposes.

Effects of stressors in the Great Lakes: Reference conditions are critical

- Measure variation at appropriate spatial scale.
- Acquire information on both response and explanatory variables.
- Explain natural variation from reference conditions, and use variation in response variables to identify stress.
- Understand that community structure data using this approach are empirical only.
- Use mechanistic models to understand linkages.
- Determine degree of stress by scale of response and actual species lost.

We have adopted a similar approach for defining degrees of impact, using a multivariate approach with 3 probability ellipses (Figure 6-9). Sites inside the smallest ellipse (90% probability) would be considered unstressed, sites between the smallest and next ellipse (99% probability) would be considered potentially stressed, sites between the 99% probability and the largest ellipse (99.9% probability) would be considered stressed, and sites located outside the 99.9% ellipse would be designated as severely stressed.

The process of determining whether an invertebrate community is impaired involves

1) sampling the community and measuring the predictor variables at the site of interest,
2) running the discriminant model with the reference database and test site to predict the expected community at the test site, and
3) comparing the test sites to reference community sites.

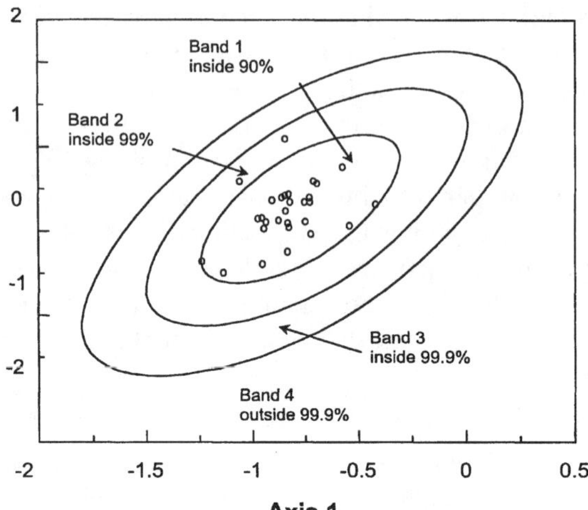

Figure 6-9 Probability ellipses calculated for reference sites to indicate potential stress (90%), stress (99%), and severe stress (99.9%) in invertebrate community of test sites

Once a site is predicted to have a specific community structure, the actual structure is compared to the communities at the equivalent reference sites. A data file is constructed that includes the species counts for the appropriate reference sites. This is ordinated so that a matrix is calculated for both reference and test sites. The sites can then be plotted in ordination space, showing the 3 ordination dimensions that synthesize the biological attributes of the sites (Figure 6-9). Note: Probability ellipses are calculated for the reference sites only. The location of the test sites relative to the reference sites can then be determined. The overall assessment of stress is based on site locations on the first versus second and first versus third axes. It is also possible to assess which species may be responsible for the change in biological state and which environmental variables are associated with the biological response by examining the species and environmental vectors. The results of principle axis correlation with the species matrix provides an improved interpretation of the biological response.

Similarly, performing principle axis correlation between all the environmental variables, not just predictor variables, provides some indication of the causation in the biological response.

Case Study 8: Amphibian Population Declines

Since the early 1990s, amphibian biologists have become aware of several examples of declines in amphibian populations at various locations worldwide. For example, in the U.S., the Cascades frog (*Rana cascadae*) has declined in the southern portion of its range in a manner linked with introduced species and habitat destruction (Fellers and Drost 1993). Perhaps more disturbing are declines in areas with no obvious human encroachment, because these declines could signal some previously undetected environmental degradation (Wake 1991). For example, several species of frogs and toads in Costa Rica have disappeared, following severe population declines in 1987 (Pounds et al. 1999).

Although population declines appear to signify a global trend, it is often impossible to conduct investigations at the global scale. Therefore, populations must be studied intensively at the local level. Understanding amphibian population declines requires understanding amphibian population dynamics and their natural fluctuations. Pounds et al. (1997) use null models to compare population trends with expected natural variability. Their results suggest that amphibian declines in the Monteverde Cloud Forest reserve of Costa Rica are significantly different from what would be expected using conservative estimates of natural variability. Other long-term studies indicate that populations are fluctuating within the range of natural variability (Meyer et al. 1998). Long-term studies of an amphibian community in the southeastern U.S. highlight the difficulty of determining significant population trends (Pechmann et al. 1991; Semlitsch et al. 1996).

There is certainly an overwhelming need for long-term data on population abundance and habitat characteristics for amphibian species in many different regions (Blaustein, Wake and Sousa 1994). However, only a few such datasets exist. Amphibians may represent biological indicator species, sensitive to degradation of environmental quality. Amphibians have permeable skin as adults. Their typical biphasic life history, in which adults are terrestrial and eggs and larvae aquatic, exposes them to potential risks both on land and in water. If amphibians are sensitive species, then their population declines may indicate degradation in environmental quality.

Stressor Attributes and Variability

Some amphibian population declines have been linked to anthropogenic influences. Habitat destruction, increases in the intensity of ultraviolet (UV)-B radiation, chemical pollution, climate change, pathogens, human consumption, and introduced species may all contribute to declining populations. The prevalence of these factors appears to be increasing with human impacts on the environment (e.g., Peters and Lovejoy 1992; Kerr and McElroy 1993; McKenzie et al. 1999). Additionally, several anthropogenic stressors may interact in a single situation. For example, studies have found synergistic effects between UV radiation and a pathogen

(Kiesecker and Blaustein 1995), UV radiation and chemical pollutants (Hatch and Burton 1998; Zaga et al. 1998), and pH and metal toxicity (Horne and Dunson 1995a, 1995b).

Selecting Endpoints to Determine Impairment and Their Uncertainty

Ultimately, information on population fluctuation over time is most desirable. To this end, Blaustein, Wake, and Sousa (1994) recommend mark–recapture studies, noting the difficulties of surveying species that are reclusive or highly dispersed in nature. Following suggestions by Connell and Sousa (1983), they suggest that a population is extinct when no adults are present over 1 complete turnover in the population. Reductions in the spatial range of a population or species might also be indicative of impairment, and a likely consequence of habitat destruction or fragmentation. Experiments may be able to look at one or a few of these issues, but experiments cannot practically or ethically involve entire populations. Therefore, experiments are most useful when they are combined with long-term data on the population and when they are based on realistic exposures. Blaustein, Wake, and Sousa (1994) recommend monitoring amphibian populations on the landscape scale. In population monitoring, lack of long-term studies and sampling difficulties add uncertainty and confound the issue of separating natural fluctuation from trends.

Manipulative experiments have been used in several studies to assess the impact of environmental stressors on amphibian populations. Manipulative experiments usually are limited to a representative sample of a population of animals but can be useful in elucidating the mechanism of stressor effects. Experimental studies have used a wide array of endpoints to assess significant impacts on both larval- and adult-stage animals. The goal of these experiments is usually to investigate some combination of stressors at realistic exposures and to determine whether a statistically significant effect exists. Uncertainty in experiments can result from the problem of using small-scale laboratory or field experiments to explain effects in nature. Some examples of endpoints that have been used to study the effects of stressors include enzyme levels (Blaustein, Hoffman et al. 1994); survival (Blaustein, Hoffman et al. 1994; Hatch and Burton 1998); growth, time to hatching or metamorphosis (Mahaney 1994); bioaccumulation (Russell et al. 1995; Hopkins et al. 1998); and behavior (Kutka 1994; Kiesecker 1995; Marco and Blaustein 1999).

Separating Effects of Anthropogenic Stress and Natural Variability

Studies have used a combination of experimental and observational approaches to assess amphibian population declines and their causes. Often, both approaches are necessary to determine the cause of a population decline: Manipulative experiments provide for causal inference, while observations link experimental results to the population in nature. Pounds et al. (1997) developed comparative statistical models to determine whether changes in population abundance are significant events or are due to natural variability. Results suggest that, in the amphibian assemblages of the Monteverde Cloud Forest, population declines are significant and are associated with climate change (Pounds et al. 1999).

Some studies with long-term datasets have concluded that the populations studied are not in decline but are fluctuating within a natural range in numbers (Meyer et al. 1998). Other studies have highlighted the difficulties of analyzing long-term data (Pechmann et al. 1991; Semlitsch et al. 1996). In a 16-year study of an amphibian community in the southeastern U.S., the number of breeding females in a population varied by up to 3 orders of magnitude (Semlitsch et al. 1996).

Experimental work suggests that embryo mortality may be linked to population declines in some situations (Blaustein, Hoffman et al. 1994). In field experiments in the Pacific Northwest, amphibian species with lower levels of the photorepair enzyme photolyase, such as the Cascades frog *Rana cascadae,* were more susceptible to UV radiation than were species with higher levels of this enzyme, such as *Hyla regilla* (Blaustein, Hoffman et al. 1994). However, there are exceptions to the UV sensitivity hypothesis for amphibians; for example, the red-legged frog *Rana aurora* does not appear to be sensitive to UV radiation but is declining in some parts of its range (Blaustein et al. 1996).

Spatial and Temporal Scales

Ideally, long-term studies of populations over their entire range are most informative. Specifically, landscape-scale monitoring of populations over at least 1 turnover in a population are ideal (Blaustein, Wake, and Sousa 1994). However, amphibian populations in any given location may not be stable in either space or time. Confounding the consideration of spatial scale is the possibility that amphibians may exist as source–sink populations, in which a population may be present at 1 location but dependent on immigration from another location to persist (Blaustein, Wake, and Sousa 1994). Also, the complex life history that is typical of amphibians means that effective monitoring should take place in both terrestrial and aquatic habitats. Because rates of birth, death, and immigration may vary over time, populations may not be characterized by a stable age distribution.

Conclusions

Associating a biological change in a natural system with the specific influences of any single variable, or stressor, requires us to demonstrate that

1) the affected process is sensitive to the stressor;
2) a feasible mechanism exists to explain the effect, separating stressor-induced change from background fluctuations in the factor that was changed; and
3) confounding variables can be identified and their influences assessed.

These requirements, coupled with the complexity of environmental variability and interactions among multiple stressors, contribute to the challenge of identifying the causes of biological change. Nevertheless, identifying the effects of stressors in complicated natural environments is not intractable. A number of common strategies and considerations are highlighted from the 8 case studies discussed here:

- Defining correlative associations is a first step in linking cause and effect, but it is not sufficient.
- A multifaceted body of work is usually necessary to identify the role of the stressor. Unreplicated studies are not convincing. The successful body of work also takes advantage of the synergy between observational information from the field and experimental studies to elucidate mechanisms.
- Plausible mechanisms must exist to explain how a stressor causes a change. Mechanistic models can help in understanding linkages. As observations are increasingly supported by theoretical explanation, experimental demonstration of a plausible mechanism, and mechanistic models, they become increasingly convincing.
- Choosing sensitive response variables is critical to identifying stressor responses in complicated environments. Experimental studies can greatly aid in such choices.
- Stressors and most environmental variables must be characterized at scales relevant to each particular variable.
- Extensive sampling of reference sites can be critical in identifying influences of natural variability. Oversampling at times or in localities unaffected by a stressor can be the key to understanding effects of that stressor.
- Detection of small effects or complicated trends requires a design with more power than does detection of large effects or simple trends.
- Removal or reduction of 1 key stressor can provide opportunities to identify interactions and to separate stressor effects from influences of other environmental factors. In recent decades, opportunities to study recovery from a stress have been used to characterize effects of stressors.
- Long-term monitoring is critical to most successful identifications of stress effects. But if monitoring is too elaborate, it may not be sustainable.
- During recovery, stressor effects may be defined by unidirectional trends; responses to environmental factors can be identified by nondirectional dynamic variability. A study must be sustained for long enough to identify these differences in trends.
- Both temporally intensive studies at few locations and spatially extensive studies at few times can be valuable approaches, but intensity at 1 scale almost always requires tradeoffs at another scale.
- Design decisions often require tradeoffs between statistical power, precision, and biological realism.

CHAPTER 7

Stressor Interactions in Ecological Systems

William H. Clements, Samuel N. Luoma, Jeroen Gerritsen, Audrey Hatch, Paul Jepson, Trefor Reynoldson, Ronald M. Thom

Although it is generally recognized that most aquatic ecosystems are affected by multiple stressors, the historical emphasis in water quality regulation has been on single polluting agents. In the United States, this early focus on single chemicals and related public health issues led to widespread primary treatment before the Clean Water Act was adopted in 1971. The Clean Water Act itself was motivated by both public health and overall integrity of waterbodies, in part because of widespread and frequent fish kills. Early implementation of the Act included construction of sewage treatment plants and cleanup of many industrial discharges to reduce biological oxygen demand (BOD) and release of hydrocarbons and metals. While the stressors coming from discharges before the 1970s were indeed complex, there was no widespread attempt to characterize their complexity, their interactions, or the effects of environmental variability on the response to stressors. The reason was that the sources, their effects, and the causal links between them were clear even to the casual observer. Making cleanup contingent on understanding complexity and interactions among stressors during this early period could have set back the improvements in water quality by a decade or more. As these acute stressors were removed or reduced, more complex and chronic stressors became apparent and were included in the overall effort. These included removal of ammonia for both chemical oxygen demand and ammonia toxicity, reduction of P in areas where fresh waters were subject to eutrophication, and removal of Cl that had been added in water treatments to kill pathogens.

It is widely assumed in the U.S. that implementation of the Clean Water Act has resulted in incremental refinements in source control and significant improvements in water quality. Unfortunately, there have been relatively few rigorous attempts to document these changes (Hart 1994). Each successive control or remediation could have been employed as an experimental opportunity to test a causal relationship

Ecological Variability: Separating Natural from Anthropogenic Causes of Ecosystem Impairment.
D.J. Baird and G.A. Burton, Jr., editors.
ISBN 1-880611-43-0

between a particular stressor and associated ecological effects. One area in which this approach could be especially useful is in understanding potential interactions among stressors, where the effectiveness of eliminating a single stressor in a system receiving multiple stressors could be monitored. This requires long-term monitoring and study. Few studies have monitored both water quality and ecological endpoints following improvements in waste treatment and water quality, thus few examples have rigorously documented successful remediation. Removal of at least some pollutants, remediation of some habitat disturbances, and recovery of ecosystems are ongoing in the developed world, so opportunities for better understanding multiple stressor effects will probably continue to present themselves. Understanding the state of knowledge with regard to multiple stressors is an important starting point for such studies.

Multiple Perturbations of Natural Systems: An Ecological Perspective

Ecologists have long recognized the importance of disturbance in structuring natural communities (Connell 1978; Sousa 1984). A large body of theoretical and empirical evidence supports the idea that all communities are subjected to varying levels of natural disturbance and that disturbance regimes influence community structure and the life-history characteristics of component species. Although most of this research has focused on large-scale physical disturbances (e.g., hurricanes, floods, volcanoes), information derived from the study of natural disturbance may be applicable to ecotoxicological research. For example, factors that influence resistance and resilience of communities to large-scale natural disturbance may also influence community responses to anthropogenic disturbances (Rapport et al. 1985).

Like the historical emphasis on single stressors in toxicology, ecologists have generally limited their investigations to single disturbance events. However, recently there has been increased interest in understanding how ecological systems respond to multiple perturbations. Because anthropogenic disturbances will be superimposed on existing natural disturbance regimes, multiple perturbations will become increasingly common. It is unlikely that the outcome of multiple perturbations can be predicted from the study of single stressors. Paine et al. (1998) present a theoretical model of ecological responses to single and multiple perturbations that has important implications for the study of contaminants and other anthropogenic stressors (Figure 7-1). Although all communities are subjected to and have the ability to recover from natural disturbance (Figure 7-1A), a sequential disturbance, especially one that occurs during the recovery phase, may reset the community to an alternative stable state (Figure 7-1B).

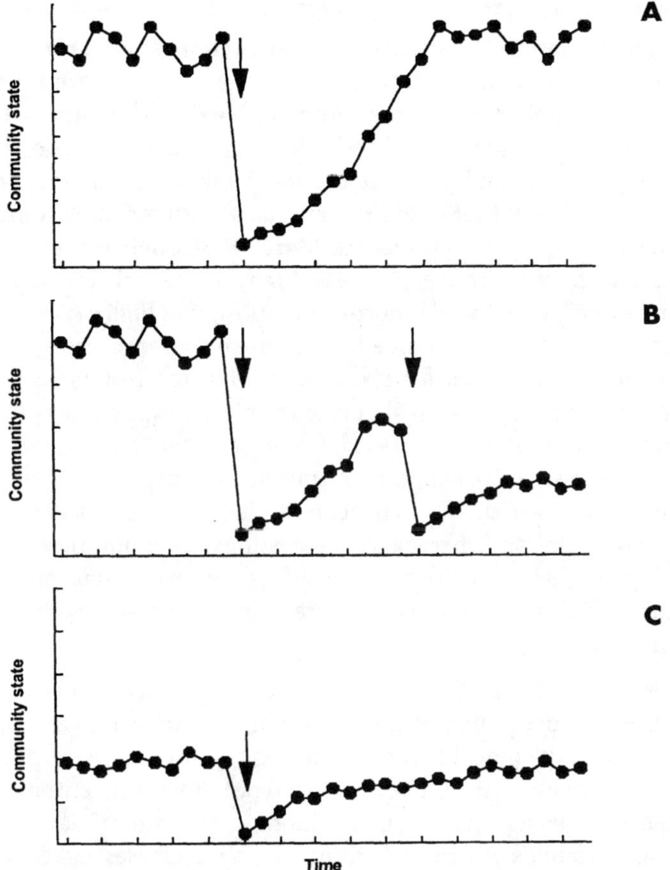

Figure 7-1 Hypothetical effects of single and multiple perturbations (indicated by arrows) on communities. A) Single perturbation is followed by gradual recovery. B) Second perturbation occurs before recovery, and community achieves altered steady state. C) Perturbation occurs on an already disturbed system, and system fails to achieve full recovery. (Modified and redrawn from Paine et al. 1998.)

Interactive Effects of Stressors on Biota

An important question with regard to multiple stressors is whether interactions among stressors result in effects that are fundamentally different from effects of individual stressors?

Toxicity of Chemical Mixtures: An Introduction

Pollution in ecosystems can rarely be attributed to single substances, with impacts being linked to multiple-substance exposure. Thus the potential for interactions among pollutants has received much attention in aquatic toxicology. Under controlled conditions in the laboratory, methods for detecting and quantifying synergistic and antagonistic interactions are well defined. Yet, there is no consensus among toxicologists about how multiple-contaminant exposures differ from single-toxicant exposures. Different pollutant mixtures probably have different interactive effects (Kraak et al. 1994). Antagonism has been reported in bioaccumulation of some metal pairs (e.g., Cd and Zn; Oakden et al. 1984; Simkiss and Mason 1984; Devineau and Triquet 1985; Fischer 1988). Additive effects, at least for some combinations of metals, are also cited (Voyer et al. 1982; de March 1988), and greater-than additive effects also can be found (Kraak et al. 1994). Many of the early investigations of intercontaminant effects involved short-term exposure to high concentrations. Competition for membrane carriers, effects on membrane processes, and failure to induce detoxification processes, for example, could distort results from such experiments, compared to what might occur in animals exposed over the long term in a contaminated ecosystem. Kraak et al. (1994) investigated all combinations of Cu, Cd, and Zn on a sensitive endpoint (filtration rate), presumably using lower concentrations than were employed in acute studies. They found all combinations of antagonism, additivity, and effects at concentrations below additivity, depending upon the mixture. Their conclusion characterizes the current state of knowledge on metal mixtures: "Effects of equitoxic mixtures cannot be predicted from effects of individual metals."

Additivity does predict the effects of some mixtures and is probably a pragmatic approach for interactive pollutant effects until more is known. Swartz et al. (1995) developed an additivity model to predict the toxicity of mixtures of polycyclic aromatic hydrocarbons (PAHs). They predicted porewater concentrations in sediment bioassays using equilibrium partitioning; then they predicted toxicity of dissolved concentrations of each PAH from existing databases and octanol–water partitioning. They summed the toxic units (TUs) predicted to be present in pore waters from several field-collected sediments and compared those to actual toxicity observed in tests of those sediments. The consistency in technique between model development and model verification and the use of factors to correct for bioavailability were probably important ingredients in demonstrating the effectiveness of the additivity model. An additivity model is also logical for PAHs because it might be expected that the mechanisms of toxicity are similar among these related compounds. Long et al. (1998) demonstrated broad agreement in co-occurrence of pollutants in sediments and effects in bioassays, using an additivity model. In earlier work, they had defined "effects-range median" (ERM) to identify concentrations of individual pollutants that are frequently (> 50% incidence) associated with adverse effects in (mostly) sediment toxicity tests, across a large database of such tests. To test multiple-pollutant effects, they divided the concentration of each contaminant by its ERM (similar to the TU approach), summed the TUs, and then divided by the number of contaminants to obtain a "mean ERM quotient." The ERM quotient was compared to a database of 1089 sediment toxicity tests with amphipods. The proportion of sediments classified as highly toxic began to increase with the ERM quotient, progressively, as the quotient exceeded ~ 0.2. The highest likelihood of observing toxicity (> 80%) existed when the ERM quotient exceeded 1.0. Thompson et al. (1999) used a similar approach to evaluate sediment toxicity tests from San Francisco Bay, California, USA and also found an ERM quotient threshold of ~ 0.2. Like other aspects of the co-occurrence approach, the ERM quotient is a purely statistical, nonmechan-istic evaluation of sediment toxicity thresholds. This does not consider bioavailability, and the resulting quotient is misleading because it dilutes high toxicity values with low toxicity values, which is undesirable. Nevertheless, statistical tests can provide broad insights. In this case, the onset of increased incidence of toxicity at ERM values << 1.0 at least raise the hypothesis that greater-than additive effects (synergism) might broadly characterize some multiple-pollutant interactions. If so, then the additivity model, and certainly evaluations based upon tests of individual chemicals, will underestimate effects in nature where multiple pollutants co-occur.

Mechanistic understanding might hold the greatest promise in drawing generalizations about the nature of interactive, multiple-pollutant effects. Antagonism might be expected, for example, where chemicals compete for membrane transporters, so it could be important to know the concentration dependence of such competition. Greater-than additive interactions might be expected, for example, in circumstances where the presence of 1 pollutant facilitates the uptake of another or where detoxification mechanisms for 1 pollutant are impeded by another. For example, some organic chemicals may influence the activity of metallothioneins, or some metals may interfere with the microsomal oxidase detoxification properties. A limited literature suggests that inhibition of detoxification mechanisms could characterize at least some interactions between metals and organic contaminants (Chadwick et al. 1978; Fair and Sick 1983; George 1990), but a convincing body of work at realistic exposure concentrations has not developed. Recent studies also show that metals react strongly with ligands in some types of carbamate pesticides. Because the carbamates are lipophilic, the result is a rapid acceleration of metal accumulation into the tissues of algae (Phinney and Bruland 1997) and their consumers (bivalves). Greater metal exposure might be expected where such pesticides are present in natural waters, but no body of evidence exists. Toxic mechanisms and toxicity test–based evaluations of multiple-chemical effects are not sufficient, however. They must be accompanied by a full evaluation of processes in ecosystems and by field-testing. Some authors have speculated that effects in nature appear to be the result of the joint action of several metals (Ward and Young 1984), but further direct study of those conjectures did not occur. On the other hand, bioaccumulation data from nature, reflecting lifetime exposures of bivalves, show little statistical evidence of interaction among metals (Luoma and Bryan 1982). Perhaps more important, Stewart (1999) evaluated the ecosystem processing of Cd in an experimental lake as concentrations of Cu, Zn, Pb, and Ni were increased, using realistic concentrations of these elements. Cd removal from the water column was fastest in the absence of the other metals, resulting in the lowest Cd bioaccumulation by the mussels. Although higher bioaccumulation occurred in all treatments with metal mixtures as a result of the ecosystem-level effect, Cd bioaccumulation declined progressively as other metals were added in higher concentration among the mixture treatments alone. So, in this case, organism-level and ecosystem-level influences of metal mixtures were both present.

Simultaneous Exposure to Multiple Stressors

In a laboratory study, toxicity is defined by a pre-selected adverse effect termed an "endpoint." Variables that might confound the endpoint response are controlled. In ecosystems, effects of contaminants are manifested in a more complex fashion. The presence of multiple stressors can confound simple determination of the influences of a single stressor. A response may co-occur with more than 1 stressor, so specification of cause and effect is difficult. It is likely that complexities beyond simple co-occurrence are also common. For example, studies of energetics are used to assess the response of mussels (*Mytilus edulis*) to stresses in their environment (Bayne and Widdows 1978). This basic knowledge was used to establish the energetic cost of different body burdens of pollutants (PAH or tributyltin [TBT]) to growth potential (scope for growth [SfG]) in mussels, both in experiments and in relatively simple field circumstances (Widdows and Donkin 1989; Widdows, Donkin, Brinsley, Evans, Salkeld et al. 1995). Widdows et al. (1996) applied this approach to Venice Lagoon, Italy, which is subjected to a mixture of stresses common to many urbanized marine ecosystems. SfG in mussels from Venice Lagoon was negatively correlated with both pollutant burdens and positively correlated with particulate organic material concentrations. However, a closer inspection of the data showed that the inhibitory effects of toxicants were masked at a site subject to nutrient addition. Thus, the SfG expected from the pollutant concentrations was offset, apparently, by the positive effects of very high food availability. Experimental studies also have shown reduced manifestation of metal stress in well-fed animals, compared to those given low food levels (Chandini 1988). Thus organisms may overcome at least some level of stress on energetics if adequate food is available.

Numerous examples also exist of the opposite effect: that pre-exposure to 1 stress reduces the ability to resist another. Moller et al. (1994) showed increases in Cd toxicity to snails outside optimal salinity ranges. Meador (1993) showed increased toxicity of TBT to marine amphipods as laboratory holding time increased and lipid levels in the animals declined. Lemly (1996) showed that fish subjected to winter stress syndrome were of increased sensitivity to Se. Pre-exposure to pollutants or an accumulated pollutant burden in tissues (de Zwaan et al. 1995) reduced the ability of mussels to survive exposure to air (Eertman et al. 1995; Viarengo et al. 1995) or anaerobic conditions. There are a surprising number of these examples of stress-on-stress in the experimental literature. Such results are highly relevant to nature, where populations can be living in moderately contaminated conditions that, alone, may not be sufficiently stressful to cause adverse effects. But if an environmental stress is superimposed (a common occurrence in all environments), the population may expire. If such responses are common, moderate contamination could be exerting unexpected effects in many circumstances in nature that are, as yet, unattributed to specific causes.

Tolerance of Populations to Contaminants: Uncoupling Exposure and Response in Nature

Populations chronically exposed to contaminants may exhibit enhanced tolerance relative to unexposed or naîve populations (Luoma 1977). In such circumstances, effects on the population will be decoupled from measures of environmental toxicity determined in the absence of the pre-exposure. So the biological property of genetic or physiological elasticity in a species confers an ability to resist exposures to a stress that normally would eliminate the population. Physiological acclimation and/or selection for a tolerant genotype are demonstrated, but an important question is whether such compensation occurs at a cost (some aspect of the biological process is weakened) and how much cost accompanies compensation. A corollary is what will be the response when additional stress (e.g., natural disturbance) is imposed on the population. A few examples of this complicated chain of events have been discussed in the literature (reviewed by Weis and Weis 1989).

Although increased tolerance has been demonstrated in populations previously exposed to contaminants, few studies have examined resistance at higher levels of biological organization. The most common explanations for increased tolerance at the population level include physiological acclimation and genetic changes that favor resistant genotypes. We argue that these intraspecific mechanisms may also account for resistance of communities and ecosystems. In addition, because communities and ecosystems consist of large numbers of interacting species, it is likely that other mechanisms, unique to these systems, will contribute to tolerance. For example, increased tolerance at the community level may result from replacement of sensitive species by tolerant species. Research at the Arkansas River Superfund site (Leadville, Colorado, USA) shows that upstream reference sites are dominated by several species of mayflies (Ephemeroptera), organisms reported to be highly sensitive to heavy metals, whereas downstream sites are dominated by metal-tolerant chironomids (Diptera) and caddisflies (Trichoptera) (Clements 1994).

Quantifying Effects of Multiple Stressors in the Field

Here we ask what types of field studies can best detect interactions among stressors and allow us to separate and rank the relative importance of individual stressors in systems that receive multiple disturbances (natural and/or anthropogenic). If multiple stressor responses are common in nature, then single-variable tests, such as analysis of a biomarker in isolation or along a surmised gradient, or studies that exclude variables other than pollutants, could be insensitive to all but the most extreme influences of contamination. Preponderance of evidence approaches will be similarly insensitive if designs are too simplistic. Combinations of persistent and intensive study of exposure and response in the field, study of critical ecosystem-specific and organism-specific processes, as well as iteration with experimental studies are useful (and perhaps necessary) strategies to discern interactions among stressors. As our understanding of mechanisms responsible for changes at lower levels of organization improves, responses to complex stressors become more predictable. This improved mechanistic understanding could lead to a similar degree of understanding for responses at higher levels of biological organization. Below, we discuss 3 examples in which researchers have attempted to identify and quantify the relative importance of individual stressors in systems receiving complex stressors. The first example demonstrates how intensive field studies identified multiple stressors and how a management plan resulted in mitigation of these stressors. The second example describes a series of field experiments designed to identify the relative importance of water quality and substrate quality on benthic macroinvertebrates in a metal-polluted stream. The final example illustrates the difficulty of sorting out the direct and indirect influences of global climate change on populations.

Cascade Effects from Ferry Terminals on Seagrass Meadows In Puget Sound

Ferry terminals in Puget Sound, Washington, USA extend out from the shoreline to cross over intertidal and shallow subtidal estuarine habitats. Because of deteriorating terminals and dramatically increasing passenger numbers, most terminals are scheduled for replacement and expansion. We investigated the direct and indirect impacts from terminals and ferry operations on eelgrass, for the purpose of redesigning terminals to minimize or avoid impacts on the eelgrass (*Zostera marina* L.), a submerged aquatic macrophyte abundant in temperate estuaries (Thom et al. 1997). Although shading by the terminals was initially believed to be the primary stressor to eelgrass, shading did not fully explain why the meadow either was absent or highly fragmented in the 30- to 50-m region surrounding the terminal.

The study, which coupled extensive field investigations at 3 terminals with laboratory investigations, identified 6 reasons for the present distribution of eelgrass:

1) Initial construction disturbance, including pile driving and grounding of construction vessels
2) Maintenance disturbance, scouring of the bottom during annual activities to replace damaged pilings and dock components
3) Light reduction by the terminal, direct shade cast on the bottom
4) Propeller scour from frequently enhanced bottom current velocities, removal of eelgrass recruits
5) Light reduction from the propeller bubble plume created during ferry dockings and departures, further reducing light at the deepest edge of the eelgrass meadow where light was already near limiting

6) Enhanced bioturbation, where foraging by seastars on barnacles and mussels attached to pilings created shell piles under the terminal that attracted and supported enhanced populations of Dungeness crab. The enhanced crab populations, along with enhanced seastar populations, increased bioturbation disturbance through burial and feeding activities at the edge of the eelgrass meadow.

Sorting out the stressors, their strengths, and their interrelationships was the basis for mitigation plans. For example, control measures were included in construction plans to minimize damage during construction and maintenance activities. To eliminate propeller scouring and bubble plume impacts on the eelgrass, it was recommended that docks be made longer to move ferry slips further offshore. An added benefit of lengthening the terminals was that vehicle-holding capacity of the terminals could be achieved while the terminals were kept narrow. Thus the terminal did not have to be made as wide over the area occupied by eelgrass. Other suggestions included incorporating transparent material into walkways, reducing the number of pilings by switching from wood to concrete, and coating the underside of the terminals with bright white paint—all intended to reduce the shading stress. The reduction in pilings effectively reduced the surface area for barnacles and mussels, thus lowering the carrying capacity of the site for bioturbators such as crab and sea stars. The effect of these actions will be determined during a 10-year monitoring program.

Our study demonstrates an attempt to systematically sort out multiple stressor effects. There was little doubt that light reduction is a major stressor on the eelgrass, but this stress is catastrophic and occurs in concordance with the construction of the terminals. Cascading (e.g., bioturbators) and ancillary (e.g., bubble plume) stressors act at various levels and scales. The frequency and intensity of these stressors vary in time and space, affecting the rates and probability of recovery of the eelgrass meadow damaged initially during construction and in areas of the meadow not damaged by construction. The bioturbators likely retard the recruitment of meadows back into unshaded areas they formerly occupied and may also actually be further damaging the meadow. The capacity of the animals to do this depends upon other factors (e.g., reproduction) that control their population levels. About every 10 to 20 minutes during the day, coinciding with ferry runs, the bubble plumes reduce light reaching the eelgrass. The spatial scale is dependent on the size of the vessel and on its operation. Changes in the frequency of runs and in the size and operation of the vessel would likely result in varying eelgrass meadow distribution.

Long-term Cycles of Giant Kelp Forests in Southern California and Influences of Human Disturbances

The kelp forest described in Chapter 6 is an example of the effect of a natural random stressor on a system that is already subjected to nonlethal anthropogenic stress. Extinction of the otter population, introduction of sewage particulates, and toxicity contributed to the decline of the kelp. As the human population exploded in southern California between about 1930 and 1970, effluent discharge rates increased concordantly, as did the impacts to the already altered kelp forests. The proximate cause of the extinction of the kelp on the Palos Verdes peninsula, however, was the major El Niño event in 1957 and 1958. This stress, which included greater water temperatures, lower nutrients, and increased wave energies in the region, resulted in physiological stress as well as physical damage. We cannot conclude that kelp would have eventually suffered the same fate in the absence of this anomalous climatic event. However, the El Niño dramatically hastened its decline.

Metal Effects on Stream Ecosystems

A series of field experiments was conducted to estimate the relative influence of heavy metals in water and substrate quality on benthic macroinvertebrates in the Animas River, Colorado, USA. Our initial experiments tested the hypothesis that benthic invertebrates can colonize the Animas River when provided with clean substrate. To investigate the influence of metal-contaminated substrate on macroinvertebrates, we measured colonization of benthic communities on substrate collected from the Animas River in Elk Creek (the reference stream). Finally, to investigate the effects of Animas River water on benthic macroinvertebrate assemblages, we conducted in situ toxicity tests with benthic communities collected from Elk Creek.

Colonization of clean substrate by benthic macroinvertebrates was greatly reduced in the Animas River compared to Elk Creek. Abundance and species richness of aquatic insects were significantly greater in Elk Creek, and community composition differed between sites. Substrates placed in Elk Creek were colonized by a diverse assortment of benthic macroinvertebrates and dominated by metal-sensitive mayflies, whereas metal-tolerant caddisflies and chironomids were more abundant in the Animas River (Figure 7-2A). The second experiment showed that aquatic insects colonized both clean and contaminated substrate placed in the reference stream; however, there were differences in community composition between substrate types (Figure 7-2B). For example, total abundance of mayflies was greater on clean substrates; however, this difference was primarily due to the lack of colonization by *Drunella doddsi* on Animas River substrate. Finally, in the third experiment, results of the in situ studies showed that water in the Animas River was highly toxic to several species of benthic macroinvertebrates (Figure 7-2C). In particular, abundance of the mayflies and stoneflies was significantly reduced in Animas River cages.

The simple experimental approach employed in this study allowed us to distinguish the relative importance of heavy metals in water and contaminated substrate. The absence of metal-sensitive organisms, especially mayflies, from clean substrate placed in the Animas River suggested a lack of organisms for colonization. Subse-

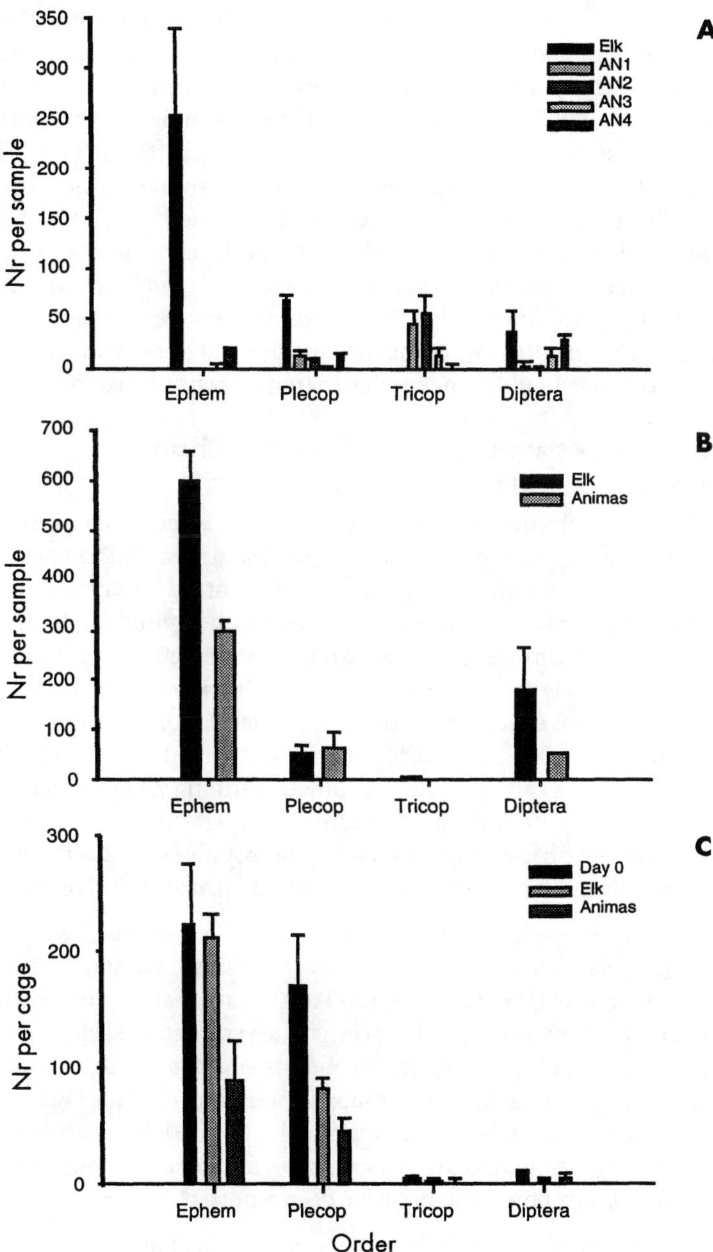

Figure 7-2 Results of field experiments quantifying effects of heavy metals in water and substrate quality. A) Colonization of clean substrate placed in Elk Creek (reference stream) and Animas River, Colorado, USA. B) Colonization of clean and contaminated substrate in Elk Creek. C) Survival of benthic invertebrates in Elk Creek and Animas River cages. Day 0 refers to abundances at start of the in situ experiment.

quent experiments showed that the lack of colonization resulted from both high levels of metals and poor substrate quality; however, the relative importance of these stressors differed among species. Experiments comparing the colonization of clean and contaminated substrate placed in Elk Creek indicated that most taxa were able to colonize both substrate types. Although poor substrate may restrict the abundance of some taxa, most species could tolerate Animas River substrate. In contrast to these results, total abundance, species richness, and abundance of most dominant species were significantly reduced following short-term (4-d) exposure in the Animas River. These findings suggest that elevated metal concentration in water was the primary factor limiting benthic macroinvertebrates in the Animas River. Therefore, improvements in water quality and associated reductions in metal levels in the Animas River should allow recolonization by benthic macroinvertebrates.

Interactions between Global Climate Change and Other Stressors

One of the most likely interactions expected in aquatic ecosystems is between global climate change and other natural or anthropogenic stressors. The magnitude and rate of global environmental changes in the 20th century is unprecedented, and current rates of global warming may be an order of magnitude faster than any other change in the global climate in the past 2 million years (Lubchenco et al. 1991). Global climate models predict a 2 to 4 °C rise in air temperature within the next century, with greatest effects expected in alpine habitats. In addition to well-documented changes in temperature patterns, changes in regional rainfall and moisture patterns may also result from climate warming (Peters and Lovejoy 1992; Pounds et al. 1999). Although most research on the effects of increased temperature has focused on terrestrial ecosystems and agroecosystems, moderate increases in temperature will also have profound effects on the structure and function of aquatic communities.

Anthropogenic depletion of ozone in the lower stratosphere and the subsequent increase in ultraviolet (UV) radiation (280 to 400 nm) reaching the earth's surface has been of global environmental concern for several decades. Negative effects of UV-B radiation, including mutagenesis, photochemical damage, and physiological stress have been observed in many groups of aquatic organisms (Vincent and Roy 1993). Environmentally relevant background levels of UV-B radiation are lethal to some aquatic organisms, and dramatic changes in primary production, community composition, and trophic structure have been reported.

Another important global environmental problem that has received considerable attention is acidification of aquatic environments. Although most research has focused on the northeastern U.S. and Canada, recent studies have demonstrated that aquatic ecosystems in other regions of the U.S. are susceptible to acid precipitation (Harte and Hoffman 1989; Bradford et al. 1991). For example, lower pH associated with atmospheric acidification may account for observed declines in amphibian populations in the western states (Harte and Hoffman 1989).

These global anthropogenic stressors may affect biological systems in many ways, both directly and indirectly. For example, direct effects of climate warming could include altered temperature tolerance or altered geographical range. Species unable to adapt their physiology or behavior to cope with the speed of predicted climate change will be at greater risk of extinction. Climate change has recently been linked to range shifts in tropical amphibian assemblages. Pounds et al. (1999) associate severe population declines or extinctions of 20 amphibian species in the Monteverde Cloud Forest in Costa Rica with climate warming. Population crashes were significantly correlated with an increase in the number of dry days per year and decreasing mist frequency, in a manner that could not be explained by direct habitat alteration (Pounds et al. 1999). Indirect effects could involve changes in species interactions or changes in the spread of disease (Lubchenco et al. 1991). For example, an experimental field study in Oregon, USA demonstrated a synergistic interaction between the pathogen *Saprolegnia* and ambient levels of UV radiation (Kiesecker and Blaustein 1995). In this outdoor experiment, the presence of fungus enhanced the effects of UV radiation on hatching success in the Cascades frog (*Rana cascadae*) and the Western toad (*Bufo boreas*).

Studies of climate change have the difficult challenge of teasing apart the effects of climate change from the effects of other stressors. These studies often must use correlational information and observations to determine whether effects are due to climate change or direct human disturbance (Parmesan 1996; Pounds et al. 1999). Climate change could interact synergistically with other stressors such as pollution and habitat degradation (Peters and Lovejoy 1992). Increasing temperature may alter stress tolerance, making biota more sensitive to the effects of other stressors. Increasing temperature may also interact with chemical contaminants directly because some contaminants are more toxic at higher temperatures (e.g., Boone and Bridges 1999; Fisher et al. 1999). Effects of natural disturbances may be altered or enhanced in combination with climate change. For example, climate change could increase the effects of El Niño events on marine phytoplankton, which are particularly vulnerable to changes in ocean temperature and currents (Peters and Lovejoy 1992). The kelp forest example described in Chapter 4 demonstrates the complex effects of a natural stressor on a system that is already subjected to nonlethal anthropogenic stress. Elimination of the otter population, introduction of sewage particulates, and direct toxicity contributed to the decline of the kelp on Palos Verdes peninsula. However, the proximate cause of the extinction of the kelp was the major El Niño storm event in 1957 and 1958. This stress, which includes greater water temperatures, lower nutrients, and increased wave energies in the region, resulted in physiological stress as well as physical damage. Increased intensity and frequency of El Niño storm events are expected consequences of global climate warming. Identifying the influences of these disturbances on ecological systems will require an understanding of their direct effects on populations as well as their interactions with other anthropogenic stressors.

Climate change may also affect the outcome of species interactions such as predation and competition. Changes in ocean temperature similar to those expected by climate change–altered predator–prey interactions in the rocky intertidal community (Sanford 1999). Other subtle effects of climate warming on species interactions and behavior are predicted. For example, forest-floor and pond-breeding amphibians in neotropical regions may be affected by climate warming in several ways (Donnelly and Crump 1998). As the climate becomes warmer and drier, ephemeral ponds will hold water for shorter periods of time each year. This could result in more species using the same pond at the same time, thus increasing competition. In addition, the shorter length of time could mean that fewer males successfully mate, ultimately reducing genetic diversity (Donnelly and Crump 1998).

In summary, because of its global scope, climate change will add stress to situations where natural and anthropogenic perturbations already exist (Figure 7-1C). It is unlikely that the stressor–response models developed for single stressors can be used to predict these interactions. Ultimately, our understanding of the biological effects of global climate change will result from studies investigating both direct and indirect effects.

Influence of Novel Stressors on Communities and Ecosystems: Global Atmospheric Pollutants

The addition of a natural or anthropogenic disturbance event on a population or community already subjected to a chronic stressor may have important consequences for recovery. Paine et al. (1998) suggest that disturbance in chronically stressed systems may impede recovery and force the system to a new, stable state (Figure 7-1C). One very likely scenario for multiple perturbations involves systems currently subjected to widespread global perturbations such as global warming, UV-B radiation, and acidification.

Because both pristine and polluted systems will be affected by these global anthropogenic impacts, we need to evaluate potential interactions between effects of contaminants and these novel stressors. Organisms that inhabit polluted environments often develop increased tolerance, compared to organisms in unpolluted habitats. However, acclimating or adapting to 1 set of environmental stressors may increase an organism's sensitivity to novel stressors (Wilson 1988). We hypothesize that communities from chemically stressed environments will be at greater risk to increased temperature, UV-B radiation, and acidification than will communities from pristine environments. Support for this hypothesis is provided by the results of experimental studies in which communities from metal-polluted and unpolluted sites were exposed to novel stressors (acidification, UV-B radiation, and stonefly predation). In all instances, communities from the metal-polluted sites were more sensitive to novel stressors (Kiffney et al. 1997; Courtney 1998; Courtney and Clements 1998; Clements 1999). An understanding of the potential effects of novel stressors on disturbed systems is essential for predicting ecological responses. This understanding can only be achieved by integrating long-term monitoring with laboratory and field experiments.

Influence of Natural Heterogeneity on Responses to Multiple Stressors

Natural variation in structure and function is a fundamental property of all aquatic ecosystems. Natural variability can also complicate assessments of ecological integrity, especially in situations where natural gradients are superimposed on gradients of anthropogenic stressors (Clements and Kiffney 1995). While ecologists have long acknowledged the influence of natural variation on ecosystems, the traditional focus in ecology has been on small-scale studies conducted within a single habitat (Wiley et al. 1997). Consequently, large-scale comparative studies are rare, and ecologists often lack an appreciation for the influences of variability at a larger spatial scale (e.g., landscape level) on the structure and function of ecosystems.

As a result of recent technological advances in remote sensing and geographic information systems (GIS), ecologists are now developing a better understanding of the direct and indirect influences of landscape-level characteristics (Poff 1997; Richards et al. 1997). An emerging paradigm in stream ecology is that landscape attributes operate as filters (sensu Southwood 1977) that determine life-history characteristics, community composition, and trophic structure (Poff 1997). We suggest that recognition of the importance of these landscape filters, especially within the context of environmental heterogeneity, offers an unprecedented opportunity to improve our ability to predict the effects of anthropogenic disturbance on ecosystems.

The influence of natural heterogeneity on life-history characteristics, patterns of species diversity, and ecosystem processes has been examined by stream ecologists (Poff and Ward 1990; Matthaei et al. 1996). For example, streams characterized by high levels of natural variability are often dominated by opportunistic species (Poff and Ward 1990; Palmer 1995; Moser and Minshall 1996), resulting in greater resilience to natural disturbance (Matthaei et al. 1996). In other words, evolutionary history and the frequency of disturbance in an ecosystem may affect its ability to withstand additional perturbations.

Despite an appreciation for the importance of environmental heterogeneity, there is no information concerning how natural variability influences responses to contaminants. Although there is some empirical support for the hypothesis that effects of contaminants vary among ecosystems (Poff and Ward 1990; Howarth 1991; Kiffney and Clements 1996; Medley and Clements 1998), there have been no attempts to identify specific watershed characteristics responsible for this variation. Most of the previous research in ecotoxicology has examined the role of local abiotic factors, such as water hardness, pH, and total organic C, in contaminant availability and toxicity. In contrast, there has been little research investigating how landscape-level attributes and natural variability influence the sensitivity of ecosystems to contaminants. Effects of anthropogenic disturbances may vary among different types of ecosystems or among similar ecosystems in different locations. If some ecosystems are inherently more fragile than others, identifying characteristics that increase sensitivity and the mechanisms responsible for ecosystem recovery are important areas of research. Rapport et al. (1985) suggest that ecosystems in unstable environments may be "preadapted" to moderate levels of anthropogenic stress. Howarth (1991) speculated that ecosystems with fewer opportunistic species, lower diversity, and closed element cycles would be sensitive to contaminants. Yount and Niemi (1990) investigated the patterns of recovery of a large number of aquatic ecosystems subjected to a diverse assortment of natural and anthropogenic stressors (floods, drought, biocides, toxic spills). The most striking generalization from these studies was the rapid recovery observed in many lotic ecosystems. Rapid recovery of lotic ecosystems resulted from the life-history characteristics of resident species, the presence of upstream refugia, and the high flushing rate of streams. They suggest that because of natural variation of streams, resident species have flexible life-history characteristics and are adapted to fluctuating conditions.

It is possible that these same characteristics that influence resilience of streams to natural disturbance will also influence responses to contaminants. Specifically, we hypothesize that the effects of anthropogenic disturbance will be greater on stable ecosystems, compared to naturally variable systems (Figure 7-3). Furthermore, we predict that recovery from anthropogenic disturbance will take longer in stable systems than in highly variable ecosystems. These predictions have important implications for comparing effects of anthropogenic disturbances among ecosystem types and across latitudinal gradients. For example, we would expect that the effects of a toxic spill would be much greater on a highly stable coral reef than on a highly variable estuarine system. We also would expect that ecosystems in tropical environments would be more susceptible to anthropogenic stressors than would their temperate counterparts. Support for the hypothesis that contaminant effects are greater in stable systems is provided by microcosm experiments and field studies conducted in metal-polluted streams (Kiffney and Clements 1996). Results of these investigations indicate that the effects of heavy metals were greater on benthic communities in stable, high-elevation streams compared to those in more unpredictable, low-elevation streams (Figure 7-4).

Summary

Any discussion of complex stressors, and especially of the complexity of anthropogenic stressors, begs the question of the magnitude and importance of such complexity with respect to practical management and remediation. In both the practical

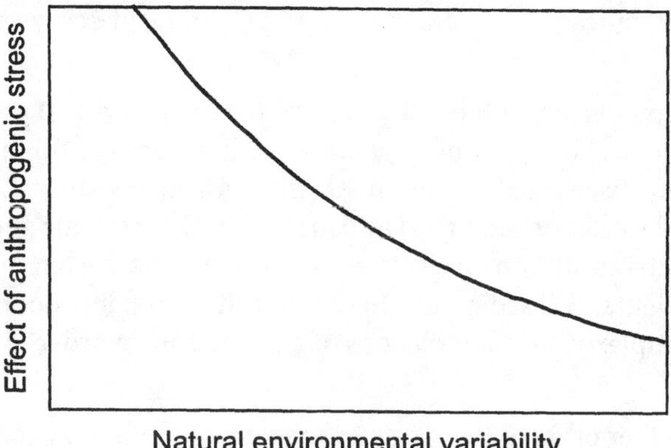

Figure 7-3 Predicted influence of natural environmental variability on responses to anthropogenic stressors

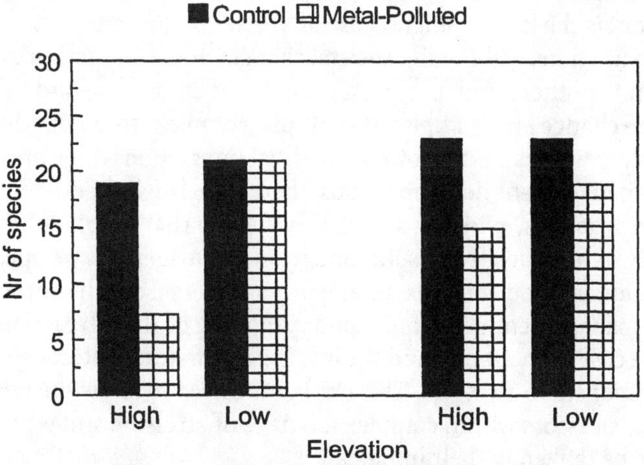

Figure 7-4 Influence of elevation on responses of benthic communities to heavy metals in field (left) and in experimental streams (right), showing number of species in stable, high-elevation streams and in variable, low-elevation streams (after Kiffney and Clements 1996)

(engineering) and the political sense, society typically manages recognized deleterious stressors as a whole but bases that management on assessment of a single contaminant, usually by its source rather than by its characteristics as a stressor. Conversely, many scientists relish complexity and interactions among stressors because they are intellectually exciting; however, they have focused primarily on simpler systems, using simplistic assessment tools with broad assumptions. The science has now advanced to a point where complexity can be accurately assessed, leading toward significant advances in our understanding of ecosystem responses to disturbances.

Despite improvements in water quality in the U.S., severe residual problems remain, especially in biological integrity of large water bodies. Some of these problems are due to historically discharged contaminants that remain in sediments. However, the most widespread causes of biological impairment in U.S. streams and rivers today are runoff and alteration from agricultural and other intensive land-use changes (pesticides, nutrients, and sediment). In this case, it is probably not necessary to elucidate the complexity and interactions of the stressors in order to manage the problem.

We may ask whether other anthropogenic stressors are commonly complex and multiple, or whether a single anthropogenic stressor is clearly dominant in causing unacceptable degradation, and whether the answer matters for managing and remediating sources of the stressors. In order to answer these questions, we need some systematic approach for separating and quantifying effects of individual stressors in complex systems. On a global scale, habitat alteration (loss, conversion, or degradation) is the largest single cause of species decline and extinction. Stemming these extinctions will require societal changes and exploitation of natural resources. On the other hand, pest management for agriculture and the influence of global climate change are examples of multiple, complex stressors, which will require our clear understanding of the potential interactions if we are to make appropriate management decisions. Thus, there is no single answer. In some media and for some problems, evidence and history suggest that simple source removal or remediation will have positive results, and from a management perspective, there is little need to understand complex, interacting stressors. For other problems, enlightened management will require understanding of the interactions among stressors, the cost of tolerance, and the influence of natural heterogeneity on ecological responses to stressors. This implies the need to know the relative influence of single stressors within complex mixtures of stressors and whether individual stressors can be ranked in their importance.

CHAPTER 8

Identifying Watershed Stressors Using Database Evaluations Linked with Field and Laboratory Studies: A Case Example

G. Allen Burton Jr., Scott D. Dyer, Susan M. Cormier, Glenn W. Suter II, Elaine J. Dorward-King

While there have been many new and exciting developments in the field of ecotoxicology in the past couple of years, most of the tools and principles for evaluating stressor effects have existed for many years. Despite this, it is rare to see an assessment of impact that systematically identifies an ecologically significant stressor or group of stressors with a high degree of certainty. However, while the tools and approaches do exist, it is still a challenge to separate and rank the effects of multiple stressors (both natural and anthropogenic). This challenge is not insurmountable, as is evidenced by the numerous case examples presented in this book, showing how multiple stressors were assessed in a wide range of ecosystems.

The following example is based on actual case studies and will highlight approaches and considerations that apply to many assessments of ecosystem impairment. Two types of assessments are commonly conducted: 1) evaluations that primarily use existing databases and 2) field and laboratory assessments in which new site data are collected (Figure 8-1). This example shows how both can be used effectively in a large, basin-wide survey and in a narrowly focused site-specific survey, either separately or in a more effective tiered or iterative process, in which uncertainties are identified and then systematically addressed in a cost-effective manner. While we discuss specific methods that have been used effectively in these assessments, there is a wide range of assessment tools that are equally effective if used properly, particularly as components of a holistic assessment process (Chapter 1; Foran and Ferenc 1999). Each assessment tool has its own inherent strengths and weaknesses, which are amplified by the study design and the user's expertise.

Ecological Variability: Separating Natural from Anthropogenic Causes of Ecosystem Impairment.
D.J. Baird and G.A. Burton, Jr., editors. © 2001 Society of Environmental Toxicology and Chemistry (SETAC).
ISBN 1-880611-43-0

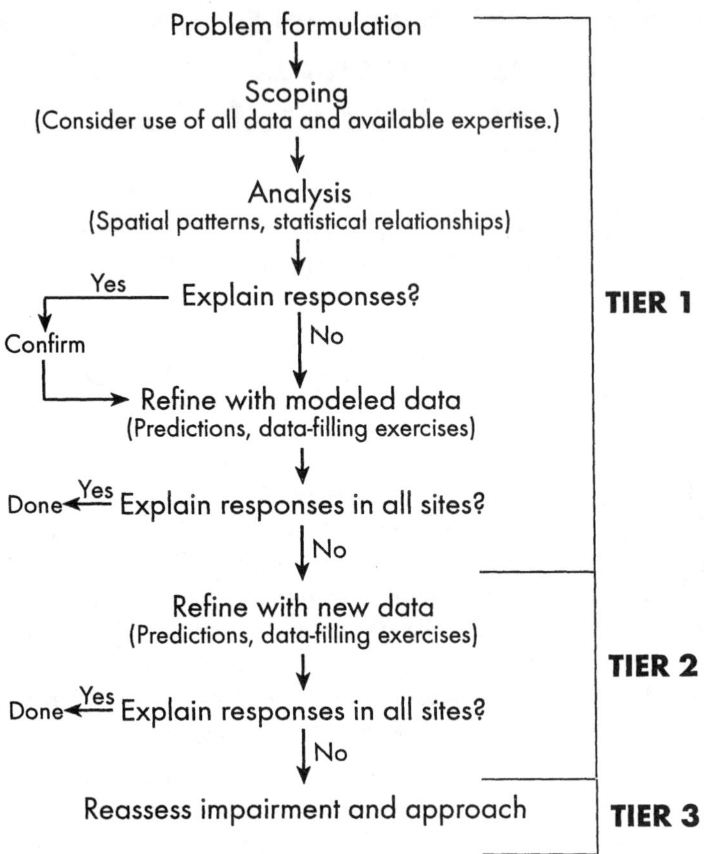

Figure 8-1 Tiered approach to assessing causes of ecosystem impairment

The decision logic used in this case study is illustrated in Figure 8-1. First, in Tier 1, the problem formulation and scoping steps were conducted. This included a statistical analysis to identify potential causal associations between the endpoint effects and the environmental metrics in existing datasets for the entire watershed. Second, in Tier 2, the hypothesized causal relationships were tested using field and laboratory experiments. These experiments allowed the assessors to eliminate all but one of the hypothesized causes of each endpoint effect at a particular location. If the experiments provide ambiguous results, additional evidence can be obtained in a Tier 3 evaluation and can be used in a weight-of-evidence analysis. Although the case study was not designed using the method of causal analysis proposed by Suter et al. (in press) and used in U.S. Environmental Protection Agency (USEPA 2000) guidance, it fits within that framework.

Tier 1: Initial Approach

In Tier 1, there is a scoping process of potential causes of water quality impairment. Its purpose is to identify the principle factors associated with impact or to reduce the plethora of potential stressors to a manageable and interpretable number that can be evaluated in subsequent tiers.

Problem Formulation

Regional USEPA offices regularly monitor biology, habitat, and water and sediment chemistry for each river basin within their jurisdiction. For our hypothetical river, analysis of the 1997 monitoring results indicated that there was an increased occurrence of fish with fin erosions (beyond the expected, compared to reference sites) and a reduced number of species (also compared to reference sites) in several areas along the river. Causes were neither obvious nor located in restricted portions of the river but appeared scattered throughout, although fin erosions were concentrated more in the lower portion of the river. There was a need to identify the causal factors that might be involved in producing these adverse effects within the local ecosystem.

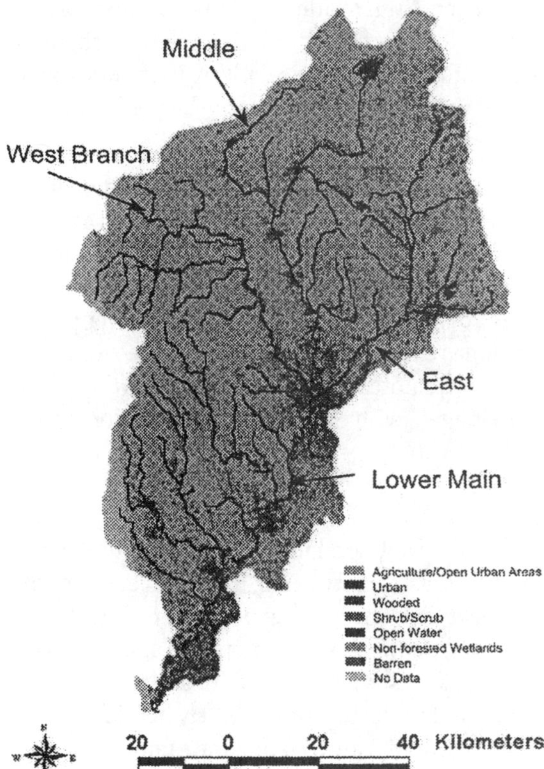

Figure 8-2 Land-use map of hypothetical river site used in case study

Basin description

The river is a typical mid–North American river, draining an approximately 1700-km^2 area that includes agriculture, forest, and urban land uses (Figure 8-2). The upper portion of the watershed comprises 3 main branches: west, middle, and east. The west branch drains agricultural land, including poultry farms, range livestock, and row crops such as corn and soybeans, which cover much of the landscape. The population density is low (< 50 persons per km^2) throughout the basin, with the exception of 2 municipalities of 5,000 and 10,000 people. Sources of river water are runoff, municipal wastewater treatment plants (WWTPs), and septic tanks. Because of row crop agricultural practices, headwater reaches are heavily managed, leaving little riparian width where reaches have seasonally intermittent flow, whereas perennial flowing downstream reaches have extensive riparian zones (100 m to 1000 m). Instream habitat (cover, riffle/pool sequence, dams, channelization, etc.) has not been altered. Of particular note is that this branch has been designated a Scenic River by a national conservation group and has a noted sport fishery.

The middle branch is much like the west branch: dominated by row crop farming and low population density, with nonpoint source runoff and municipal WWTPs providing the majority of annual river flow. Unlike the west branch, forest cover in the middle branch is restricted to stream banks and tributary ravines. Even so, the perennial flowing reaches have not been modified, other than at a gravel-mining site. Much of the middle branch constitutes a noted sport fishery.

Land use in the east branch consists of agricultural and urban areas. The population density ranges widely, from < 50 persons per km^2 to > 500 persons per km^2. Discharges to the east branch consist of municipal and industrial WWTPs and nonpoint source runoff from both agricultural and urban lands. Both streamside and instream habitat have been highly managed. Streamside riparian habitats range from extensive riparian buffers to impervious surfaces bordering the river channel. Historical channelization pierced the groundwater table, resulting in groundwater supplying the majority of the perennial river flow. A coldwater fishery is managed throughout the east branch.

Below the confluence of the 3 branches, the lower watershed is dominated by urban land use and forest. The population density is high all along the main stem and low along the minor western tributaries. Discharges along the main stem range widely: utility thermal discharge, small to large WWTPs, gravel mining, and nonpoint. Historical pulp and metal-plating plant discharges were located in the lower portion of the watershed. Extensive riparian zones exist throughout most of the western tributaries, whereas few riparian zones were noted in urban settings. Further, flow within the lower branch was highly managed, with extensive channelization and low-head dams.

Reference condition

A key to success in diagnosing the causes of biological impairment is an understanding of the ecological meaning of the relationship between fish with fin erosions and numbers of fish species. For instance, is it appropriate to expect fin erosions to be near zero regardless of river size? Further, do the number of species and types of species occur equally in headwater reaches versus large rivers? To address this, calibration curves that provide an expected value may be extremely helpful to dimension the degree of impairment. Figures 8-3 and 8-4 illustrate nonlinear regressions for ecoregion reference sites and the river for percentage fish with fin erosions and number of fish species versus drainage area. In Figure 8-3, the percent-

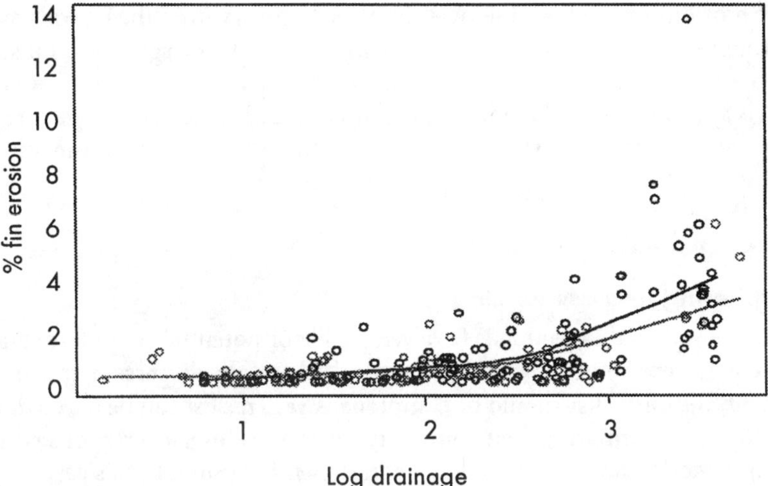

Figure 8-3 Nonlinear regressions and scatter plot of % fin erosions versus drainage area for ecoregion reference sites (gray) and example river (black). Increased number of fin erosions per drainage area (drainage area > log 3 km^2) for study river indicates degree of ecosystem impairment.

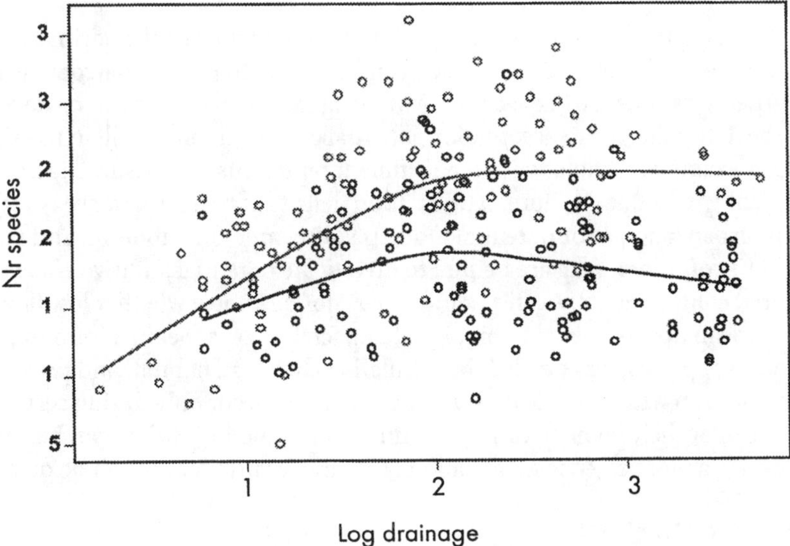

Figure 8-4 Nonlinear regressions of number of fish species versus drainage area for ecoregion reference sites (gray) and study river (black). Decreased number of fish species per drainage area compared to reference sites indicates general depression in species richness.

age of fin erosions throughout the river drainage basin is within the bounds expected from reference sites. However, where the river is at its largest (> 1000 km^2), the number of erosions exceeds the expectation. A map of the residuals from the expected is a helpful method to identify impaired sites. Unlike percentage fin erosions, the number of fish species is depressed throughout much of the watershed (Figure 8-4), indicating a generalized phenomenon.

Scoping and Analysis of Watershed

Data gathering and assembling

The principle goal in this section is to develop a list of potential candidate sources of impairment. Given the size of the river's watershed and the disperse nature of the impairment, the initial list should be comprehensive. This list can be pared down to suspect causes by comparing their coincidence with the biological responses, and/or a conceptual model can be created that justifies their inclusion. In this case study, all available data regarding industrial and municipal discharge locations, along with biological, chemical (water, sediment, tissues), instream habitat, and land-use data, were assembled into a geographic information system (GIS). Types of data were not uniform throughout the watershed. For instance, 200 sites for fish, invertebrates, and instream habitat were available, whereas chemistry samples for water, sediment, and fish tissue residues were available for fewer sites. Land-use classification data on a 1:250,000 scale were readily available from a national environmental agency (Figure 8-5).

The impacts to the fishery were observed throughout much of the basin. Because a large number of potential causes may exist, simple methods that narrow the focus for further iterations are needed. Proxies and aggregate measures can be developed from the data collected to account for potential unknowns and to fill in missing data. For example, cumulative percent effluent represents unmeasured factors coincident with effluents, human population density may serve as a proxy for percent urban land use or potential industrial discharge contribution, and toxic units (TUs) or mmoles/kg are helpful to investigate potential additive effects of measured contaminants. At this point, it may not be known whether low flow, high flow, or mean flow is related to biological impact. Hence, a mean, minimum, and maximum approach was needed for calculating chemical, habitat, and proxy events for comparison with biological data. Many stressors occur only during certain times of the year, or their potency varies over time. Consequently, if data can be parsed temporally, a more extensive evaluation of their causative nature can be determined.

Initial focusing efforts

The spatial coincidence of discharge location, proxies, chemical concentrations, and land-use characteristics with biological data can be observed using several maps (Figure 8-5).

Figure 8-5 Intensity maps of selected historical water chemistry, instream habitat, and biological data. Diameter of dot is proportional to intensity of factor. (BOD = biological oxygen demand, TSS = total suspended solid)

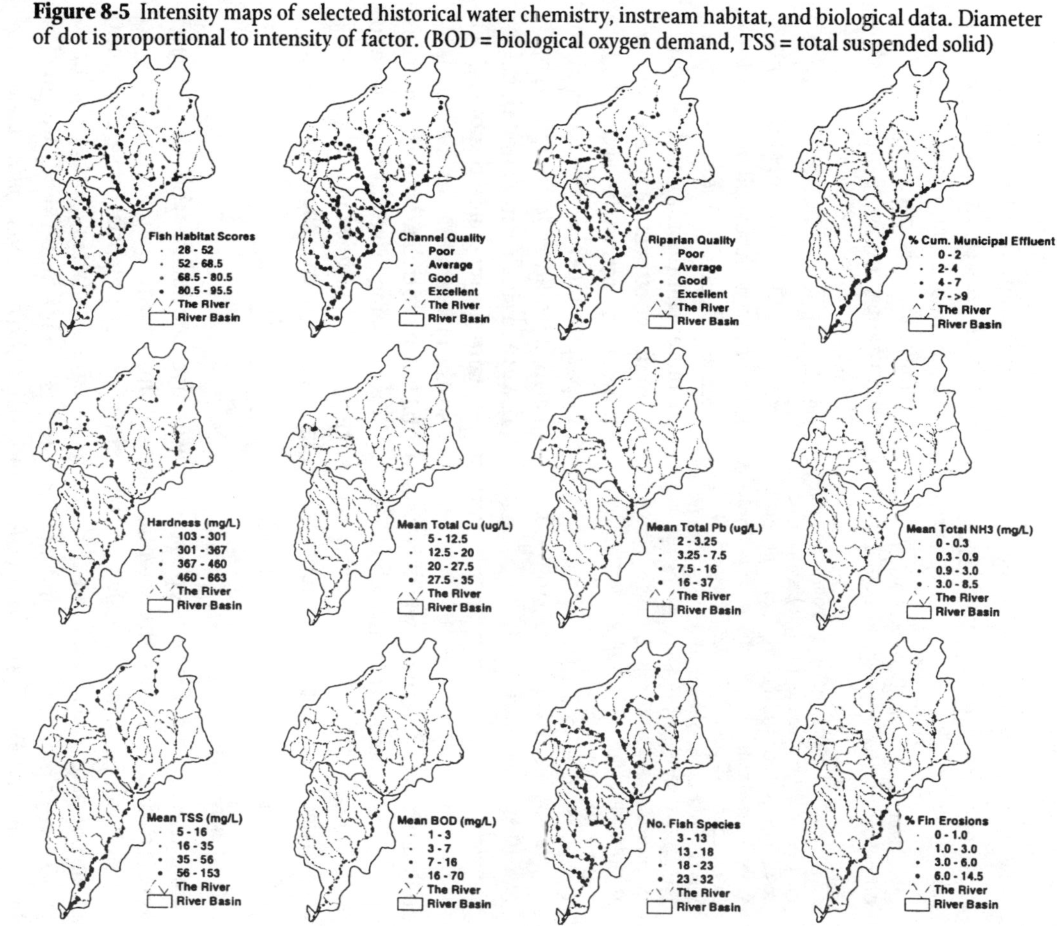

Practical note on spatial aggregation

The appropriate diagnosis of attribution of stressors may depend on the spatial resolution of field data available in a retrospective analysis. Data from different organizations may be compiled into a single system. Unfortunately, it is often the case that organizations interested in tracking chemistry may not coordinate sampling locations and sampling periods for biology. For instance, water and sediment chemistry data are most often collected near bridges, whereas biological and habitat data are often found near major stressor sources such as discharges. Hence, reconciliation of different sampling locations into a single site or segment basis may be appropriate, if not critical, for complex stressor evaluation. If a segment approach is used, segments should not cross tributaries or discharges.

Based on an appropriate spatial aggregation method, a correlation matrix of proxies, chemical concentrations, habitat, and land use to biological data is made (Figure 8-6). For instance, total suspended solids (TSSs) showed significant positive relationships with biological oxygen demand (BOD), Pb, Zn, and ammonia but not with several other factors such as Cd, Cu, hardness, TUs (from metals and ammonia), and percent cumulative effluent at mean and low flows. Considering the widespread impact (number of fish species) in the basin, it would be serendipitous to discover a single correlation that explains the biological response. Even so, the greater utility of the correlation matrix is to determine which environmental factors correlate with each other. Correlation plots also identify which relationships are being driven by extremes and not by consistent trends. If a multivariate statistic (e.g., multiple linear regression) is used to derive a potency value for these factors, then covariance should be considered.

Initial attribution

Initial attribution is the last step in the retrospective focusing exercise. The purpose here is to diagnose or list the potential causes of impact. Inevitably, multivariate statistics will be needed to assign potency values for each potential cause. Examples of appropriate statistics include multiple linear regression (where partial r^2 values may serve as potency indicators), nonlinear regression methods (locally weighted least squares [LOESS], regression trees), and ordination methods coupled with regression.

Refine with Modeled Data: Results of First Iteration

The goal of the refinement step is to synthesize available data into a format that allows for the construction of a conceptual model that explains the biological impairments. In this case, a series of maps, correlations, and multivariate statistical methods were used to simplify the long list of potential stressors to a more focused, smaller list. Visual observation of the mapped environmental factors indicated that the number of species appears to be smallest in urban areas and greatest in rural areas (both forested and agricultural). The correlation matrix (not shown) indicated that percent urban land use was highly significantly correlated with population

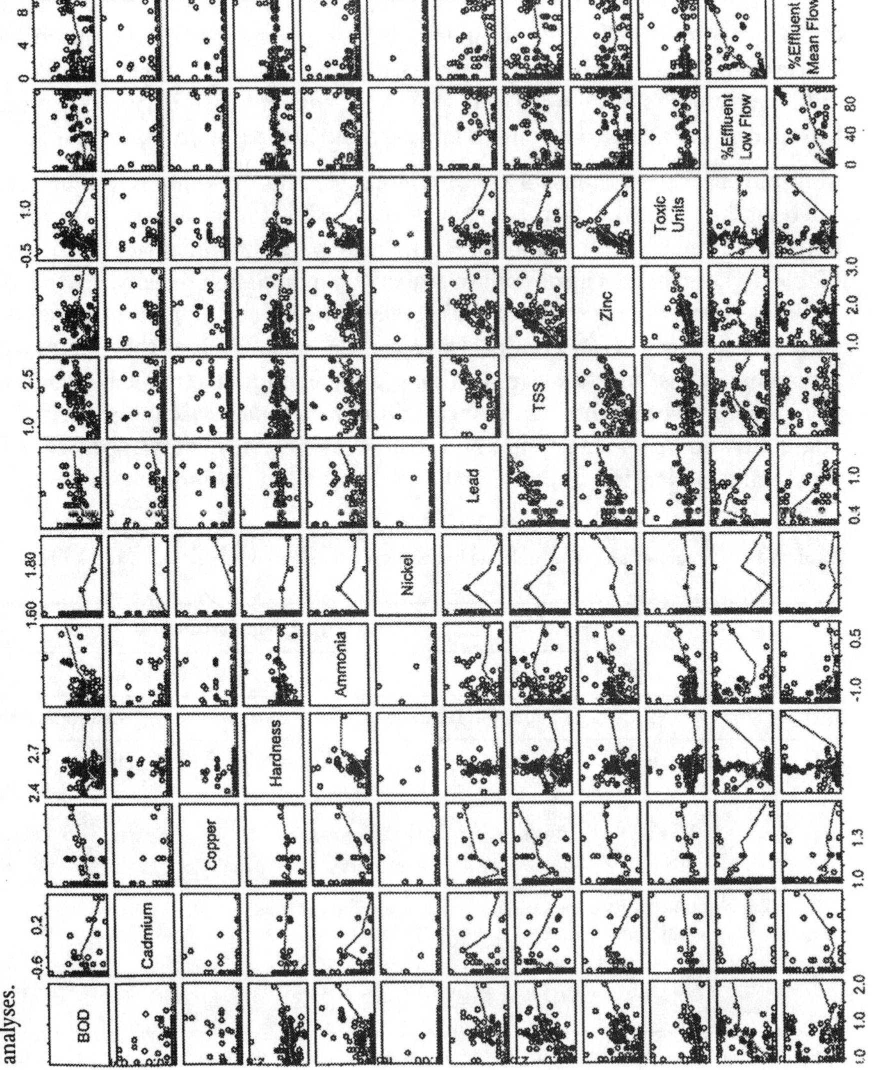

Figure 8-6 Correlation matrix illustrating potential interrelationships between various water chemistry factors and % cumulative effluent at mean and low flows; helpful for determining proxy parameters and interpreting multivariate analyses.

density (i.e., they were proxies for each other). TSSs were highly correlated with each of these factors as well. Multiple linear regressions were used on raw number-of-species data as well as on residuals from the habitat calibration. Because few sediment sample sites ($n = 10$) were available, sediment chemistry versus number of fish species was assessed through a nonparametric correlation (Spearman).

Multiple linear regressions for raw species and habitat-calibration residuals indicated that the following factors might be responsible for the observed impacts: population density, urban land use, cumulative percent municipal effluent, and TSS (Table 8-1). Again, it is important to realize that population density, % urban land use, and TSS were each significantly correlated to each other, hence may serve as a surrogate for each factor. The Spearman correlation analysis of the 10 sites with sediment chemistry data suggested that sediment ammonia ($r = -0.66$) concentration might also be a contributory factor for reduced species richness. Percent cumulative effluent at mean flow and sediment substrate quality appeared to be the most influential factors addressing the variation in fin erosions.

Table 8-1 Summary of stepwise multiple linear regressions for species richness and % fish with fin erosions

	Dependent variable[a]			
	Species richness		Fin erosions (%)	
Step	Raw data	Calibration residuals	Raw data	Calibration residuals
1	Drainage area 0.30	Urban land (%) −0.18	Effluent mean flow (%) 0.44	Effluent mean flow (%) 0.16
2	QHEI 0.45	Effluent mean flow (%) −0.25		Substrate −0.24
3	Population density −0.60			
4	Effluent mean flow (%) −0.67			

[a] Partial r^2 values represent cumulative coefficients of determination per step. Only sites where location-specific matches occurred for each parameter were employed in regressions. Signs in front of r^2 denote positive or negative associations with dependent variables.
QHEI = Qualitative Habitat Evaluation Index.

Based on these analyses, there are 2 plausible explanations of the depressions in species richness and increased fin erosions:

1) Sediment toxicity from solids loadings from nonpoint and point sources—Supporting this explanation were the high partial r^2 values in the multiple linear regressions for land use and population density, the correlation of TSS with land use and population density, and the high coefficient of determination for number of species versus sediment ammonia concentration.

2) Unknown toxicants from nonpoint runoff and WWTPs—Supporting evidence includes high r^2 for proxies of urban land use and cumulative percent municipal wastewater effluent at mean flow.

A conceptual model further describing each explanation is shown in Figure 8-7. Note that the schematic illustrates how several factors may also play a part in the depressed number of species and increased number of fin erosions. While the focusing effort has decreased some uncertainties and provided a conceptual model for impairment, significant uncertainties remain. Hence, appropriate diagnosis will require further refinement.

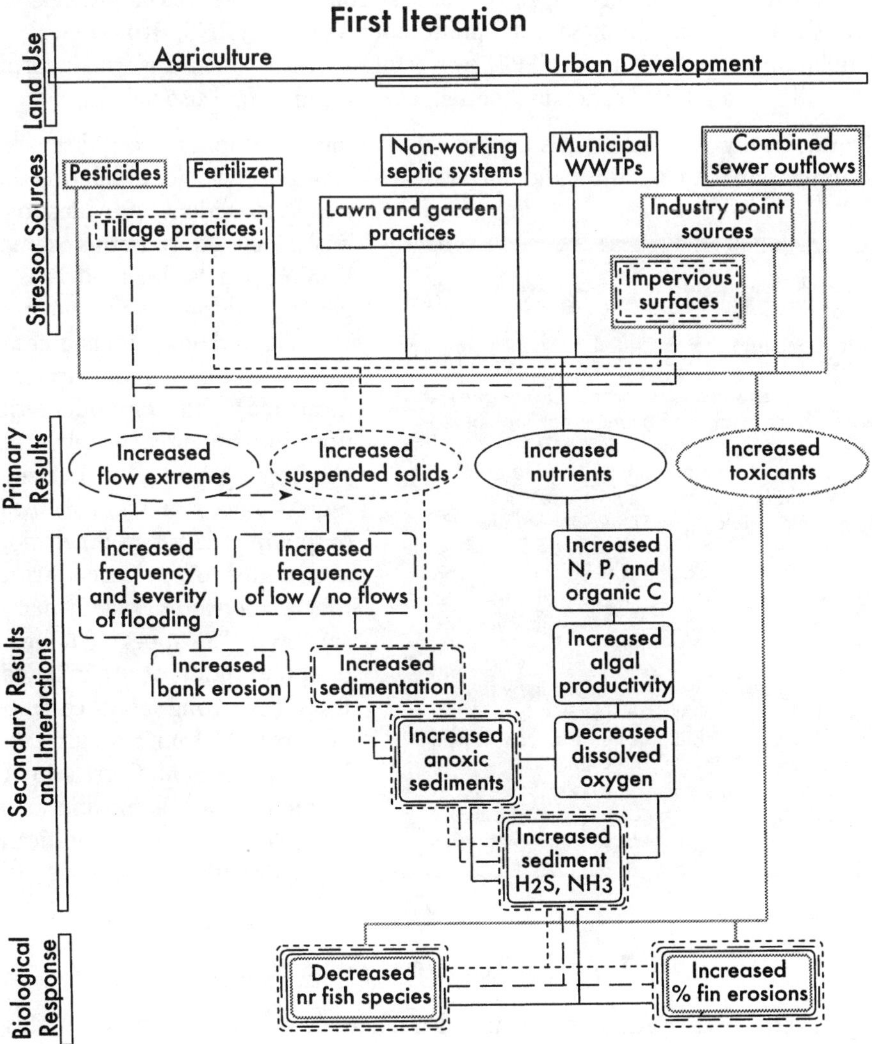

Figure 8-7 Schematic of primary stressors and their potential relationships in governing reduced number of fish species and increased fin erosions

The acceptability of a diagnosis is dependent on the strength of linking effects to the appropriate stressor and source. In this instance, ambient water and sediment chemistries, coupled with proxies for WWTP and nonpoint discharges in urban areas, provide correlative evidence of causal links. Refinement of this conclusion will require gaining additional chemistry data for WWTPs and nonpoint runoff. Contacting WWTPs for effluent chemistry and toxicity data is appropriate, as is obtaining a hydrological model (e.g., Better Assessment Science Integrating Point and Nonpoint Sources [BASINS; www.epa.gov/ost/basinsv2.htm], Rout [Dyer and Caprora 1997], Geography-Referenced Exposure Assessment Tool for European Rivers [GREAT-ER; Feijtel et al. 1997]) that can route chemical discharges in a watershed. The same applies for nonpoint models (e.g., BASINS, Hydrological Simulation Program Fortran [HSPF], universal soil loss). Models that can incorporate sedimentation and estimate sediment concentrations are also helpful.

Hydrologically oriented models that take into account multiple sources, discharge, accumulation, and loss throughout a basin are highly beneficial for evaluating the weight of evidence provided in the initial focus effort. In this example, BASINS was used to model suspended solids and ammonia from WWTPs and nonpoint sources. Model results were compared to measured results for model verification. Subsequently, model outputs per river segment were analyzed by regression. A regression of BASINS output for TSS versus fin erosions and number of species confirmed that relationships exist. Primary sources of TSS appeared to be from nonpoint runoff of urban lands. Even so, a strong spatial correspondence of TSS from agricultural lands was evident. Correlation of ambient water column ammonia from WWTPs did not significantly correlate with impairment.

Conclusions from Tier 1 analysis of existing data

Conclusions of the Tier 1 analysis were mixed. While a statistical model indicated that suspended solids correlated strongly with biological impairment, a conceptual mechanism was rather tenuous. The lack of data about nonpoint source loading on potential toxicants, high flow effects, matching biology and chemical data for all reaches, and sediment toxicity add uncertainty to any conclusions regarding primary stressors and causality. Further, the BASINS model was not applicable for estimating sediment ammonia concentrations.

Hence, demonstrating causality would require experimental approaches and addition field monitoring efforts. Data layers and model outputs in the GIS guide the selection of future field-site collection efforts and provide useful mechanisms to predict biological impairment based on new data.

Tier 2: Refined Approach

Tier 1 retrospective analyses may not adequately resolve study questions. In this case, site-specific assessments must be designed.

Refine Assessment with New Data

The previous database assessment concluded that a field evaluation was needed to better determine which stressors were important and what their relationship was to beneficial use impairment. Water quality problems (fin erosions) were elevated below a large urban area. The above assessment suggested that either urban non-point source runoff or municipal WWTP effluents might be a source of reduced fish species richness and of fin erosions, with TSS and sediment ammonia having a significant relationship with these impairments. However, the database was inadequate for determining causality or specific stressor sources. Therefore, TSS and sediment ammonia may simply be surrogates or indicators of other stressors or co-varying factors. Other potential stressors include metals, synthetic organics (e.g., polycyclic aromatic hydrocarbons [PAHs], polychlorinated biphenyls [PCBs], pesticides), pathogens, temperature, habitat, and, indirectly, nutrients (Table 8-2). These may exist at sublethal levels that, when present in combination, impair the fish populations (see also Chapter 1). In addition, the database was unable to adequately assess whether other aquatic populations, such as invertebrates, were adversely affected. It is difficult to directly measure a stressor causing fin erosions, and a wide range of stressors may directly or indirectly cause fin erosions. Therefore, it is more practical to measure the potential for fin erosions by measuring toxicity and/or bioaccumulation in tissues in order to link the presence of metals or synthetic organics to organism stress and potential for fin erosions. These stressors may occur in both nonpoint source runoff from urban and agricultural areas and in effluents; thus, they may dominate at high and low flow, respectively.

Table 8-2 Possible stressor media and components: Dominant exposure routes in test stream

In situ exposure	Possible stressor source	Stressor media components	Possible stressor
Base flow	PS, GW	Water, food	Metals, organics, nutrients, pathogens, temperature, habitat
High flow	NPS	Water, suspended solids	Flow, metals (urban or natural), organics, nutrients, pathogens
Sediment	PS, NPS, GW	Sediment, pore water, food	Metals, organics, nutrients, pathogens, habitat

PS = point source inputs (National Pollutant Discharge and Elimination System [NPDES] discharges, septic systems, feedlot discharges).
GW = groundwater inputs.
NPS = Nonpoint source inputs (dry and wet deposition, runoff).

For purposes of the field evaluation, both nonpoint source urban and agricultural runoff and WWTP were targeted as being potentially responsible for the beneficial use impairment. The sampling design was a paired, upstream–downstream approach allowing for *t*-test type statistical comparisons. Two WWTP effluents were evaluated, one in a predominately urban-use area and one in an agricultural area. Both were sampled during both low and high flow periods to separate nonpoint source runoff from effluent-based effects. The test hypotheses were these:

1) Aquatic organisms exposed to nonpoint source runoff during high flow and depositional sediments during low flow will be adversely affected.
2) The WWTP effluent will not cause adverse effects in aquatic organisms.

If adverse responses are observed, then more detailed studies (Tier 3) can be conducted to determine whether TSS and ammonia are the primary stressors. Both will be measured during the initial (Tier 2) field evaluations. First, both test sites have the potential for habitat- and flow-related stressors that must be assessed to determine whether they are comparable between upstream and downstream stations. In the watershed area dominated by agriculture, pesticides, nutrients, and pathogens may be stressors during high flow periods in the spring. In the watershed dominated by urban activities, runoff and potential stressors will be more complex, with possible loadings of metals, pesticides, nutrients, pathogens, and toxic nonpolar organics such as PAHs. Metal and organic pollutants in nonpoint source runoff may occur either as dissolved or TSS-associated fractions; therefore, the TSS could exert both a physical and/or a toxicity stress.

Based on these realities, the initial experimental design captured these possible stressors and exposure periods (Table 8-2). Once the relative contribution of stress associated with surface water (low or high flow) or sediment is defined, then knowledge of the loadings likely from the upstream inputs will help focus the chemical analyses and follow-up laboratory or field stressor identification (Tier 3; USEPA 2000). For example, if increased mortality is associated with high flow periods, then stressors such as suspended solids and pesticides are more likely if upstream loadings are primarily from farming practices. Targeted chemical analyses and toxicity identification evaluations (TIEs) will help discern whether pesticides are the primary stressor.

The 2 study areas had a large WWTP located just downstream from an urban or agricultural area. Concentrations of chemicals in the WWTP effluents routinely met National Pollutant Discharge Elimination System (NPDES) permit levels, and toxicity was not found in standard short-term chronic toxicity tests. Given the low concentrations of metals observed in the effluents and the high hardness of the receiving waters, it was very unlikely that metals were stressors. The assessment focused on low flow and high flow evaluations above and below the WWTP outfalls and with upstream or ecoregion reference sites. Testing was conducted during late spring and late summer to coincide with pesticide applications and low flow periods, respectively. The ecosystem compartments monitored included surface water (low and high flow) and sediments (depositional and large-grain from low and high energy stream sections). A tiered, iterative assessment approach was used to more effectively direct a limited budget for the project. Information from the watershed (basin-wide) assessment identified questions requiring field data. The Tier 2 assessment with minimal chemical analyses allowed for more focused sampling and targeted chemistry in Tier 3, thereby reducing costs. The findings from these assessments then were used for site-specific management decisions and to refine the watershed model to better answer the basin-wide question of point source impacts.

Design of the Field Study

Station locations

Sites that were used in the field study included

1) upstream and/or ecoregion reference (suggested by USEPA as having good water quality and healthy fish and benthic macroinvertebrate communities),
2) upstream WWTP,
3) near-field downstream, and
4) far-field downstream.

Rationale for station locations

An ecoregion reference was needed because the upstream reference water received nonpoint source inputs of solids, nutrients, and chemicals. Effects observed at the upstream WWTP station were compared to the near-field downstream site to determine the effects of the outfall and were compared to the ecoregion reference to determine effects of upper watershed inputs. The far-field downstream site was compared to the near-field site to assess whether stressor effects were declining downstream.

Assessment tools

In Tier 2, the field study was designed to use these primary assessment tools:

1) in situ toxicity testing,
2) laboratory toxicity testing,
3) indigenous biota characterizations,
4) habitat evaluations, and
5) simultaneous physical–chemical profiling.

Each of these components added unique strengths to the assessment process, and each is briefly described below.

Rationale for in situ testing

Species respond differently to chemicals, so their combined use increases the likelihood of detecting any existing chemical stress. Although impact on fish was identified as the system impairment of interest, invertebrates tend to be equally or more sensitive to stressors. Thus the use of invertebrates may aid in prediction of effects on fish. These species occupy different niches important in most aquatic ecosystems. Exposure in situ reduces laboratory extrapolation uncertainties and sampling-induced artifacts and provides a more realistic exposure to potentially fluctuating levels of stressors. The 3 exposures quantify the degree of stress associated with water, suspended solids, and surficial sediments, thereby helping to determine the source of primary stressors.

In situ testing consisted of caged exposures of 3 species (early life stages of Cladocera, fish, and amphipod) for periods ranging from 2 to 14 days (see Chapter 1). The chamber exposures were optimized to compartmentalize overlying water, overlying water only with reduced suspended solids, and surficial sediments. The measurement endpoint for the exposed organisms was survival.

Rationale for laboratory sediment toxicity testing

Collection of the upper few cm of sediment provides information on potential sediment toxicity that may affect benthic and water column organisms through resuspension or diffusion and bioturbation. Such occurrences may not be detected with in situ surficial sediment exposures if contaminants are more then several mm from the sediment surface. The 2 assays are standardized and have been shown to be reliable.

The second assessment tool used was laboratory sediment toxicity testing. The USEPA whole sediment toxicity test methods were used, consisting of 10-d amphipod and midge exposures in static renewal systems (USEPA 1999a). The measurement endpoints included survival (amphipod and midge) and growth (midge only).

Rationale for characterization of indigenous benthic invertebrate community

The indigenous community provides essential information on the biological quality of receiving water. While this tool may be robust because of natural variability and insensitivity of some resident populations to stress, it is a proven tool that provides a ground-truth for the assessment process. Ohio and other states have effectively used both the benthic macroinvertebrate and fish communities as biological criteria. Fish communities were not assessed because of their ability to easily migrate between the 3 stations. Downstream fish surveys had already revealed degraded communities with elevated fin erosions.

The third essential assessment tool used in this Tier 2 study was a characterization of the indigenous benthic macroinvertebrate community. The benthic macroinvertebrate community was quantitatively sampled using the Rapid Bioassessment Protocol (RBP; USEPA 1999b). Measurement endpoints included taxonomic identification to lowest practical level and subsequent analyses using 10 indexes. Other indexes, metrics, and statistical analysis methods may be used.

Rationale for characterization of habitat at each sampling station

QHEI provides a crude quantification of habitat quality and its potential role as a stressor or as an environment capable of supporting high-quality aquatic communities.

Another critical assessment component is characterization of habitat at each sampling station, to ascertain its role in stress and its ability to support high-quality biological communities. The habitat quality was characterized using the Qualitative Habitat Evaluation Index (QHEI) approach (Ohio EPA [OEPA] 1987; USEPA 1989). Habitat is measured through best professional judgment scoring of numerous habitat characteristics, with subsequent summation of scores. While the QHEI was designed for assessing fish habitat, it provides a useful indication of habitat characteristics and relative similarity that is important for benthic invertebrate communities.

The final essential assessment component is the characterization of the physical and chemical conditions that exist during the exposure period. This component obviously can vary markedly in its level of detail, depending on the sampling frequency (minutes to years), media measured (water, sediment, tissue), parameters (major ions and conventional NPDES limits to trace-level contaminants), and their level of detection (mg/L to ng/L). Decisions on what level of detail to use are often dictated by available resources. Because of the expense of thorough, comprehensive characterizations, a high level of detailed study should be limited to Tier 3–level studies. In our current Tier 2 study, water and sediment quality were assessed for the common and critical stressors or indicators, including continuous measurements of dissolved oxygen, pH, temperature, and conductivity. Measurements made at test initiation and termination and during first flush conditions (high flow) included suspended solids, ammonia, P, hardness, and pesticides (immunoassay for triazines).

> ***Rationale for characterization of physical and chemical conditions***
>
> Use of datasondes allows for continual monitoring of key water quality parameters to show changing exposure conditions. The remaining parameters are easily measured at key times. The immunoassay for triazines provides information on the occurrence of the herbicide atrazine, which is widely used and can act as a crude but inexpensive surrogate for other pesticides.

Explanation of Responses: Tier 2 Survey Results, Conclusions, and Actions

The Tier 2 evaluation resulted in the following findings about in situ exposures, laboratory sediment toxicity, indigenous community, habitat, and physical–chemical factors.

In situ exposures

No effects were found at low flow in the water column (all stations), suggesting continuous point source inputs were not primary stressors. Slight toxicity was recorded in surface sediments at the depositional site (far-field). This suggests that low levels of organic chemicals originating from either a point source or an non-point source have recently entered the stream, but the lack of toxicity in deeper sediments lowers the probability of historical inputs or ground water being a current stressor. Greater toxicity was observed in late summer in the surficial sediment exposures during warmer, low flow conditions. This suggested a biologically related phenomenon such as a seasonal increase in sediment ammonia. Supporting chemistry showed porewater ammonia concentrations were near threshold levels. No effects were observed in surficial sediments of the other 3 stations (ecoregion reference, upstream, and near-field). No fine-grained sediments existed at any site other than the far-field sites, so contamination sources could occur over a wide range of the upstream watershed and not be detected in larger-grained sediments. Moderate mortality occurred in the water column exposures during late summer high flow, and high mortality occurred in the spring (all stream stations, including the ecoregion reference). Less mortality, however, occurred during high flow in chambers without suspended solids (all stations). This suggests that suspended solids (and possibly contaminants adsorbed to them) are contributing to organism stress. The in situ results (Table 8-3) did not identify the primary stressors; however, they did add to the weight of evidence indicating that the most significant toxicity occurs during high flow conditions when suspended solids are elevated and when peak pesticide application periods occur. In addition, sediment ammonia concentrations are known to often peak during late summer periods, and their elevated levels are likely the primary cause of the sediment toxicity. The ammonia may be originating from both nonpoint and point sources.

Table 8-3 Summary of in situ toxicity testing

		Mortality		
	In situ exposure	Spring	Late summer	Likely stressors
1	Base flow	–	+	Ammonia
2a	High flow	+++	++	Pesticides
2b	High flow with reduced suspended solids	+	+	Suspended solids
3	Sediment	–	++	Ammonia

Laboratory sediment toxicity

Low toxicity was observed at far-field sites. Sediments were collected from the field by ponar dredge down to a depth of 10 cm. This information, in combination with the in situ toxicity data, suggests that deeper historical sediments were not contributing to toxicity.

Indigenous community

RBP metrics revealed a fair community at the depositional far-field site and fair to good communities at the other sites. Ratings also were calculated using the Invertebrate Community Index (ICI; OEPA 1987) and compared to historical data. New ICI values were lower at the far-field sites than historical values that were measured using artificial substrates (Hester-Dendy's). A possible explanation is that artificial substrates separated organisms from contaminated surficial sediments and allowed colonization from upstream refugial drift. Given the robust nature of these data, they did not help discern site differences.

Habitat

Habitat at all sites was fair in quality. An adequate level of shelter and organic matter was present to support good aquatic communities; however, siltation and embeddedness were evident. These results suggest habitat is being affected similarly by the physical presence of suspended solids throughout the watershed and is not responsible for site differences.

Physical–chemical factors

The only parameters measured showing possible threshold exceedances were ammonia, suspended solids, and Atrazine. Ammonia levels were high in sediment pore waters at the far-field sites. Atrazine was detected in spring high flow water and in sediment extracts. Other toxicants may exist but were not analyzed. Suspended solids were elevated during high flow conditions.

Conclusions from Tier 2 Assessment

Far-field sites had sediment toxicity with potential ammonia problems. Suspended solids are causing mortality during high flow periods. Some chemicals may be associated with solids (with elevated toxicity during spring pesticide runoff periods). Warmer water temperatures and elevated pathogen loadings during high flow may also be contributing to fin erosions but were not adequately assessed in Tier 2.

Actions from Tier 2 Assessment

These results and conclusions were provided to the watershed assessor and linked to the results of the Tier 1 database study. Considerations of causal factors from both Tier 1 and Tier 2 studies are evaluated by an approach described in Table 8-1 (see also Chapter 1). Results from Tiers 1 and 2 showed that nonpoint sources exert adverse impacts on the receiving system through the addition of suspended solids. Plausible mechanisms for impairment of receiving water biota include production of ammonia in sediments during low flow and physical effect during high flow periods. While pathogens may present another factor worth investigating, the assessor has enough information to recommend control of nonpoint source runoff. Causality will be confirmed or refuted based on the response of the instream biology to changes in runoff. If the watershed does recover to an expected state, then another iteration may be necessary, perhaps investigating pathogens.

Tier 3: Reassess Causes of Impairment and Approaches Used

Once the iterative process is conducted, or immediately following a Tier 2–type field survey, it may be apparent that a Tier 3 study is needed to adequately characterize the primary stressors, their sources, and/or their magnitude, frequency, and duration. A number of synoptic-type studies can be designed, consisting of varying degrees of laboratory and field evaluations. These should be focused at reducing uncertainty and answering specific questions while considering the various stressors and their environmental interactions (resulting in additivity, independence, antagonism, or synergism; see also Foran and Ferenc 1999). At this level of study, it is important from a resource allocation perspective to scale the design of the study to the size of the problem being addressed. A wide variety of useful assessment options exist for Tier 3–type studies to help conclusively identify and prioritize the system stressors. Many of these approaches are discussed in Chapter 1, with more details in Chapters 2 through 7.

Summary and Conclusions

Databases used for retrospective studies are usually deficient, containing limited data on synthetic organics, sediments, tissue residues, toxicity, habitat, and biological communities. However, as shown in this chapter, existing water quality data can be used to determine which questions need to be asked and where and what kind of additional data need to be collected. It is important to remember that when significant correlations and relationships are revealed in the data analyses, they do not usually constitute causation, and additional stressors may exist. For example, in the database evaluation, P concentrations were highly correlated with impairment to the fish communities (elevated fin erosions), yet this nutrient was not directly the cause of the fin erosions. In this system, P may in fact be a useful surrogate for other chemicals or pathogens that are the primary cause of the fin erosions. Through carefully designed synoptic studies (laboratory and field), these unknowns can be resolved.

Another reality is that it may be difficult to identify all the possible detrimental interactions. Many of the widely used pesticides (e.g., pyrethroids) and pharmaceuticals are rarely analyzed and are difficult to measure. These may be primary stressors in some systems and may act in additive and synergistic fashions with other stressors. However, by using a simple weight-of-evidence design (such as described above), the stressor class, source, proxies, and/or critical exposure pathway can be ascertained and thereby managed. As our ability to understand and measure stressor mixture effects grows, so will our ease in detecting stress and assigning ranking or importance levels to each.

Ideally, experiments like those discussed above could be designed to eliminate confounding variables and to give a clean result that eliminates individual hypothesized causes. However, experiments do not always provide clear results, and in some cases, the test endpoint (e.g., mortality) cannot be confidently extrapolated to the assessment endpoint (e.g., fin erosion). In such cases, experimental results along with results of site-specific observations and information from the literature are combined in a weight-of-evidence analysis. Because an endpoint effect such as loss of fish species richness may have different causes at different locations in a watershed, a weight-of-evidence analysis would require information about the specific reach in which the effects occur and must be explained. This information would include the occurrence and intensity of different putative causes, the occurrence and frequency or magnitude of effects, and the spatial and temporal associations of causes and effects. In addition, the weight-of-evidence analysis would require information from the literature about the effects potentially induced by each putative cause. For example, does ammonia or suspended sediment cause fin erosion? A subsequent Tier 3 evaluation could effectively design a laboratory and field evaluation, using caged fish with various exposures to ammonia and suspended solids, to assess this question.

This 2-tiered case example showed a statistically significant relationship between the presence of ammonia and suspended solids and the erosion of fins and loss of fish species richness. But some degree of biological variance is still unexplained and could be due to other associated factors such as pesticide or PAH exposures. The degree of certainty of any analysis will always be a function of how many iterations the assessor is able to effectively conduct (see Figure 8-1), thereby increasing the weight of evidence of the conclusions.

Abbreviations

alkw	alkalinity of water
AOC	area of concern
BASINS	Better Assessment Science Integrating Point and Nonpoint Sources
BCF	bioconcentration factor
BCFV	plant–air volumetric bioconcentration factor
BEAST	BEnthic Assessment of SedimenT
BOD	biological oxygen demand
BSAF	biota–sediment accumulation factor
CATS	Contaminants in Aquatic and Terrestrial ecoSystems
CCU	cumulative criterion unit
CERCLA	Comprehensive Environmental Response Compensation Liability Act
CF	conventional management practice
CG	California Gulch
CI	condition index
CLASSIC	Community Level Aquatic System Studies Interpretation Criteria
C_s	concentration in sediment
CV	coefficient of variation
DDE	delay-differential equation
DDE	dichlorodiphenyldichloroethylene
DDT	dichlorodiphenyltrichloroethylene
DEB	dynamic energy budget
DFA	discriminant function analysis
DIN	dissolved inorganic N
DNA	deoxyribonucleic acid

Ecological Variability: Separating Natural from Anthropogenic Causes of Ecosystem Impairment.
D.J. Baird and G.A. Burton, Jr., editors.
ISBN 1-880611-43-0

DOC	dissolved organic carbon
dw	dry weight
EAC	ecologically acceptable concentration
EMAP	Environmental Monitoring and Assessment Program
ERM	effects-range median
FIFRA	Federal Insecticide, Fungicide, and Rodenticide Act
FL	lipid content in organism
GCTE	Global Change in Terrestrial Ecosystems
GIS	geographical information system
GREAT-ER	Geography-Referenced Exposure Assessment Tool for European Rivers
GW	groundwater inputs
HARAP	Higher-tier Aquatic Risk Assessment for Pesticides
HC5	hazardous concentration for 5% of a species
HCH	hexachlorocyclohexane
HCN	Health Council of the Netherlands
HSPF	Hydrological Simulation Program Fortran
IBGP	International Biosphere Geosphere Program
IBM	individual-based model
ICI	Invertebrate Community Index
IF	integrated management practice
IFEM	Integrated Fate and Effects Model
IJC	International Joint Commission
IUCN	International Union for Conservation of Nature and Natural Resources
K_{oa}	octanol–air partition coefficient
K_{ow}	octanol–water partition coefficient

K_{pw}	plant–water partition coefficient
K_{wa}	dimensionless Henry's Law constant
LBB	lethal body burden
LC50	lethal concentration to 50% of a population
LMTD	Leadville Mine Drainage Tunnel
LOESS	locally weighted least squares
LOI	loss on ignition
LTRE	life-table response experiment
mRNA	messenger ribonucleic acid
MVP	minimum viable population
n	number of subjects
NAPAP	National Acid Precipitation Assessment Program
NEPA	National Environmental Policy Act
NOEC	no-observed-effect concentration
NPDES	National Pollutant Discharge Elimination System
nr	number
NRDA	Natural Resource Damage Assessment
OECD	Organization for Economic Cooperation and Development
OEPA	Ohio Environmental Protection Agency
PAH	polycyclic aromatic hydrocarbon
PCB	polychlorinated biphenyl
PDE	partial-differential equation
PEC	probable or predicted effect concentration
PENC	probable no-effect concentration
pK_a	dissociation constant
pp	density of plant tissue

PVA	population viability analysis
pw	density of water
QHEI	Qualitative Habitat Evaluation Index
RAPD	random amplified polymorphic DNA
RBP	Rapid Bioassessment Protocol
RCRA	Resource Conservation Recovery Act
R-EMAP	Regional Environmental Monitoring and Assessment Program
RFLP	restriction fragment length polymorphism
RIVPACS	River Invertebrate Prediction and Classification System
SD	standard deviation
SE	standard error
SEL	severe effect level
SfG	scope for growth
SOC	sediment organic C
SOM	sediment organic matter
SPM	suspended particulate matter
SW	seawater
TBT	tributyltin
TE	time to extinction
TEG	triethylene glycol
TIE	toxicity identification evaluation
TOC	total organic C
TSCA	Toxic Substances Control Act
TSS	total suspended solid
TU	toxic unit

UNEP-IETC	United Nations Environmental Program – International Environmental Technology Center
USEPA	U.S. Environmental Protection Agency
USGS	U.S. Geological Survey
UV	ultraviolet
WWTP	wastewater treatment plant

References

Abrams PA, Menge BA, Mittelbach GG, Spiller D, Yodzis P. 1996. The role of indirect effects in food webs. In: Polis GA, Winemiller KO, editors. Food webs: Integration of patterns and dynamics. New York NY, USA: Chapman & Hall. p 371–395.

Aebischer NJ. 1990. Assessing pesticide effects on non-target invertebrates using long term monitoring and time series modeling. *Funct Ecol* 4:369–373.

Ahlborg UG, Brouwer A, Fingerhut MA, Jacobson JL, Jacobson SW, Kennedy SW, Kettrup AA, Koeman JH, Poiger H, Rappe C, Safe SH, Segal RF, Tuomisto I, Van den Berg M. 1992. Impact of polychlorinated dibenzo-*p*-dioxins, dibenzofurans and biphenyls on human and environmental health with special emphasis on the application of the toxic equivalency factor concept. *Eur J Pharmacol* 228:179–199.

Akçakaya HR, Atwood JL. 1997. A habitat-based metapopulation model of the California gnatcher. *Conserv Biol* 11:422–434.

Allred PM, Giesy JP. 1985. Solar radiation-induced toxicity of anthracene to *Daphnia pulex*. *Environ Toxicol Chem* 4:219–226.

Andrén O, Lindberg T, Boström U, Clarholm M, Hansson AC, Johansson G, Lagerlöf J, Paustian K, Persson J, Petterson R, Schnürer J, Sohlenius B, Wivstad M. 1990. Organic carbon and nitrogen flows. *Ecol Bull* 40:85–125.

Antonovics J, Bradshaw AD, Turner RG. 1971. Heavy metal tolerance in plants. *Ad Ecol Res* 7:1–85.

Armitage PD, Gunn RJM, Furse MT, Wright JF, Moss D. 1987. The use of prediction to assess macroinvertebrate response to river regulation. *Hydrobiologia* 144:25–32.

Ashby P. 1999. The risks to wildlife from pesticides. In: Kedwards TJ, Thomas MB, editors. Challenges in applied population biology. Aspects of applied biology 53. Wellsbourne UK: Assoc of Applied Biologists. p 1–7.

Atwell L, Hobson KA, Welch HE. 1998. Biomagnification and bioaccumulation of mercury in an arctic marine food web: Insights from stable nitrogen isotope analysis. *Can J Fish Aquat Sci* 55:1114–1121.

Axtmann EV, Cain DJ, Luoma SN. 1997. The effect of tributary inflows on the distribution of trace metals in fine-grained bed sediments and benthic insects of the Clark Fork River, Montana. *Environ Sci Technol* 31:750–758.

Axtmann EV, Luoma SN. 1991. Large-scale distribution of metal contamination in the fine-grained sediment of the Clark Fork River, Montana. *Appl Geochem* 6:75–88.

Bailey RG. 1998. Ecoregions: The ecosystem geography of the oceans and continents. New York NY, USA: Springer-Verlag.

Bailey HC, Miller JL, Miler MJ, Wiborg LC, Deanovic L, Shed T. 1997. Joint acute toxicity of diazinon and chlorpyrifos to *Ceriodaphnia dubia*. *Environ Toxicol Chem* 16:2304–2308.

Baird DJ, Maltby L, Greig-Smith PW, Douben PET, editors. 1996. Ecotoxicology: Ecological dimensions. London UK: Chapman & Hall.

Baker AJM. 1987. Metal tolerance. *New Phytol* 106(Suppl):93–111.

Barbour MT, Gerritsen J. 1996. Subsampling of benthic samples: A defense of the fixed-count method. *J N Am Benthol Soc* 15:386–391.

Barbour MT, Gerritsen J, Griffith GE, Frydenborg R, McCarron E, White JS, Bastian ML. 1996. A framework for biological criteria for Florida streams using benthic macroinvertebrates. *J N Am Benthol Soc* 15:185–211.

Barbour MT, Gerritsen J, Snyder BD, Stribling JB. 1999. Rapid bioassessment protocols for use in streams and wadeable rivers: Periphyton, benthic macroinvertebrates and fish. 2nd ed.

Ecological Variability: Separating Natural from Anthropogenic Causes of Ecosystem Impairment.
D.J. Baird and G.A. Burton, Jr., editors. © 2001 Society of Environmental Toxicology and Chemistry (SETAC).
ISBN 1-880611-43-0

Washington DC, USA: U.S. Environmental Protection Agency (USEPA) Office of Water. EPA/ 841-B-99-002.

Barnthouse L. 1993. Population-level effects. In: Suter II GW, editor. Ecological risk assessment. Boca Raton FL, USA: Lewis. p 247–274.

Barret GW, Rosenberg R. 1981. Stress effects on natural ecosystems. Chichester UK: Wiley.

Bartell SM, Gardner RH, O'Neill RV. 1992. Ecological risk estimation. Chelsea MI, USA: Lewis. 252 p.

Baveco JM, de Roos AM. 1996. Assessing the impact of pesticides on lumbricid populations: An individual-based modelling approach. *J Appl Ecol* 33:1451–1468.

Bayne BL, Brown DA, Burnes K, Dixon DR, Ivanovici A, Livingstone DR, Lowe DM, Moore MN, Stebbing ARD, Widdows J. 1985. The effects of stress and pollution on marine animals. New York NY, USA: Praeger.

Bayne BL, Widdows J. 1978. Physiological ecology of two populations of *Mytilus edulis* L. *Oecologia* 37:137–162.

Begon M, Mortimer M. 1986. Population ecology: A unified study of animals and plants. Oxford UK: Blackwell Scientific.

Belanger SE. 1997. Literature review and analysis of biological complexity in model stream ecosystems: Influence of size and experimental design. *Ecotoxicol Environ Saf* 36:1–16.

Bengtsson J, Martinez N. 1996. Causes and effects in food webs: Do generalities exist? In: Polis GA, Winemiller KO, editors. Food webs: Integration of patterns and dynamics. London UK: Chapman & Hall. p 179–184.

Bennett RS. 1994. Do behavioural responses to pesticide exposure affect wildlife population parameters? In: Kendall RJ, Lacher Jr TE, editors. Wildlife toxicology and population modeling: Integrated studies of agroecosystems. Boca Raton FL, USA: Lewis.

Benton MJ, Guttman SI. 1990. Effect of allozyme genotype on the survivorship of mayflies (*Stenonema femoratum*) exposed to copper. *J N Am Benthol Soc* 9:271–276.

Bentzen E, Lean DRS, Taylor WD, Mackay D. 1996. Role of food web structure on lipid and bioaccumulation of organic contaminants by lake trout (*Salvelinus namaycush*). *Can J Fish Aquat Sci* 53:2397.

Bergeron JM, Crews D, McLachlan JA. 1994. PCBs as environmental estrogens: Turtle sex determination as a biomarker for environmental contamination. *Environ Health Persp* 102:780–781.

Berglund O. 1999. The influence of ecological processes on the accumulation of persistent organochlorines in aquatic ecosystems [Ph.D. thesis]. Lund, Sweden: Lund Univ. ISBN 91-7105-114-7.

Betin C, Oehlmann J, Stroben E. 1996. TBT-induced imposex in marine neogastropods is mediated by an increasing androgen level. *Helgol Meeresunters* 50:299–317.

Black DE, Gutjahr-Gobell R, Pruell RJ, Bergen B, Mills L, McElroy AE. 1998. Reproduction and polychlorinated biphenyls in *Fundulus heteroclitus* (Linnaeus) from New Bedford Harbor, Massachusetts, USA. *Environ Toxicol Chem* 17:1405–1414.

Blaustein AR, Hoffman PD, Hokit DG, Kiesecker JM, Walls SC, Hays JB. 1994. UV repair and resistance to solar UV-B in amphibian eggs: A link to population declines? *Proc Natl Acad Sci* 91:1791–1795.

Blaustein AR, Hoffman PD, Kiesecker JM, Hays JB. 1996. DNA repair activity and resistance to solar UV-B radiation in eggs of the red-legged frog. *Conserv Biol* 10:1398–1402.

Blaustein AR, Wake DB, Sousa WP. 1994. Amphibian declines: Judging stability, persistence, and susceptibility of populations to local and global extinctions. *Conserv Biol* 8:60–71.

Boone MD, Bridges CM. 1999. The effect of temperature on the potency of carbaryl for survival of tadpoles of the green frog (*Rana clamitans*). *Environ Toxicol Chem* 18:1482–1484.

Bordin G, McCourt J, Cordeiro Raposo F, Rodriguez AR. 1997. Metallothionein-like metalloproteins in the Baltic clam *Macoma balthica*: Seasonal variations and induction upon metal exposure. *Mar Biol* 129:453–463.

Borgmann U, Whittle DM. 1991. Contaminant concentration trends in Lake Ontario lake trout (*Salvelinus namaycush*): 1977 to 1988. *J Great Lakes Res* 17(3):368–381.

Bowling JW, Leversee GJ, Landrum PF, Giesy JP. 1983. Acute mortality of anthracene-contaminated fish exposed to sunlight. *Aquat Toxicol* 3:79–90.

Bradford DF, Swanson C, Gordon MS. 1991. Effects of low pH and aluminum on two declining species of amphibians in the Sierra Nevada, California. *J Herpetol* 26:369–377.

Bradley RA, Bradley DW. 1993. Wintering shorebirds increase after kelp (Macrocystis) recovery. *The Condor* 95:372–376.

Brick CM, Moore JN. 1996. Diel variation of trace metals in the Upper Clark Fork River, Montana. *Environ Sci Technol* 30:1953–1960.

Briggs GG, Bromilow RH, Evans AA. 1982. Relationships between lipophilicity and root uptake and translocation of non-ionized chemicals by barley. *Pesticide Sci* 13:495–504.

Briggs GG, Bromilow RH, Evans AA, Williams M. 1983. Relationships between lipophilicity and the distribution of of non-ionized chemicals in barley shoots following uptake by the roots. *Pesticide Sci* 14:492–500.

Brock TCM, Budde BJ. 1994. On the choice of structural parameters and endpoints to indicate responses of freshwater ecosystems to pesticide stress. In: Hill IA, Heimbach F, Leeuwangh P, Matthiessen P, editors. Freshwater field tests for hazard asseement of chemicals. Boca Raton FL, USA: Lewis. p 19–56.

Brock TCM, Crum SJM, Van Wijngaarden R, Budde BJ, Tijink J, Zupelli A, Leeuwangh P. 1992. Fate and effects of the insecticide Dursban 4E in indoor Elodea-dominated and macrophyte-free freshwater model ecosystems: I. Fate and primary effects of the active ingredient chlorpyrifos. *Arch Environ Contam Toxicol* 23:69–84.

Brock TCM, Roijackers RMM, Rollon R, Bransen F, Van der Heyden L. 1995. Effects of nutrient loading and insecticide application on the ecology of Elodea-dominated freshwater microcosms: II. Responses of macrophytes, periphyton and macroinvertebrate grazers. *Arch Hydrobiol* 134:53–74.

Brock TCM, Van den Bogaert M, Bos AR, Van Breukelen SWF, Reiche R, Terwoert J, Suykerbuyk REM, Roijackers RMM. 1992. Fate and effects of the insecticide Dursban 4E in indoor Elodea-dominated and macrophyte-free freshwater model ecosystems: II. Secondary effects on community structure. *Arch Environ Contam Toxicol* 23:391–409.

Brock TCM, Vet JJRM, Kerkhofs MJJ, Lijzen J, Van Zuilekom WJ, Gijlstra R. 1993. Fate and effects of the insecticide Dursban 4E in indoor Elodea-dominated and macrophyte-free freshwater model ecosystems: III. Aspects of ecosystem functioning. *Arch Environ Contam Toxicol* 25:160–169.

Broman D, Naf C, Rolff C, Zebuhr Y, Fry B, Hobbie J. 1992. Using ratios of stable nitrogen isotopes to estimate bioaccumulation and flux of polychlorinated dibenzo-*p*-dioxins (PCDDs) and dibenzofurans (PCDFs) in two Baltic food chains from the northern Baltic. *Environ Toxicol Chem* 11:331–345.

Brussaard L, Bouwman LA, Geurs M, Hassink J, Zwart KB. 1990. Biomass, composition and temporal dynamics of soil organisms of a silt loam soil under conventional and integrated management. *Neth J Agric Sci* 38: 283–302.

Brussaard L, Van Veen, JA, Kooistra MJ, Lebbink G. 1988. The Dutch programme on soil ecology of arable farming systems. I. Objectives, approach and some preliminary results. *Ecol Bull* 39:35–40.

Bryan GW, Gibbs PE, Burt GR. 1988. A comparison of effectiveness of tri-*n*-butyltin chloride and five other organotin compounds in promoting the development of imposex in the dog-whelk, *Nucella lapillus*. *J Mar Biol Assoc UK* 68:733–744.

Bryan, GW, Gibbs PE, Hummerstone LG, Burt GR. 1986. The decline of the gastropod *Nucella lapillus* around south-west England: Evidence for the effect of tributyltin from antifouling paints. *J Mar Biol Assoc UK* 66:611–640.

Buckland ST, Anderson DR, Burnham KP, Laake JL. 1993. Distance sampling: Estimating abundance of biological populations. Dordrecht, Netherlands: Kluwer Academic.

Buekema JJ, DeBruin W. 1977. Seasonal changes in dry weight and chemical composition of the soft parts of the tellinid bivalve *Macoma balthica* in the Dutch Wadden Sea. *Neth J Sea Res* 11:42–55.

Burn AJ. 1992. Interaction between cereal pests and their predators and parasites. In: Greig-Smith PW, Frampton GK, Hardy AR, editors. Pesticides, cereal farming and the environment. London UK: HMSO. p 110–131.

Burton Jr GA. 1991. Assessing the toxicity of freshwater sediments. *Environ Toxicol Chem* 10:1585–1627.

Burton Jr GA. 1992. Sediment collection and processing: Factors affecting realism. In: Burton Jr GA, editor. Sediment toxicity assessment. Boca Raton FL, USA: Lewis. p 37–66.

Burton Jr GA, Hickey CW, DeWitt TH, Morrison DJ, Roper DS, Nipper MG. 1996. In situ toxicity testing: Teasing out the environmental stressors. *SETAC News* 16(5):20–22.

Burton Jr GA, Moore L. 1999. An assessment of storm water runoff effects in Wolf Creek, Dayton, OH, USA. Final Report. City of Dayton, OH.

Burton Jr GA, Pitt R. 2001. Handbook for assessing stormwater effects: A toolbox of approaches and methods for watershed managers, scientists and engineers. Boca Raton FL, USA: CRC/Lewis (in press).

Burton Jr GA, Rowland CD, Greenberg MS, Lavoie DR, Nordstrom JF, Eggert LM. 2002. A tiered, weight-of-evidence approach for evaluating aquatic ecosystems. In: Munawar M, editor. Sediment quality assessment: Approaches, insights and technology for the 21st century. Leiden, Netherlands: Backhuys. Ecovision World Monograph Series (in press).

Bush CP, Weis JS. 1983. Effects of salinity on fertilization success in two populations of killifish, *Fundulus heteroclitus*. *Biol Bull* 164:406–417.

Cairns J, Loos J. 1967. Changed feeding rate of *Brachydanio rerio* (Hamilton-Buchanan) resulting from exposure to sublethal concentraions of zinc, potassium dichromate and alkyl benzene sulfonate detergent. *Proc Pennsylvania Acad Sci* 40:47–52.

Calabrese E, Baldwin L. 1999. Reevaluation of the fundamental dose-response relationship. *BioScience* 49:725–732.

Calow P, Sibly RM. 1990. A physiological basis of population processes: Ecotoxicological implications. *Funct Ecol* 4:283–288.

Calow P, Sibly RM, Forbes VE. 1997. Risk assessment on the basis of simplified life-history scenarios. *Environ Toxicol Chem* 16:1983–1989.

Campbell PJ, Arnold DJS, Brock, TCM, Grandy NJ, Heger W, Heimbach F, Maund SJ, Streloke M. 1999. Guidance document on Higher-tier Aquatic Risk Assessment for Pesticides (HARAP). Brussels, Belgium: SETAC-Europe Pr.

Campfens J, Mackay D. 1997. Fugacity-based model of PCB bioaccumulation in complex aqautic food webs. *Environ Sci Technol* 31:577–583.

Carey J, Cook P, Giesy J, Hodson P, Muir D, Owens JW, Solomon K. 1998. Ecotoxicological risk assessment of the chlorinated organic chemicals. Proceedings from the Pellston Workshop, 24–29 July 1994. Pensacola FL, USA: SETAC Pr.

Carignan R, Rapin F, Tessier A. 1985. Sediment porewater sampling for metal analysis: A comparison of techniques. *Geochim Cosmochim Acta* 49:2493–2497.

Caswell H. 1989. Matrix population models - construction, analysis, and interpretation. Sunderland MA, USA: Sinauer Assoc.

Caswell H. 1996. Analysis of life table response experiments II. Alternative parameterizations for size- and stage-structured models. *Ecol Model* 88:73–82.

Caswell H, Nisbet RM, de Roos AM, Tuljapurkar S. 1996. Structured-population models: Many methods, a few basic concepts. In: Tuljapurkar S, Caswell H, editors. Structured-population models in marine, terrestrial and freshwater systems. Dordrecht, Netherlands: Kluwer. 656 p.

Chadwick RW, Faeder EJ, King LC, Copeland MF, Williams K, Chuang LT. 1978. Effect of acute and chronic Cd exposure on lindane metabolism. *Ecotox Environ Safety* 2:301–316.

Chandini T. 1988. Effects of different foods (Chlorella) concentrations on the chronic toxicity of cadmium to survivorship, growth and reproduction of *Echinisca triserialis* (Crustacea: Cladocera). *Environ Pollut* 54:139–154.

Chapin III FS, Walker BH, Hobbs RJ, Hooper DU, Lawton JH, Sala OE, Tilman D. 1997. Biotic control over the functioning of ecosystems. *Science* 277:500–504.

Chappie DJ, Burton Jr GA. 2000. Applications of aquatic and sediment toxicity testing in situ. *J Soil Sediment Contam* 9:219–246.

Chesson PL. 1986. Environmental variation and the coexistence of species. In: Diamond J, Case TJ, editors. Community ecology. New York NY, USA: Harper & Row. p 240–256.

Chesson PL, Case TJ. 1986. Overview: Nonequilibrium community theories: Chance, variability, history and coexistence. In: Diamond J, Case TJ, editors. Community ecology. New York NY, USA: Harper & Row. p 229–239.

Clark KE, Gobas FAPC, Mckay D. 1990. Model of organic chemical uptake and clearance by fish from food and water. *Environ Sci Technol* 24:1203–1213.

Clements WH. 1994. Benthic invertebrate community responses to heavy metals in the Upper Arkansas River Basin, Colorado. *J N Am Benthol Soc* 13:30–44.

Clements WH. 1999. Metal tolerance and predator-prey interactions in benthic macroinvertebrate stream communities. *Ecol Appl* 9:1073–1084.

Clements WH, Carlisle DM, Lazorchak JM, Johnson PC. 2000. Heavy metals structure benthic communities in Colorado mountain streams. *Ecol Appl* 10:626–638.

Clements WH, Kiffney PM. 1995. The influence of elevation on benthic community responses to heavy metals in Rocky Mountain streams. *Can J Fish Aquat Sci* 52:1966–1977.

Cloern JE. 1996. Phytoplankton bloom dynamics in coastal ecosystems: A review with some general lessons from sustained investigation of San Francisco Bay, California. *Rev Geophys* 34:127–168.

Cohen JE. 1978. Food webs and niche space. Princeton NJ, USA: Princeton Univ Pr.

Cole LC. 1954. The population consequences of life history phenomena. *Quart Rev Biol* 29:103–137.

Connell JH. 1978. Diversity in tropical rainforests and coral reefs. *Science* 199:1302–1310.

Connell JH, Sousa WP. 1983. On the evidence needed to judge ecological stability or persistence. *Am Nat* 121:789–824.

Connolly JP, Tonelli R. 1985. Modelling kepone in the striped bass food chain of the James River estuary. *Estuar Coastal Shelf Sci* 20:349–336.

Couillard YP, Campbell GC, Pellerin-Massicotte J, Auclair JC. 1995. Field transplantation of a freshwater bivalve, Pyganodon grandis, across a metal contamination gradient. II. Metallothinionein response to Cd and Zn exposure, evidence for cytotoxicity, and links to effects at higher levels of biological organization. *Can J Fish Aquat Sci* 52:703–715.

Courtney LA. 1998. Influence of past exposure to heavy metals on aquatic insect assemblage responses to chemical stressors: Metals and acidic pH [MS thesis]. Fort Collins CO, USA: Colorado State Univ.

Courtney LA, Clements WH. 1998. Effects of acidic pH on benthic macroinvertebrate communities in stream microcosms. *Hydrobiologia* 379:135–145.

Cousins SH. 1996. Food webs: From the Lindeman paradigm to a taxonomic general theory of ecology. In: Polis G, Winemiller K, editors. Food webs: Integration of patterns and dynamics. New York NY, USA: Chapman & Hall.

Cox DR. 1986. Some general aspects of the theory of statistics. *Internat Stat Rev* 54:117–126.

Crane M, Maltby L. 1991. The lethal and sublethal responses of *Gammarus pulex* to stress: Sensitivity and variation in an in situ bioassay. *Environ Toxicol Chem* 10:1331–1339.

Craven P, Wahba G. 1979. Smoothing noisy data with spline functions. *Numer Math* 31:377–403.

Crowder LB. 1988. Food web interactions in lakes. In: Carpenter SP, editor. Complex interactions in lake communities. New York NY, USA: Springer-Verlag. p 141–160.

Crump ML, Hensley FR, Clark KL. 1992. Apparent decline of the golden toad: Underground or extinct? *Copeia* 1992:413–420.

Culp JM, Podemski CL, Cash KJ, Lowell RB. 1996. Utility of field-based artificial streams for assessing effluent effects on riverine ecosystems. *J Aquat Ecosystem Health* 5:117–124.

Culp JM, Podemski CL, Cash KJ, Lowell RB. 2000. A research strategy for using stream microcosms in ecotoxicology: Integrating single population and community experiments with field data. *J Aquat Ecosys Stress Recovery* 7:167–176.

Cummins KW. 1973. Trophic relations of aquatic insects. *Ann Rev Entomol* 18:183–206.

Cuppen JGM, Gylstra R, Van Buesenkom S, Budde BJ, Brock TCM. 1995. Effect of nutrient loading and insecticide application on the ecology of Elodea-dominated microcosms. III. Responses of macrinvertbrate detritivores, breakdown of plant litter and final conclusions. *Arch Hydrobiol* 134:157–177.

Daniels RE, Allan JD. 1981. Life table evaluation of chronic exposure to a pesticide. *Can J Fish Aquat Sci* 38:485–494.

Davison W, Zhang H. 1994. In situ speciation measurements of trace metal components in natural waters using thin-film gels. *Nature* 367:546–548.

Dayton PK, Currie V, Gerrodette T, Keller BD, Rosenthal R, Ven Tresca D. 1984. Patch dynamics and stability of some California kelp communities. *Ecol Monogr* 54:253–289.

Dayton PK, Tegner MJ, Parnell PE, Edwards PB. 1992. Temporal and spatial patterns of disturbance and recovery in a kelp forest community. *Ecol Monogr* 62:421–445.

DeAngelis DL. 1992. Dynamics of nutrient cycling and food webs. London UK: Chapman & Hall.

DeAngelis DL. 1996. Indirect effects: concepts and approaches from ecological theory. In: Baird, DJ, Maltby L, Greig-Smith PW, Douben PET, editors. ECOtoxicology: Ecological dimensions. London UK: Chapman & Hall. p 25–41.

DeCaprio AP. 1997. Biomarkers: Coming of age for environmental health and risk assessment. *Environ Sci Technol* 31:1837–1848.

Decho AW, Luoma SN. 1996. Flexible digestion strategies and trace metal assimilation in marine bivalves. *Limnol Oceanogr* 41:568–572.

De Groot M. 1986. Probability and statistics. New York NY, USA: Addison Wesley.

de March BGE. 1988. Acute toxicity of binary mixtures of five cations Cu, Cd, Zn, Mg, and K) to the freshwater amphipod *Gammarus lacustris* (Sars): Alternative descriptive models. *Can J Fish Aquat Sci* 45:625–633.

Depledge MH, Lundebye AK. 1996. Physiological monitoring of contaminant effects in individual rock crabs, *Hemigrapsus edwardsi*: The ecotoxicolgical significance of variability in response. *Comp Biochem Physiol* 113C:277–282.

De Roos AM. 1988. Numerical methods for structured population models: The Escalator Boxcar Train. *Numer Methods Partial Diff Equ* 4:173–195.

De Roos AM, Diekmann O, Metz JAJ. 1992. Studying the dynamics of structured population models: A versatile technique and its application to Daphnia. *Am Nat* 139:123–147.

De Roos AM, McCauley E, Wilson WG. 1991. Mobility versus density limited predator-prey dynamics on different spatial scales. *Proc R Soc Lond (B)* 256:117–122.

De Ruiter PC, Neutel AM, Moore JC. 1994. Modeling food webs and nutrient cycling in agro-ecosystems. *TREE* 9(10) 378–383.

De Ruiter PC, Neutel AM, Moore JC. 1995. Energetics, patterns of interaction strengths, and stability in real ecosystems. *Science* 269:1257–1260.

De Ruiter PC, Neutel AM, Moore JC. 1998. Biodiversity in soil ecosystems: The role of energy flow and community stability. *Appl Soil Ecol* 10:217–228.

De Ruiter PC, Van Veen JA, Moore JC, Brussaard L, Hunt HW. 1993. Calculation of nitrogen mineralization in soil food webs. *Plant Soil* 157:263–273.

Devineau J, Triquet CA. 1985. Patterns of bioaccumulation of an essential trace element (zinc) and a pollutant metal (cadmium) in larvae of the prawn *Palaemon serratus*. *Mar Biol* 86:139–143.

De Zwaan A, Cortesi P, Cattani O. 1995. Resistance of bivalves to anoxia as a response to pollution-induced environmental stress. *Sci Total Environ* 171:121–125.

Di Toro DM, Mahony JD, Hansen DJ, Scott KJ, Hicks MB, Mayr SM, Redmond MS. 1990. Toxicity of cadmium in sediments: The role of acid volatile sulfide. *Environ Toxicol Chem* 9:1487–1502.

Diamond ML. 1995. Application of a mass balance model to assess in-place arsenic pollution. *Environ Sci Technol* 29:29–42.

Donkin P, Widdows J. 1986. Ecotoxicology: Scope for growth as a measurement of environmental pollution and its interpretation using structure-activity relationships. *Chem Ind* Nov:732–737.

Donnelly MA, Crump ML. 1998. Potential effects of climate change on two neotropical amphibian assemblages. *Clim Change* 39:541–561.

Drake GJ, Moore JN. 1997. Long-term variability of trace metal concentrations in fine-grained sediment of the upper Clark Fork River, Montana [MS thesis]. Missoula MT, USA: Univ Montana.

Droop MR. 1973. Some thoughts on nutrient limitation in algae. *J Phycol* 9:264–272.

Droop MR. 1974. The nutrient status of algal cells in continuous culture. *J Mar Biol* 54:825–855.

Droop MR, Mickelson M.J, Scott JM, Turner MF. 1982. Light and nutrient status of algal cells. *J Mar Biol Assoc UK* 62:403–434.

Dunning JB, Stewart DJ, Danielson BJ, Noon BR, Root TL, Lamberson RH, Stevens EE. 1995. Spatially explicit population models: Current forms and future uses. *Ecol Appl* 5:3–11.

Dyer SD, Caprora RJ. 1997. A method for evaluating consumer product ingredient contributions to surface and drinking water: Boron as a test case. *Environ Toxicol Chem* 16:2070–2081.

Earll RC. 1992. Commonsense and the precautionary principle: An environmentalist's perspective. *Mar Pollut Bull* 24:182–186.

[EC] European Commission. 1996. Technical guidance document in support of commission directive 93/67/EEC on risk assessment for new notifed substances and the commission regulation (EC) 1488/94 on risk assessment for existing substances. Part II: Environmental risk assessment. Luxemburg: Office for the Official Publications of the European Community. XI/919/94-EN.

Eertman RHM, Groenink CLFMG, Sandee B, Hummel H, Smaal AC. 1995. Response of the blue mussel *Mytilus edulis* L. following exposure to PAHs or contaminated sediment. *Mar Environ Res* 39:169–174.

Eldridge PM, Jackson GA. 1993. Benthic trophic dynamics in California coastal basin and continental-slope communities inferred using inverse analysis. *Mar Ecol Prog Ser* 99:115–135.

Elliott NG, Ritz DA, Swain R. 1985. Interaction between copper and zinc accumulation in the barnacle *Elminius modestus* Darwin. *Mar Environ Res* 17:13–17.

Ellison AM. 1996. An introduction to Bayesian inference for ecological research and environmental decision-making. *Ecol Appl* 6:1036–1046.

Engle VD, Summers JK, Gaston GR. 1994. A benthic index of environmental condition of Gulf of Mexico estuaries. *Estuaries* 17:372–384.

Faber JH. 1991. Functional classification of soil fauna: A new approach. *Oikos* 62(1):110–117.

Fair PA, Sick LV. 1983. Accumulation of naphthalene and cadmium after simultaneous ingestion by the black sea bass, *Centropristis striata*. *Arch Environ Contam Toxicol* 12:551–557.

Fairweather PG. 1991. Statistical power and design requirements for environmental monitoring. *Austral Mar Freshw Res* 42:555–67.

Farr JA. 1977. Impairment of antipredator behavior in *Palaemonetes pugio* by exposure to sublethal doses of parathion. *Trans Am Fish Soc* 106:287–290.

Fietjel TC, Borije G, Matthies M, Young S, Morris G, Gandolfi C, Hansen B, Fox K, Koch V, Schroder R, Cassani G, Schowanek D, Rosenblom J, Niessen H. 1997. Development of a geography-referenced regional exposure assessment tool for European rivers: GREAT-ER. *Chemosphere* 34:235.

Fellers GM, Drost CA. 1993. Disappearance of the Cascades frog (*Rana cascadae*) at the southern end of its range, California, USA. *Biol Conserv* 65:177–181.

Ferson S, Ginzburg L, Slivers A. 1989. Extreme event risk analysis for age-structured populations. *Ecol Model* 47:175–183.

Fischer H. 1988. *Mytilus edulis* as a quantitative indicator of dissolved cadmium. Final study and synthesis. *Mar Ecol Prog Ser* 48:163–174.

Fisher NS, Hooke S. 1997. Effects of dietary siler on egg production in copepods. The 5th International Conference on Transport, Fate and Effects of Silver in the Environment [extended abstracts]. Toronto, Ontario, Canada.

Fisher R. 1992. Sediment interstitial water toxicity evaluations using *Daphnia magna* [MS thesis]. Dayton OH, USA: Wright State Univ.

Fisher SW, Hwang H, Atanasoff M, Landrum PF. 1999. Lethal body residues for pentachlorophenol in zebra mussels (*Dreissena plymorpha*) under varying conditions of temperature and pH. *Ecotoxicol Environ Saf* 43:274–283.

Fisk AT, Norstrom RJ, Cymbalisty CD, Muir DCG. 1998. Dietary accumulation and depuration of hydrophobic organochlorines: Bioaccumulation parameters and their relationship with the octanol/water partition coefficient. *Environ Toxicol Chem* 17(5):951–961.

Fjeld E, Haugen TO, Vollestad LA. 1998. Permanent impairment in the feeding behavior of grayling (*Thymallus thymallus*) exposed to methylmercury during embryogenesis. *Sci Total Environ* 213:247–254.

Foley P. 1994. Predicting extinction times from environmental stochasticity and carrying capacity. *Conserv Biol* 8:124–137.

Foran JA, Ferenc SA. 1999. Multiple stressors in ecological risk and impact assessment. Pensacola FL, USA: SETAC Pr.

Forbes VE. 2000. Is hormesis an evolutionary expectation? *Funct Ecol* 14:12–24.

Forbes VE, Calow P. 1999. Is the per capita rate of increase a good measure of population-level effects in ecotoxicology? *Environ Toxicol Chem* 18:1544–1556.

Forbes VE, Forbes TL. 1993. Ecotoxicology in theory and practice. London UK: Chapman & Hall.

Forget J, Pavillon JF, Beliaeff B, Bocquene G. 1999. Joint action of pollutant combinations (pesticides and metals) on survival (LC50 values) and acetylcholinesterase activity of *Tigriopus brevicornis* (Copepoda, harpacticoida). *Environ Toxicol Chem* 18:912–918.

Fox GA. 1993. What biomarkers told us about the effects of contaminants on the health of fish-eating birds in the Great Lakes: The theory and a literature review. *J Great Lakes Res* 19:722–736.

Fry DM, Toone CK. 1981. DDT-induced feminization of gull embryos. *Science* 213:222–224.

Furse MT, Moss D, Wright JF, Armitage PD. 1984. The influence of seasonal and taxonomic factors on the ordination and classification of running-water sites in Great Britain and on the prediction of their macro-invertebrate communities. *Freshw Biol* 14:257–280.

Gearing JN. 1989. The role of aquatic microcosms in ecotoxicological research as illustrated by large marine systems. In: Levin SA, Harwell MA, Kelly JR, Kimball KD, editors. Ecotoxicology: Problems and approaches. New York NY, USA: Springer-Verlag. p 411–470.

George SG. 1990. Biochemical and cytological assessments of metal toxicity in marine animals. In: Furness RW, Rainbow PS, editors. Heavy metals in the marine environment. Boca Ration FL, USA: CRC. p 123–132.

Gerritsen J, Jessup B, Leppo EW, White J. 1999. Development of a biological index for Florida lakes. Tallahassee FL, USA: Florida Department of Environmental Protection. Review draft.

Gibbs PE. Bryan GW. 1986. Reproductive failure in populations of the dog-whelk, *Nucella lapillus*, caused by imposex induced by tributyltin from anti-fouling paints. *J Mar Biol Assoc UK* 66:767–777.

Gibbs PE, Pascoe PL, Burt GR. 1988. Sex change in the female dog whelk, *Nucella lapillus*, imposex induced by tributyltin from antifouling paints. *J Mar Biol Assoc UK* 68:715–731.

Giddings JM. 1980. Types of aquatic mesocosms and their research applications. In: Giesy JP, editor. Microcosms in ecological research. Washington DC, USA: U.S. Department of Energy, Technical Information Center. DOE Symposium Series 52. p 248–266.

Giddings JM. 1981. Laboratory tests for chemical effects on aquatic population interactions and ecosystem properties. In: Hammons AS, editor. Methods for ecological toxicology: A critical review of laboratory multi-species tests. Ann Arbor MI, USA: Ann Arbor Science. p 25–91.

Giddings J, Brock T, Heger W, Heimbach F, Maund S, Norman S, Ratte T, Schaefers C, Stein B, Streloke M. 2002. Community Level Aquatic System Studies Interpretation Criteria (CLASSIC). Pensacola FL, USA: SETAC Pr (in press).

Gillman MP, Hails R. 1997. An introduction to ecological modelling: Putting practice into theory. Oxford UK: Blackwell Science.

Gillman MP, Silvertown J. 1997. Population extinction and the uncertainty of measurement. Proceedings of the JNN/BES Symposium on the role of genetics in conserving small populations.

Gilpin M. 1996. Forty-eight parrots and the origins of population viability analysis. *Conserv Biol* 10:1491–1493.

Gilpin ME, Soule ME. 1986. Minimum viable populations: The processes of species extinctions. In: Soule ME, editor. Conservation biology: The science of scarcity and diversity. Sunderland MD, USA: Sinauer Assoc. p 19–34.

Gimeno S, Gerritsen T, Bowmer T, Komen H. 1996. Feminization of male carp. *Nature* 384:221–222.

Ginzburg L, Ferson S, Akçakaya R. 1990. Reconstructibility of density dependence and the conservative assessment of extinction risks. *Conserv Biol* 4:63–70.

Ginzburg LR, Slobodkin LB, Johnson K, Bindman AG. 1982. Quasi-extinction probabilities as a measure of impact on population growth. *Risk Anal* 2:171–182.

Gleason TR, Bengtson DA. 1996a. Growth, survival and size-selective predation mortality of larval and juvenile inland silversides, *Menidia beryllina* (Pisces; Atherinidae). *J Exp Mar Biol Ecol* 199:165–177.

Gleason TR, Bengtson DA. 1996b. Size-selective mortality of inland silversides: Evidence from otolith microstructure. *Trans Am Fish Soc* 125:860–873.

Gobas FAPC. 1993. A model for predicting bioaccumulation of hydrophobic organic chemicals in aquatic food webs: Application to Lake Ontario. *Ecol Model* 69:1–17.

Goldingay R, Possingham H. 1995. Area requirements for viable populations of the Australian gliding marsupial *Petaurus australis*. *Biol Conserv* 73:161–167.

Gomes MC, Haedrich RL. 1992. Predicting community dynamics from food web structure. In: Rowe GT, Pariente W, editors. Deep-sea food chains and the global carbon cycle. *NATO ASI Ser* 360:277–293.

Goodland R, Daly H. 1996. Environmental sustainability: Universal and nonnegotiable. *Ecol Appl* 6:1002–1017.

Grant A. 1998. Population consequences of chronic toxicity: Incorporating density dependence into the analysis of life table response experiments. *Ecol Model* 105:325–335.

Gray Jr LE, Kelce WR. 1996. Latent effects of pesticides and toxic substances on sexual differentiation of rodents. *Toxicol Ind Health* 12:515–531.

Gray JS, Bewers JM. 1996. Towards a scientific definition of the Precautionary Principle. *Mar Pollut Bull* 32:768–771.

Gray MA, Metcalfe CD. 1997. Induction of testis-ova in Japanese medaka *Oryzias latipes* exposed to p-nonylphenol. *Environ Toxicol Chem* 16:1082–1086.

Greenberg M, Rowland C, Burton Jr GA, Hickey C, Stubblefield W, Clements W, Landrum P. 1998. Isolating individual stessor effects at sites with contaminated sediments and waters. Proceedings from Annual Meeting, Society of Environmental Toxicology and Chemistry. Pensacola FL, USA. Nr 166. p 25.

Greig-Smith P. 1983. Quantitative plant ecology. 3rd ed. Berkeley CA, USA: Univ California Pr.

Greig-Smith PW, Frampton GK, Hardy AR, editors. 1992. Pesticides, cereal farming and the environment. London UK: HMSO.

Griffiths BS, Ritz K, Bardgett RD, Cook R, Christensen S, Ekelund F, Sørensen S, Bååth E, Bloem J, De Ruiter PC, Dolfing J, Nicolardot B. 2000. Ecosystem response of pasture soil communities to fumigation-induced microbial diversity reductions: An examination of the biodiversity-ecosystem functioning relationship. *Oikos* 90(2):274–294.

Grothe DR, Dickson KL, Reed-Judkins DK, editors. 1996. Whole effluent toxicity testing: An evaluation of methods and prediction of receiving system impacts. Pensacola FL, USA: SETAC Pr.

Grue CE, Gibert PL, Seeley ME. 1997. Neurophysiological and behavioural changes in non-target wildlife exposed to organophosphate and carbamate pesticides: Thermoregulation, food consumption, and reproduction. *Am Zool* 37:369–388.

Grue CE, Shipley BK. 1984. Sensitivity of nestling and adult starlings to dicrotophos, an organophosphate pesticide. *Environ Res* 35:454–465.

Gu B, Schelske CL, Brenner M. 1996. Relationship between sediment and plankton isotope ratios and primary productivity in Florida lakes. *Can J Fish Aquat Sci* 53:875–883.

Guillette Jr LJ, Crain DA, Rooney AA, Pickford DB. 1995. Organization versus activation: The role of endocrine-disrupting contaminants (EDCs) during embryonic development in wildlife. *Environ Health Persp* 103(suppl 7):157–164.

Guillette Jr LJ, Gross TS, Masson TR, Matter JM, Percival HF, Woodward AR. 1994. Developmental anomalies of the gonad and abnormal sex hormone concentrations in juvenile alligators from contaminated and control lakes in Florida. *Environ Health Persp* 102:680–688.

Gurney WSC, Nisbet, M. 1998. Ecological dynamics. New York NY, USA: Oxford Univ Pr.

Gustafsson O, Haghseta F, Chan C, MacFarlane J, Gschwend PM. 1997. Quantification of the dilute sedimentary soot phase: Implications for PAH speciation and bioavailability. *Environ Sci Technol* 31:203–209.

Guttman SI. 1994. Population genetic structure and exotoxicology. *Environ Health Persp* 102(suppl):97–100.

Hairston Jr NG, Hairston Sr NG. 1993. Cause-effect relationships in energy flow, trophic structure, and interspecific interactions. *Am Nat* 142:379–411.

Hairston Jr NG, Li KT, Easter Jr SS. 1982. Fish vision and the detection of planktonic prey. *Science* 218:1240–1242.

Hairston Jr NG, Munns Jr WR. 1984. The timing of copepod diapause as an evolutionary stable strategy. *Am Nat* 123:733–751.

Hairston Jr NG, Olds EJ, Munns Jr WR. 1985. Bet-hedging and environmentally cued diapause strategies of diaptomid copepods. *Internationale Vereinigung fur Theoretische Angewandte Limnologie* 22:3170–3177.

Hairston Jr NG, Pastorok RA. 1975. Response of Daphnia population size and age structure to predation. *Verh Internat Verein Limnol* 19:2898–2905.

Hairston Jr NG, Walton WE, Li KT. 1983. The causes and consequences of sex-specific mortality in a freshwater copepod. *Limnol Oceanog* 28:935–947.

Halley JM, Iwasa Y. 1998. Extinction rate of a population under both demographic and environmental stochasticity. *Theor Popul Biol* 53:1–15.

Halley JM, Thomas CFG, Jepson PC. 1996. A model for the spatial dynamics of linyphiid spiders in farmland. *J Appl Ecol* 33:471–492.

Hamilton WD. 1967. Extraordinary sex ratios. *Science* 156:477–488.

Hanratty MP, Stay FS. 1994. Field evaluation of the Littoral Ecosystem Risk Assessment Model's predictions of the effects of Chlorpyrifos. *J Appl Ecol* 31:439–453.

Hansen F, Forbes VE, Forbes TL. 1999a. Effects of 4-n-nonylphenol on life-history traits and population dynamics of a polychaete. *Ecol Appl* 9:482–495.

Hansen F, Forbes VE, Forbes TL. 1999b. Using elasticity analysis of demographic models to link toxicant effects on individuals to the population level: An example. *Funct Ecol* 13:157–162.

Hanski I. 1991. Single-species metapopulation dynamics: Concepts, models and observations. *Biol J Linn Soc* 42:17–38.

Hanski I. 1994. A practical model of metapopulation dynamics. *J Anim Ecol* 63:151–162.

Hanski I. 1996. Metapopulation ecology. In: Rhodes OE, Chesser RK, Smith MH, editors. Population dynamics in ecological time and space. Chicago IL: USA: Univ Chicago Pr. p 13–43.

Hanski I. 1997. Metapopulation dynamics: From concepts and observations to predictive models. In: Hanski I, Gilpin M, editors. Metapopulation biology: Ecology, genetics and evolution. San Diego CA, USA: Academic Pr. p 69–72.

Hanski I, Gilpin M. 1991. Metapopulation dynamics: A brief history and conceptual domain. *Biol J Linnean Soc* 42:3–16.

Harrison S, Quinn JF, Baughman JF, Murphy DD, Ehlich PR. 1991. Estimating the effects of scientific study on 2 butterfly populations. *Am Nat* 137:222–243.

Hart DD. 1994. Building a stronger partnership between ecological research and biological monitoring. *J N Am Benthol Soc* 13:110–116.

Harte J, Hoffman E. 1989. Possible effects of acidic deposition in a Rocky Mountain population of the tiger salamander *Ambystoma tigrinum*. *Conserv Biol* 3:149–157.

Hatch AC, Burton Jr GA. 1998. Effects of photoinduced toxicity of fluoranthene on amphibian embryos and larvae. *Environ Toxicol Chem* 17:1777–1785.

Hawker DW, Connell DW. 1989. A simple water-octanol partition system for bioconcentration investigations. *Environ Sci Technol* 23:961–965.

Hawkins CP, MacMahon JA. 1989. Guilds: The multiple meanings of a concept. *Ann Rev Entomol* 34:423–451.

[HCN] Health Council of the Netherlands, Standing Committee on Ecotoxicology. 1997. The food web approach in ecological risk assessment. Rijswijk, Netherlands: HCN. Publication Nr 1997/14E. 75 p.

Hebert CE, Shutt JL, Norstrom RJ. 1997. Dietary changes cause temporal fluctuations in polychlorinated biphenyl levels in herring gull eggs from Lake Ontario. *Environ Sci Technol* 31:1012–1017.

Hecky RE, Hesslein RH. 1995. Contributions of benthic algae to lake food webs as revealed by stable isotope analysis. *J N Am Benthol Soc* 14(4):631–653.

Heimbach F. 1987. Fate and biological effects of Baythroid in an aquatic ecosystem. Proceedings, 11th International Congress of Plant Protection, Vol. 1; Manila, Philippines; 5.-9.19.1987. p 388–392.

Heimbach F, Pflueger W, Ratte HT. 1992. Use of small artifical ponds for assessment of hazards to aquatic ecosystems. *Environ Toxicol Chem* 11:27–34.

Hendrix PF, Crossley DA Jr, Coleman DC, Parlemee RW, Beare MH. 1987. Carbon dynamics in soil microbes and fauna in conventional and no-tillage agroecosystems. *INTECOL Bulletin* 15:59–63.

Hendrix PF, Parmelee RW, Crossley Jr DA, Coleman DC, Odum EP, Groffman PM. 1986. Detritus food webs in conventional and no-till agroecosystems. *Bioscience* 36:374–380.

Heneghan PA, Biggs J, Jepson PC, Kedwards TJ, Maund SJ, Sherratt TN, Shillabeer N, Stickland T, Williams P. 1999. Pond-FX: An Internet tool for exploring invertebrate population recovery. In: Kedwards TJ, Thomas M, editors. Challenges in applied population biology. Aspects of Applied Biology 53. Wellesbourne UK: Assoc Applied Biologists. p 263–270.

Hengeveld R. 1985. Dynamics of Dutch beetle species during the twentieth century (Coleoptera: Carabidae). *J Biogeogr* 12:389–411.

Hesslein RH, Hallard KA, Ramlal P. 1993. Replacement of sulfur, carbon and nitrogen in tissue of growing broad whitefish (*Coregonus nasus*) in response to a change in diet traced by delta34S, delta13C and delta15N. *Can J Fish Aquat Sci* 50:2071–2076.

Hill AB. 1965. The environment and disease: Association or causation? *Proc Royal Soc Med* 58:295–300.

Hobson KA, Sealey SG. 1991. Marine protein contributions to the diet of Northern Saw-whet owls on the Queen Charlotte Islands: A stable-isotope approach. *The Auk* 108(2):437–440.

Holland AF, editor. 1990. Near coastal program plan for 1990: Estuaries. Narragansett RI, USA: U.S. Environmental Protection Agency (USEPA) Office of Research and Development. EPA/600/4-90/033.

Holland JM, Frampton GK, Cilgi T, Wratten SD. 1994 Arable acronyms analyzed: A review of integrated arable farming systems research in W. Europe. *Ann Appl Biol* 125:399–438.

Hommen U. 1998. Ökosystemmodelle von Modellökosystemen - Simulationen und Mikrokosmen in der Risikoanalyse von Fremdstoffen in aquatischen Ökosystemen [PhD thesis]. Shaker, Aachen, Germany: Aachen Univ Technology. 200 p.

Hommen U, Poethke HJ, Duelmer U, Ratte HT. 1993. Simulation models to predict ecological risk of toxins in freshwater systems. *ICES J Mar Sci* 50:337–247.

Hopkins WA, Mendoca MT, Rowe CL, Congon JD. 1998. Elevated trace element concentrations in southern toads, *Bufo terrestris*, exposed to coal combustion waste. *Arch Environ Contam Toxicol* 35:325–329.

Hornberger MI, Lambing JH, Luoma SN, Axtmann V. 1997. Spatial and temporal trends of trace metals in water, bed sediment, and biota of the Upper Clark Fork Basin, Montana: 1985–1995. Menlo Park CA, USA: U.S. Geological Survey (USGS). Open File Report 97-669.

Hornberger MI, Luoma S, Cain D, Parchaso F, Brown C, Bouse R, Wellise C, Thompson J. 1999. Bioaccumulation of metals by the bivalve *Macoma balthica* at a site in South San Francisco Bay between 1977 and 1997: Long-term trends and associated biological effects. Menlo Park CA, USA: U.S. Geological Survey (USGS). Open File Report 97-699. 170 p.

Hornberger MI, Luoma SN, Cain DJ, Parchaso F, Brown CL, Bouse RM, Wellise C, Thompson JK. 2000. Linkage of bioaccumulation and biological effects to changes in pollutant loads in South San Francisco Bay. *Environ Sci Technol* 34:2401–2409.

Horne MT, Dunson WA. 1994. Exclusion of the Jefferson salamander, *Ambystoma jeffersonianum*, from some potential breeding ponds in Pennsylvania: Effects of pH, temperature and metals on embryonic development. *Arch Environ Contam Toxicol* 27:323–330.

Horne MT, Dunson WA. 1995a. Effects of low pH, metals and water hardness on larval amphibians. *Arch Environ Contam Toxicol* 29:500–505.

Horne MT, Dunson WA. 1995b. The interactive effects of low pH, toxic metals, and DOC on a simulated temporary pond community. *Environ Pollut* 89:155–161.

Howarth RW. 1991. Comparative responses of aquatic ecosystems to toxic chemical stress. In: Cole J, Lovett G, Findlay S, editors. Comparative analyses of ecosystems: Patterns, mechanisms, and theories. New York NY, USA: Springer-Verlag.

Howe GE, Marking LL, Bills TD, Rach JJ, Mayer FL. 1994. Effects of water temperature and pH on toxicity of terbufos, trichlorfon, 4-nitrophenol and 2-dinitrophenol to the amphipod (*Gammarus pseudolimnaeus*) and rainbow trout (*Onchorynchus mykiss*). *Environ Toxicol Chem* 13:51–66.

Huckins JN, Manuweera GK, Petty JD. 1993. Lipid-containing semipermeable-membrane devices for monitoring organic contaminants in water. *Environ Sci Technol* 27:2489–2496.

Hugget RJ, Van Veld PA, Smith CL, Hargin Jr WJ, Vogelbein WK, Weeks BA. 1992. The effects of contaminated sediments in the Elizabeth River. In: Burton GA, editor. Sediment toxicity assessment. Boca Raton FL, USA: Lewis. p 403–430.

Hughes RM. 1995. Defining acceptable biological status by comparing with reference conditions. In: Davis WS, Simon TP, editors. Biological assessment and criteria. Tools for water resource planning and decision making. Boca Raton FL, USA: Lewis. p 31–47.

Hunt HW, Coleman DC, Ingham ER, Ingham RE, Elliott ET, Moore JC, Rose SL, Reid CPP, Morley CR. 1987. The detrital food web in a shortgrass prairie. *Biol Fertil Soil* 3:57–68.

Hunter MD, Price PW. 1992. Playing chutes and ladders: Heterogeneity and the relative roles of bottom-up and top-down forces in natural communities. *Ecology* 73:724–731.

Hurlbert SH. 1984. Pseudoreplication and the design of ecological field experiments. *Ecol Monogr* 54:187–211.

Hutchinson GE. 1957. Concluding remarks. Cold Spring Harbor. *Symp Quant Biol* 22:415–427.

Ireland DS, Burton Jr GA, Hess GG. 1996. In-situ toxicity evaluations of turbidity and photoinduction of polycyclic aromatic hydrocarbons. *Environ Toxicol Chem* 15:574–581.

Jackson GA, Eldridge PM. 1992. Foodweb analysis of a planktonic system off southern California. *Prog Oceanogr* 30:223–251.

Jak RG. 1997. Toxicant-induced changes in zooplankton communities and consequences for phytoplankton development [doctoral thesis]. Amsterdam, Netherlands: Vrije Universiteit. 144 p.

Janse JH, Traas TP. 1996. Microcosms simulated: Modelling the combined effects of nutrients and an organophosphorus insecticide. In: Kramer PRG, Jonkers DA, Van Liere L, editors. Interactions of nutrients and toxicants in the foodchain of aquatic ecosystems. Bilthoven, Netherlands: RIVM. Report 703715001. p 63–69.

Janse JH, Van Donk E, Gulati RD, 1995. Modelling nutrient cycles in relation to food web structure in a biomanipulated shallow lake. *Hydrobiologia* 304:29–38.

Jepson PC, Sherratt TN. 1996. The dimensions of space and time in the assessment of ecotoxicological risks. In: Baird DJ, Maltby L, Greig-Smith PW, Douben PET, editors. ECOtoxicology: Ecological dimensions. London UK: Chapman & Hall. p 43–54.

Jepson PC, Thacker JRM. 1990. Analysis of the spatial component of pesticide side-effects on non-target invertebrate populations and its relevance to hazard analysis. *Funct Ecol* 4:349–355.

Jobling S, Sheahan D, Osborne JA, Matthiessen P, Sumpter JP. 1996. Inhibition of testicular growth in rainbow trout *Oncorhynchus mykiss* exposed to estrogenic alkylphenolic chemicals. *Environ Toxicol Chem* 15:194–202.

Johansson C, Cain DJ, Luoma SN. 1985. Variability in the fractionation of Cu, Ag and Zn among cytosolic proteins in the bivalve *Macoma balthica. Mar Ecol Progress Series* 28:87–97.

Johnson RK. 1998. Spatiotemporal variability of temperate lake macroinvertebrate communities: Detection of impact. *Ecol Appl* 8:61–70.

Johnson RK, Goedkoop W. 1999. The 1995 national survey of Swedish lakes and streams: Assessment of ecological status using macroinvertebrates. In: Sutcliffe DW, editor. Proceedings, RIVPACS International Workshop, Jesus College, UK. Freshwater Biological Assoc.

Jones JC, Reynolds JD. 1997. Effects of pollution on reproductive behaviour of fishes. *Rev Fish Biol Fisheries* 7:463–491.

Jones LA, Hajek RA. 1995. Effects of estrogenic chemicals on development. *Environ Health Persp* 103(suppl 7):63–67.

Jung RE, Jagoe CH. 1995. Effects of low pH and aluminum on body size, swimming performance, and susceptibility to predation of green tree frog (*Hyla cinerea*) tadpoles. *Can J Zool* 73:2171–2183.

Kaag NHBM, Foekema EM, Scholten MCTh, van Straalen NM. 1997. Comparison of contaminant accumulation in three species of marine invertebrates with different feeding habits. *Environ Toxicol Chem* 16:837–842.

Kalmus H, Smith CAB. 1960. Evolutionary origin of sexual differentiation and the sex-ratio. *Nature* 186:1004–1006.

Kammenga JE, Busschers M, van Straalen NM, Jepson PC, Bakker J. 1996. Stress induced fitness reduction is not determined by the most sensitive life-cycle trait. *Funct Ecol* 10:106–111.

Kammenga JE, Korthals GW, Bongers AMT, Bakker J. 1997. Reaction norms for life-history traits as the basis for the evaluation of critical effect levels of toxicants. In: Van Straalen NM, Løkke H, editors. Ecological principles for risk assessment of contaminants in soil. London UK: Chapman & Hall. p 293–304.

Kapustka LA. 1996. Plant ecotoxicology: The design and evaluation of plant performance in risk assessments and forensic ecology. In: La Point TW, Price FT, Little EE, editors. Environmental toxicology and risk assessment, Volume 4. Conshohocken PA, USA: American Society for Testing and Materials (ASTM). ASTM STP 1262. p 110–121.

Kapustka LA, Landis WG. 1998. Ecology: The science versus the myth. *Hum Ecol Risk Assess* 4:829–838.

Kareiva PK, Sahakian R. 1990. Tritrophic effects of a simple architectural mutation in pea plants. *Nature* 345:433–434.

Kareiva P, Stark J, Wennergren U. 1996. Using demographic theory, community ecology and spatial models to illuminate ecotoxicology. In: Baird DJ, Maltby L, Greig-Smith PW, Douben PET, editors. ECOtoxicology: Ecological dimensions. London UK: Chapman & Hall. p 13–23.

Karr JR, Fausch KD, Angermeier PL, Yant PR, Schlosser LJ. 1986. Assessing biological integrity in running waters: A method and its rationale. Champaign IL, USA: Illinois Natural History Survey. Special publication 5.

Kerans BL, Karr JR. 1994. A benthic index of biotic integrity (B-IBI) for rivers of the Tennessee Valley. *Ecol Appl* 4:768–785.

Kerr JB, McElroy CT. 1993. Evidence for large upward trends of ultraviolet-B radiation linked to ozone depletion. *Science* 22:1032–1034.

Kersting K. 1994. Functional endpoints in field testing. In: Hill IA, Heimbach F, Leeuwangh P, Matthiessen P, editors. Freshwater field tests for hazard assessment of chemicals. Boca Raton FL, USA: Lewis.

Kidd KA. 1998. Use of stable isotope ratios in freshwater and marine biomagnification studies. In: Rose J, editor. Environmental toxicology: Current developments. Berkshire UK: Gordon and Breach Science. p 359–378.

Kidd KA, Hesslein RH, Ross BJ, Koczanski K, Stephens GR, Muir DCG. 1998. Bioaccumulation of organochlorines through a remote freshwater food web in the Canadian Arctic. *Environ Pollut* 102:91–103.

Kidd KA, Paterson MJ, Hesslein RH, Muir DCG, Hecky RE. 1999. Effects of pike additions on food web structure and pollutant accumulation in biota from a eutrophic and an oligotrophic lake. *Can J Fish Aquat Sci* 56:1–10.

Kidd KA, Schindler DW, Muir DCG, Lockhart WL, Hesslein RH. 1995. High concentrations of toxaphene in fishes from a subarctic lake. *Science* 269:240–242.

Kiesecker JM. 1995. pH-mediated predator-prey interactions between *Ambystoma tigrinum* and *Pseudacris triseriata*. *Ecol Appl* 6:1325–1331.

Kiesecker JM, Blaustein AR. 1995. Synergism between UV-B and a pathogen magnifies amphibian embryo mortality in nature. *Proc Natl Acad Sci* 92:11049–11052.

Kiffney PM, Clements WH. 1993. Bioaccumulation of heavy metals in benthic organisms of a montane stream in Colorado, USA. *Environ Toxicol Chem* 12:1507–1517.

Kiffney PM, Clements WH. 1996. Effects of heavy metals on stream macroinvertebrate assemblages from different elevations. *Ecol Appl* 6:472–481.

Kiffney PM, Little EE, Clements WH. 1997. Influence of ultraviolet-B radiation on the drift response of stream invertebrates. *Freshw Biol* 37:485–492.

Kiriluk RM, Servos MR, Whittle DM, Cabana G, Rasmussen JB. 1995. Using ratios of stable nitrogen and carbon isotopes to characterize the biomagnification of DDE, mirex, and PCB in a Lake Ontario pelagic food web. *Can J Fish Aquat Sci* 52:2660–2674.

Klerks PL, Levinton JS. 1989. Effects of heavy metals in a polluted aquatic ecosystem. In: Levin SA, Harwell MA, Kelly JR, Kimball KD, editors. Ecotoxicology: Problems and approaches. New York NY, USA: Springer Verlag. p 41–67.

Klok C, de Roos AM. 1996. Population level consequences of toxicological influences on individual growth and reproduction in *Lumbricus rubellus* (Lumbricidae, Oligochaeta). *Ecotoxicol Environ Saf* 33:118–127.

Klok C, de Roos AM, Broekhuizen S, Van Apeldoorn RC. 2000. Effects of heavy metals on the badger Meles meles, interactions between habitat quality and fragmentation. In: Kammenga JE, Laskowski R, editors. Demography in ecotoxicology. Sussex UK: Wiley.

Knieb RY. 1986. The role of *Fundulus heteroclitus* in salt marsh trophic dynamics. *Amer Zool* 26:259–269.

Knowlton MF, Jones JR. 1989. Comparison of surface and depth-integrated composite samples for estimating algal biomass and phosphorus values and notes on the vertical distribution of algae and photosynthetic bacteria in midwestern lakes. *Arch Hydrobiol Suppl* 83:175–96.

Köhler H-R, Belitz B, Eckwert H, Adam R, Rahman B, Trontelj P. 1998. Validation of hsp70 stress gene expression as a marker of metal effects in *Deroceras reticulatum* (Pulmonata): Correlation with demographic parameters. *Environ Toxicol Chem* 17:2246–2253.

Kolaja GJ. 1977. The effects of DDT, DDE, and their sulfonated derivatives on eggshell formation in the mallard duck. *Bull Environ Contam Toxicol* 17:697–701.

Konneman H. 1981. Fish toxicity tests with mixtures of more than 2 chemicals. A proposal for a quantitative approach and experimental results. *Toxicology* 19:223–228.

Kooijman SALM. 1987. A safety factor for LC50 values allowing for differences in sensitivity between species. *Water Res* 21:269–276.

Kooijman SALM. 1993. Dynamic energy budgets in biological systems. Cambridge UK: Cambridge Univ Pr.

Kooijman SALM, Bedaux JJM. 1996. The analysis of aquatic toxicity data. Amsterdam, Netherlands: VU Univ Pr.

Kooijman SALM, Metz JAJ. 1984. On the dynamics of chemically stressed populations: The deduction of population consequences from effects on individuals. *Ecotoxicol Environ Saf* 8:254–274.

Korthals GW, Alexiev AD, Lexmond TM, Kammenga JE, Bongers T. 1996. Long-term effects of copper and pH on the nematode community in an agroecosystem. *Environ Toxicol Chem* 15:979–985.

Kraak MHS, Lavy D, Schoon H, Toussaint M, Peeters WHM, van Straalen NM. 1994. Exotoxicity of mixtures of metals to the zebra mussel *Dreissena polymorpha*. *Environ Toxicol Chem* 13:109–114.

Kraus ML, Kraus DB. 1986. Differences in effects of mercury on predator avoidance in two populations of grass shrimp, *Palaemonetes pugio*. *Mar Environ Res* 18:77–89.

Krebs CJ. 1978. Ecology: The experimental analysis of distribution and abundance. 2nd ed. New York NY, USA: Harper & Row.

Krebs RA, Loescheke V. 1994. Costs and benefits of activation of the heat-shock response in *Drosophila melanogaster*. *Funct Ecol* 8:730–737.

Kremen C. 1992. Assessing the indicator properties of species asemblages for natural areas monitoring. *Ecol Appl* 2:203–217.

Kremen C. 1994. Biological inventory using target taxa: A case study of the butterflies of Madagascar. *Ecol Appl* 4:407–422.

Kremen C, Colwell RK, Erwin TL, Murphy DD, Noss RF, Sanjayan MA. 1993. Terrestrial arthropod assemblages: Their use in conservation planning. *Conserv Biol* 7:796–808.

Kroon de H, Plaisier A, van Groenendael J, Caswell H. 1986. Elasticity: The relative contribution of demographic parameters to population growth rate. *Ecology* 67:1427–1431.

Kucklick JR, Baker JE. 1998. Organochlorines in Lake Superior's food web. *Environ Sci Technol* 32:1192–1198.

Kuhn A, Munns Jr WR, Champlin D, McKinney R, Tagliabue M, Serbst J, Gleason T. 2001. Evaluation of the efficacy of extrapolation population modeling to predict the dynamics of *Americamysis bahia* populations in the laboratory. *Environ Toxicol Chem* 20:213–221.

Kuhn A, Munns Jr WR, Poucher S, Champlin D, Lussier S. 2000. Prediction of population-level response from mysid toxicity test data using population modeling techniques. *Environ Toxicol Chem* 19:2364–2371.

Kutka FJ. 1994. Low pH effects of swimming activity of *Ambystoma* salamander larvae. *Environ Toxicol Chem* 13:1821–1824.

Lamberson RH, McKelvey R, Noon BR, Voss C. 1992. A dynamic analysis of northern spotted owl viability in a fragmented forest landscape. *Conserv Biol* 6:505–512.

Lande R. 1993. Risks of population extinction from demographic and environmental stochasticity and random catastrophes. *Am Nat* 142:911–927.

Lande R. 1998a. Anthropogenic, ecological and genetic factors in extinction. In: Mace GM, Balmford A, Ginsberg JR, editors. Conservation in a changing world. New York NY, USA: Cambridge Univ Pr.

Lande R. 1998b. Anthropogenic, ecological and genetic factors in extinction and conservation. *Res Population Ecol* 40:259–269.

Lande R, Orzack SH. 1988. Extinction dynamics of age-structured populations in a fluctuating environment. *Proc Nat Acad Sci USA* 85:7418–7421.

Landres PB, Verner J, Thomas JW. 1988. Ecological uses of vertebrate indicator species. *Conserv Biol* 2:316–329.

Larsson P, Okla L, Woln P. 1990. Atmospheric transport of persisten pollutants governs uptake by Holarctic terrestrial biota. *Environ Sci Technol* 24(10):1599–1601.

Laughlin R. 1965. Capacity for increase: A useful population statistic. *J Anim Ecol* 34:77–91.

Lavie B, Nevo E. 1982. Heavy metal selection of phosphoglucose isomerase allozymes in marine gastropods. *Mar Biol* 71:17–22.

Lavie B, Nevo E, Zoller U. 1984. Differential ability of phosphoglucose isomerase allozyme genotypes of marine snails in nonionic detergent and crude oil-surfactant mixtures. *Environ Res* 35:270–276.

Lavigne DM. 1995. Interaction between marine mammals and their prey. In: Montevecchi WA, editor. Studies of high-latitude hoemotherms in cold ocean systems. Guelph, Ontario, Canada: International Marine Mammal Assoc. Technical Report 95–02.

Lawton JH. 1989. Food webs. In: Cherrett JM, editor. Ecological concepts: The contribution of ecology to an understanding of the natural world. Oxford UK: Blackwell Scientific. p 43–78.

Lee BG, Luoma S. 1998. Influence of microalgal biomass on absorption efficiency of Cd, Cr, and Zn by two bivalves from San Francisco Bay. *Limnol Oceanogr* 43:1455–1466.

Lefkovitch LP. 1965. The study of population growth in organisms grouped by stages. *Biometrics* 21:1–18.

Lemly D. 1996. Wastewater discharges may be most hazardous to fish during winter. *Environ Pollut* 93:169–174.

Leslie PH. 1945. On the use of matrices in population mathematics. *Biometrika* 33:183–212.

Levin L, Caswell H, Bridges T, Dibacco C, Cabrera D, Plaia G. 1996. Demographic responses of estuarine polychaetes to sewage, algal, and hydrocarbon additions: Life-table response experiments. *Ecol Appl* 6:1295–1313.

Levin SA, Harwell MA, Kelly JR, Kimball KD, editors. 1989. Ecotoxicology: Problems and approaches. New York NY, USA: Springer-Verlag.

Levins R. 1969. Some demographic and genetic consequences of environmental heterogeneity for biological control. Princeton NJ, USA: Princeton Univ Pr.

Lindeman RL. 1942. The trophic-dynamic aspect of ecology. *Ecology* 23:401–418.

Linke-Gamenick I, Forbes VE, Sibly RM. 1999. Density-dependent effects of a toxicant on life-history traits and population dynamics of a capitellid polychaete. *Mar Ecol Prog Ser* 184:139–148.

Linthurst RA, Landers DH, Eilers JM, Brakke DF, Overton WS, Meier EP, Crowe RE. 1986. Characteristics of lakes in the eastern United States. Volume I, Population descriptions and physico-chemical relationships. Las Vegas NV, USA: U.S. Environmental Protection Agency (USEPA). EPA 600/4-86/007a.

Little EE, Archeski RD, Flerov B, Kozlovskaya V. 1990. Behavioral indicators of sublethal toxicity in rainbow trout. *Arch Environ Contam Toxicol* 19:380–385.

Long ER, Field LJ, MacDonald DD. 1998. Predicting toxicity in marine sediments with numerical sediment quality guidelines. *Environ Toxicol Chem* 17:714–727.

Long LE, Saylor LS, Soule ME. 1995. A pH/UV-B synergism in amphibians. *Conserv Biol* 9:1301–1303.

Lubchenco J, Olson AM, Brubaker LB, Carpenter SR, Holland MM, Hubbell SP, Levin SA, MacMahon JA, Matson PA, Melillo JM, Mooney HA, Peterson CH, Pulliam HR, Real LA, Regal PJ, Risser PG. 1991. The sustainable biosphere initiative: An ecological research initiative. *Ecology* 72:371–412.

Lucasz LV, Koseff JR, Monismith SG, Cloern JE, Thompson JK. 1999a. Processes governing phytoplankton blooms in estuaries. I: The local production-loss overning phytoplankton blooms in estuaries. *Mar Ecol Prog Ser* 187:1–15.

Lucasz LV, Koseff JR, Monismith SG, Cloern JE, Thompson JK. 1999b. Processes governing phytoplankton blooms in estuaries II: The role of horizontal transport. *Mar Ecol Prog Ser* 187:17–30.

Lundholm CE. 1987. Thinning of eggshells in birds by DDE: Mode of action on the eggshell gland. *Comp Biochem Physiol* 88C:1–22.

Lundholm CE. 1993. Inhibition of prostaglandin synthesis in eggshell gland mucosa as a mechanism for *p,p'*-DDE-induced eggshell thinning in birds. A comparison of ducks and domestic fowls. *Comp Biochem Physiol* 106C:389–394.

Luoma SN. 1977. Detection of trace contaminant effects in aquatic ecosystems. *J Fish Res Bd Can* 34:436–439.

Luoma SN. 1996. The developing framework of marine ecotoxicology: Pollutants as a variable in marine ecosystems? *J Exp Mar Biol Ecol* 200:29–55.

Luoma SN, Bryan GW. 1982. A statistical study of environmental factors controlling concentrations of heavy metals in the burrowing bivalve *Scrobicularia plana* and the polychaete *Nereis diversicolor*. *Estuar Coastal Shelf Sci* 15:95–108.

Luoma SN, Carter JL. 1991. Effects of trace metals on aquatic benthos. In: Newman MC, McIntosh AW, editors. Metal ecotoxicology: Concepts and applications. Chelsea MI, USA: Lewis. p 261–300.

Luoma SN, Fisher NS. 1997. Uncertainties in assessing contaminant exposure from sediments: Bioavailability. In: Ingersoll C, Biddinger G, Dillon T, editors. Ecological risk assessment of contaminated sediments. Pensacola FL, USA: SETAC Pr. p 211–237.

Luoma SN, Johns C, Fisher NS, Steinberg NA, Oremland RS, Reinfelder J. 1992. Determination of selenium bioavailability to a benthic bivalve from particulate and solute pathways. *Environ Sci Technol* 26:485–491.

Luoma SN, Phillips DJH. 1988. Distribution, variability and impacts of trace elements in San Francisco Bay. *Mar Pollut Bull* 19:413–425.

Luoma SN, van Geen A, Lee BG, Cloern JE. 1998. Metal uptake by phytoplankton during a bloom in south San Francisco Bay: Implications for metal cycling in estuaries. *Limnol Oceanogr* 43:1007–1016.

Mackay D. 1982. Correlation of bioconcentration factors. *Environ Sci Technol* 16:274–278.

Mackay D. 1994. Fate models. In: Calow P, editor. Volume II, The handbook of ecotoxicology. Oxford UK: Blackwell Scientific.

Mackay D, Diamond M. 1989. Application of the QWASI (quantitative water air sediment interaction) fugacity model to the dynamics of organic and inorganic chemicals in lakes. *Chemosphere* 18:1343–1365.

Mackay D, DiGuardo A, Paterson S, Cowan CE. 1996. Evaluating the environmental fate of a variety of types of chemicals using the EQC model. *Environ Toxicol Chem* 15:1627–1637.

Mackay D, DiGuardo A, Paterson S, Kicsi G, Cowan CE. 1996. Assessing the fate of new and existing chemicals: A five-stage process. *Environ Toxicol Chem* 15:618–1626.

Macklin MG, Hudson-Edwards KA, Dawson EJ. 1997. The significance of pollution from historic metal mining in the Pennine orefields on river sediment contaminant fluxes to the North Sea. *Sci Total Environ* 194/195:391–397.

Madenjian CP, Carpenter SR, Rand PS. 1994. Why are the PCB concentrations of salmonine individuals from the same lake so highly variable? *Can J Fish Aquat Sci* 51:800–807.

Mahaney PA. 1994. Effects of freshwater petroleum contamination on amphibian hatching and metamorphosis. *Environ Toxicol Chem* 13:259–265.

Malley DF. 1996. Cadmium whole-lake experiment at the Experimental Lakes Area: An anachronism. *Can J Fish Aquat Sci* 53:1862–1870.

Maltby L. 1991. Pollution as a probe of life-history adaptation in *Asellus aquaticus* (Crustacea; Isopoda). *Oikos* 61:11–18.

Maltby L. 1994. Stress, shredders and streams: Using Gammarus energetics to assess water quality. In: Sutcliffe DW, editor. Water quality and stress indicators in marine and freshwater systems: Linking levels of organisation. Ambleside UK: Freshwater Biological Assoc. p 98–110.

Maltby L. 1999. Studying stress: The importance of organism-level response. *Ecol Appl* 9:431–440.

Maltby L, Naylor C, Calow P. 1990. Field deployment of a scope for growth assay involving *Gammarus pulex*, a freshwater benthic invertebrate. *Ecotoxicol Environ Saf* 19:292–300.

Marco A, Blaustein AR. 1999. The effects of nitrate on behavior and metamorphosis in Cascades frogs (*Rana cascadae*). *Environ Toxicol Chem* 18:946–949.

Marshall JS. 1962. The effects of continuous gamma radiation on the intrinsic rate of natural increase of *Daphnia pulex*. *Ecology* 43:598–607.

Martin M, Ichikawa G, Goetzl J, de los Reyes M, Stephenson MK. 1984. Relationships between physiological stress and trace toxic substances in the bay mussel, *Mytilus edulis* from San Francisco Bay, California. *Mar Environ Res* 11:91–110.

Matthaei CD, Uehlinger U, Meyer EI, Frutiger A. 1996. Recolonization by benthic invertebrates after experimental disturbance in a Swiss prealpine river. *Freshw Biol* 35:233–48.

Maurer D, Nguyen H, Robertson G, Gerlinger T. 1999. The infaunal trophic index (ITI): Its suitability for marine environmental monitoring. *Ecol Appl* 9:699–713.

Maxted JR, Barbour MT, Gerritsen J, Poretti V, Primrose N, Silvia A, Penrose D, Renfrow R. A framework for assessing mid-Atlantic coastal plain streams using benthic macroinvertebrates. *J N Am Benthol Soc* (in press).

May RM. 1972. Will a large complex system be stable? *Nature* 238:413–414.

May RM. 1973. Stability and complexity in model ecosystems. 2nd ed. Princeton NJ, USA: Princeton Univ Pr.

Maynard Smith J. 1978. The evolution of sex. New York NY, USA: Cambridge Univ Pr.

McGroddy SE, Farrington JW. 1995. Sediment porewater polynuclear aromatic hydrocarbons in three cores from Boston harbor, Massachusetts. *Environ Sci Technol* 29:1542–1550.

McIntosh RP. 1985. The background of ecology: Concept and theory. New York NY, USA: Cambridge Univ Pr.

McKenzie R, Connor B, Bodeker G. 1999. Increased summertime UV radiation in New Zealand in response to ozone loss. *Science* 285:1709–1711.

McLachlan MS. 1996. Bioaccumulation of hydrophobic chemicals in agricultural food chains. *Environ Sci Technol* 30(1):252–259.

McNaughton SJ. 1994. Biodiversity and function of grazing ecosystems. In: Mooney ED, Schulze HA, editors. Biodiversity and ecosystem function. London UK: Springer-Verlag. p 361–383.

Meador J. 1993. The effect of laboratory holding on the toxicity response of marine infaunal amphipods to cadmium and tributyltin. *J Exp Mar Biol Ecol* 174:227–242.

Medley CN, Clements WH. 1998. Responses of diatom communities to heavy metals in streams: The influence of longitudinal variation. *Ecol Appl* 8:631–644.

Meistrell JC, Montagne DE. 1983. Waste disposal in southern California and its effects on the rocky subtidal habitat. In: The effects of waste disposal on kelp communities. San Diego CA, USA: California Sea Grant College Program Publication. p 84–102.

Menge BA, Sutherland JP. 1976. Species diversity gradients: Synthesis of the roles of predation, competition, and temporal heterogeneity. *Am Nat* 110:351–369.

Menzie C, Henning MH, Cura J, Finkelstein K, Gentile J, Maughan J, Mitchell D, Petron S, Potocki B, Svirsky S, Tyler P. 1996. A weight-of-evidence approach for evaluating ecological risks: Report of the Massachusetts Weight-of-Evidence Work Group. *Human Ecol Risk Assess* 2:277–304.

Meyer AH, Schmidt BR, Grossenbacher K. 1998. Analysis of three amphibian populations with quarter-century long time-series. *Proc Roy Soc Lond* (B) 265:523–528.

Mikola J, Setälä H. 1998. Relating species diversity to ecosystem functioning: Mechanistic backgrounds and experimental approach with a decomposer food web. *Oikos* 83:180–194.

Mitsch WJ, Jørgensen SE. 1989. Ecological engineering: An introduction to ecotechnology. New York NY, USA: Wiley.

Moller V, Forbes VE, Depledge MH. 1994. Influence of acclimation and exposure temperature on the acute toxicity of cadmium to the freshwater snail *Potamopyrgus antipodarum* (Hydrobiidae). *Environ Toxicol Chem* 13:1519–1524.

Moore CG, Stevenson JM. 1991. The occurrence of intersexuality in harpacticoid copepods and its relationship to pollution. *Mar Poll Bull* 22:72–74.

Moore JC, De Ruiter PC. 1987a. A food web approach to assess the effects of disturbance on ecosystem structure, function and stability. In: Løkke NM. van Straalen H, editors. Ecological risk assessment of contaminants in soil. London UK: Chapman & Hall.

Moore JC, De Ruiter PC. 1987b. Compartmentalization of resource utilization within soil ecosystems. In: Brown AC, Gange VK, editors. Multitrophic interactions in terrestrial ecosystems. The 36th Symposium of the British Ecological Society. Oxford UK: Blackwell Science.

Moore JC, De Ruiter PC, Hunt HW. 1993. Influence of productivity on the stability of real and model ecosystems. *Science* 261:906–908.

Moore JC, De Ruiter PC, Hunt HW, Coleman DC, Freckman DW. 1996. Microcosms and soil ecology: Critical linkages between field studies and modelling food webs. *Ecology* 77:694–705.

Moore JC, Hunt HW. 1988. Resource compartmentation and the stability of real ecosystems. *Nature* 333:261–263.

Moore JC, Walter DE, Hunt HW. 1988. Arthropod regulation of micro- and mesobiota in belowground food webs. *Ann Rev Entomol* 33:419–439.

Mora MA, Auman HJ, Ludwig JP, Giesy JP, Verbrugge DA, Ludwig ME. 1993. Polychlorinated-biphenyls and chlorinated insecticides in plasma of Caspian terns: Relationships with age, productivity, and colony site-tenacity in the Great Lakes. *Arch Environ Contam Toxicol* 24:320–331.

Morgan MJ, Kiceniuk JW. 1990. Effect of fenitrothion on the foraging behavior of juvenile Atlantic salmon. *Environ Toxicol Chem* 9:489–495.

Morgan E, Knacker T. 1994. The role of laboratory terrestrial model ecosystems in the testing of potentially harmful substances. *Ecotoxicology* 3:213–233.

Moser DC, Minshall GW. 1996. Effects of localized disturbance on macroinvertebrate community structure in relation to mode of colonization and season. *Am MidNat* 135:92–101.

Mount DR, Hockett JR, Garrison TD, Barten KA. 1992. Summary findings from toxicity identification evaluations [abstract]. Proceedings of the Society of Environmental Toxicology and Chemistry (SETAC) 13th Annual Meeting; 1992 Sep 8–12; Cincinnati OH. Pensacola FL, USA: SETAC.

Munger C, Hare L. 1997. Relative importance of water and food as cadmium sources to an aquatic insect (*Chaoborus punctipennis*): Implications for predicting Cd bioaccumulation in nature. *Environ Sci Technol* 31:891–895.

Munshower FE. 1993. Practical handbook of disturbed land revegetation. Boca Raton FL, USA: Lewis.

Nacci DE, Coiro L, Champlin D, Jayaraman S, McKinney R, Gleason T, Munns Jr WR, Specker J, Cooper K. 1999. Adaptation of wild fish populations to persistent environmental contaminants. *Mar Biol* 134:9–17.

Nacci DE, Coiro L, Kuhn A, Champlin D, Munns Jr WR, Specker J, Cooper K. 1998. A non-destructive indicator of ethoxyresorufin-O-deethylase activity in embryonic fish. *Environ Toxicol Chem* 17:2481–2486.

Naeem S, Thompson LJ, Lawler SP, Lawton JH, Woodfin RM. 1994. Declining biodiversity may alter the performance of ecosystems. *Nature* 368:734–737.

Narbonne JF, Daubeze M, Clerandeau C, Garrigues P. 1999. Scale of classification based on biochemical markers in mussels: Application to pollution monitoring in European coasts. *Biomarkers* 4:415–424.

[NAPAP] National Acid Precipitation Assessment Program. 1991. 1990 Integrated assessment report. Washington DC, USA: The NAPAP Office of the Director.

Nelder JA, Mead R. 1964. A simplex method for function minimization. *Comput J* 7:308–313.

Neutel AM. 2001. Stability of complex food webs: Pyramids of biomass, interaction strengths and the weight of trophic loops. [PhD thesis]. Utrecht, Netherlands: Utrecht Univ. ISBN 90-393-2737-8.

Nevo E, Ben-Shlomo R, Lavie B. 1984. Mercury selection of allozymes in marine organisms: Predictions and verification in nature. *Proc Nat Acad Sci* 81:1258–1259.

Nevo E, Noy R, Lavie B, Bieles A, Muchtar S. 1986. Genetic diversity and resistance to marine pollution. *Biol J Linn Soc* 29:139–144.

Nichols FH, Thompson JK. 1985. Seasonal growth in the bivalve *Macoma balthica* near the southern limit of its range. *Hydrobiologia* 129:121–138.

Nicholson AJ. 1954. Compensatory reactions of populations to stresses and their evolutionary significance. *Austral J Zool* 2:1–8.

Nisbet RM. 1996. Delay-differential equations for structured populations. In: Tuljapurkar S, Caswell H, editors. Structured-population models in marine, terrestrial and freshwater systems. Dordrecht, Netherlands: Kluwer Academic.

Nisbet RM, Gurney WSC. 1982. Modelling fluctuating populations. New York NY, USA: Wiley.

Nisbet RM, Ross AH, Brooks AJ. 1996. Empirically based energy budget models: Theory and application to ecotoxicology. *Nonlinear World* 3:85–106.

Noonburg EG, Nisbet RM, McCauley E, Gurney WSC, Murdoch WW, de Roos AM. 1998. Experimental testing of dynamic energy budget models. *Funct Ecol* 12:211–222.

North WJ, Clendenning KA, Jones LG, Lackey LB, Leighton DL, Neushul M, Sargent MC, Scotten HL. 1964. An investigation of the effects of dischraged wastes on kelp. Sacramento CA, USA: California State Water Quality Control Board. Publication Nr 26.

North WJ, Jackson GA, Manley SL. 1986. Macrocystis and its environment, knowns and unknowns. *Aquat Bot* 26:9–26.

Norton SB, Rodier DJ, Gentile JH, Schalie van der WH, Wood WP, Slimak MW. 1992. A framework for ecological risk assessment. *Environ Toxicol Chem* 11:1663–1672.

Noss RF. 1990. Indicators for monitoring biodiversity: A heirarchical approach. *Conserv Biol* 4:355–364.

Oakden JM, Oliver JS, Flegal AR. 1984. EDTA chelation and zinc antagonism with cadmium in sediment: Effects on the behaviour and mortality of two infaunal amphipods. *Mar Biol* 84:125–130.

Odum EP. 1969. The strategy of ecosystem development. *Science* 164:262–270.

Odum EP. 1981. The effects of stress on the trajectory of ecological succession. In: Barrett GW, Rosenberg, editors. Stress effects on natural ecosystems. New York NY, USA: Wiley. p 43–46.

[OECD] Organization for Economic Cooperation and Development. 1995. Guidance document for aquatic effects assessment. Paris, France: OECD. OECD Environment monographs Nr 92.

[OEPA] Ohio Environmental Protection Agency. 1987. Biological criteria for the protection of aquatic life. Columbus OH, USA: OEPA.

Øksendal B. 1989. Stochastic differential equations. Berlin, Germany: Springer-Verlag.

Okubo A. 1980. Diffusion and ecological problems: Mathematical models. Berlin, Germany: Springer-Verlag.

Oliver AJ, Niimi AJ. 1988. Trophodynamic analysis of polychlroinated biphenyl congeners and other chlorinated hydrocarbons in the Lake Ontario ecosystem. *Environ Sci Technol* 22:388–397.

O'Neill RV. 1969. Indirect estimation of energy fluxes in animal food webs. *J Theor Biol* 22:284–290.

O'Neill RV, Gardner RH, Barnthouse LW, Suter GW, Hildebrand SG, Gehrs CW. 1982. Ecosystem risk analysis: A new methodology. *Environ Toxicol Chem* 1:167–177.

Opperhuizen A, Van der Velde EW, Gobas FAPC, Liem DAK, Van der Steen JMD, Hutzinger O. 1985. Relationships between bioconcentration in fish and steric factors of organic chemicals. *Chemosphere* 14:1871–1896.

Pahl-Wostl C. 1995. The dynamic nature of ecosystems. Chichester UK: Wiley.

Paine RR. 1988. Food webs: Road maps of interactions or grist for theoretical development. *Ecology* 69:1648–1654.

Paine RT. 1980. Food webs: Linkage, interaction strength and community infrastructure. *J Anim Ecol* 49:667–685.

Paine RT. 1992. Food-web analysis through field measurements of per capita interaction strength. *Nature* 355:73-75.

Paine RT. 1980. Food webs: Linkage, interaction strength and community infrastructure. *J Anim Ecol* 49:667–686.

Paine RT, Tegner MJ, Johnson EA. 1998. Compounded perturbations yield ecological surprises. *Ecosystems* 1:535–545.

Painter S. 1992. Regional variability in sediment background metal concentrations and the Ontario sediment quality guidelines. Burlington, Ontario, Canada: Environment Canada. NWRI Report Nr 92–85.

Palmer TM. 1995. The influence of spatial heterogeneity on the behavior and growth of two herbivorous stream insects. *Oecologia* 104:476–86.

Pape-Lindstrom PA, Lydy MJ. 1997. Synergistic toxicity of atrazine and organophosphate insecticides contravenes the response addition mixture model. *Environ Toxicol Chem* 16:2415–2420.

Parchaso F, Brown CL, Thompson JK, Luoma SN. 1997. Reproduction and secondary production in the bivalve *Potamocorbula amurensis* form San Francisco Bay. Menlo Park CA, USA: U.S. Geological Survey (USGS). Open File Report 97-420. 143 p.

Park RA. 1999a. AQUATOX for Windows: A modular toxic effects model for aquatic ecosystems. Washington DC, USA: U.S. Environmental Protection Agency (USEPA). Contract Nr 68-C4-0051, 4-13. 187 p.

Park RA. 1999b. Validation of the AQUATOX Model, Version 1.66 with data from Coralville Reservoir, Iowa. Washington DC, USA: U.S. Environmental Protection Agency (USEPA). Contract Nr 68-C4-0051, 4-13. 10 p.

Park RA. 1999c. Validation of the AQUATOX Model, Version 1.66, with data from Lake Onondaga, New York NY, USA. Washington DC, USA: U.S. Environmental Protection Agency (USEPA). Contract Nr 68-C4-0051, 4-13. 16 p.

Park RA, Firlie B, Camacho R, Sappington K, Coombs M, Mauriello KD. 1995. AQUATOX, a general fate and effects model for aquatic ecosystems. Proceedings for the Toxic Substances in Water Environments: Assessment and Control. Water Environment Federation. p (3)7–17.

Park RA, O'Neil RV, Bloomfield JA, Shugart HH, Booth RS, Goldstein RA, Mankin JB, Koonce JF, Scavia D, Adams MS, Clesceri LS, Colon EM, Dettmann EH, Hoopes J, Huff DD, Katz S, Kitchell JF, Kohberger et al. 1974. A generalized model for simulating lake ecosystems. *Simulation* 23:33–50.

Parmesan C. 1996. Climate and species' range. *Nature* 382:765–766.

Pascoe D, Kedwards TJ, Maund SJ, Muthi E, Taylor EJ. 1994. Laboratory and field evaluation of a behavioural bioassay - the *Gammarus pulex* (L.) precopula separation (GaPPS) test. *Wat Res* 28:369–372.

Paterson S, Mackay D, Bacci E, Calamari D. 1991. Correlation of the equilibrium and kinetics of the leaf air exchange of hydrophobic organic chemicals. *Environ Sci Technol* 25:866–871.

Paulsen SG, Larsen DP, Kaufmann PR, Whittier TR, Baker JR, Peck D, McGue J, Hughes RM, McMullen D, Stevens D, Stoddard JL, Lazorchak J, Kinney W, Selle AR, Hjort R. 1991. EMAP: Surface waters monitoring and research strategy, fiscal year 1991. Washington DC, USA: U.S. Environmental Protection Agency (USEPA), Office of Research and Development; Corvallis OR, USA: Environmental Research Laboratory. EPA-600-3-91–002.

Peakall DB. 1996. Disrupted patterns of behaviour in natural populations as an index of ecotoxicity. *Environ Health Persp* 104:331–335.

Pechmann JHK, Scott DE, Semlitsch RD, Caldwell JP, Vitt LJ, Gibbons JW. 1991. Declining amphibian populations: The problem of separating human impacts from natural fluctuations. *Science* 253:892–895.

Persaud D, Jaagumagi R, Hayton A. 1992. Guidelines for the protection and management of aquatic sediment quality in Ontario. Toronto, Ontario, Canada: Water Resources Branch, Ontario Ministry of Environment.

Peterman RM. 1990. The importance of reporting statistical power: The forest decline andacidic deposition example. *Ecology* 71:2024–2027.

Peters RL, Lovejoy TE, editors. 1992. Global warming and biological diversity. New Haven CT, USA: Yale Univ Pr.

Peterson BJ, Fry B. 1987. Stable isotopes in ecosystem studies. *Ann Rev Ecol Syst* 18:293–320.

Phillips G, Spoon R. 1990. Ambient toxicity assessment of Clark Fork River water toxicity tests and metal residues in brown trout organs. In: Watson V, Cook JW, editors. Proceedings, 1990 Clark Fork River Symposium, Missoula, MT, USA.

Phinney JT, Bruland KW. 1997. Trace metal exchange in solution by the fungicides Ziram and Maneb and subsequent uptake of lipophilic organic zinc, copper and lead complexes into phytoplankton cells. *Environ Toxicol Chem* 16:2046–2053.

Pimm SL. 1982. Food webs. London UK: Chapman & Hall.

Pimm SL, Lawton JH. 1977. The number of trophic levels in ecological communities. *Nature* 268:329–331.

Poff NL. 1997. Landscape filters and species traits: Towards mechanistic understanding and prediction in stream ecology. *J N Am Benthol Soc* 16:391–409.

Poff NL, Ward JV. 1990. Physical habitat template of lotic systems: Recovery in the context of historical pattern of spatiotemporal heterogeneity. *Environ Manage* 14:629–45.

Polis GA. 1991. Complex trophic interactions in deserts: An empirical critique of food-web theory. *Am Nat* 138:123–155.

Postma JF, Groenendijk D. 1999. Adaptation to metals in the midge *Chironomus riparius*: A case study in the River Dommel. In: Forbes VE, editor. Genetics and ecotoxicology. Philadelphia PA, USA: Taylor & Francis.

Potts GR, Vickerman GP. 1974. Studies on the cereal ecosystem. *Ad Ecol Res* 2:107–197.

Pounds JA, Fogden MPL, Campbell JH. 1999. Biological response to climate change on a tropical mountain. *Nature* 398:611–615.

Pounds JA, Fogden MPL, Savage JM, Gorman GC. 1997. Tests of null models for amphibian declines on a tropical mountain. *Conserv Biol* 11:1307–1322.

Press WH, Flannery BP, Teukolsky SA, Vetterling WT. 1989. Numerical recipes in Pascal. Cambridge UK: Cambridge Univ Pr.

Prinslow TE, Valiela I, Teal HN. 1974. The effect of detritus and ration size on the growth of *Fundulus heteroclitus* (L.). *J Exp Mar Biol Ecol* 16:1–10.

Pulliam HR. 1994. Incorporating concepts from population and behavioral ecology into models of exposure to toxins and risk assessment. In: Kendall RJ, Lacher Jr TE, editors. Wildlife toxicology and population modeling. Boca Raton FL, USA: Lewis.

Rapport DJ, Regier HA, Hutchinson TC. 1985. Ecosystem behavior under stress. *Am Nat* 125:617–640.

Rapport DJ, Whitford WG, Hilden M. 1998. Common patterns of ecosystem breakdown under stress. *Environ Monitor Assess* 51:171–178.

Rasmussen JB, Rowan DJ, Lean DRS, Carey JH. 1990. Food chain structure affects PCB levels in pelagic fish from Ontario lakes. *Can J Fish Aquat Sci* 47:2030–2038.

Remmert H. 1994. Minimum animal populations. Berlin, Germany: Springer-Verlag.

Ren L, Huang XD, McConkey BJ, Dixon DG, Greenberg BM. 1994. Photoinduced toxicity of three polycyclic aromatic hydrocarbons (fluoranthene, pyrene and naphthalene) to the duckweed Lemna gibba L-G-3. *Ecotoxicol Environ Saf* 28:160–171.

Renshaw E. 1990. Modelling biological population in space and time. Cambridge UK: Univ Cambridge.

Resh VH, Norris RH, Barbour MT. 1995. Design and implementation of rapid assessment approaches for water resource monitoring using benthic macroinvertebrates. *Austral J Ecol* 20:08–21.

Reynoldson TB, Bailey RC, Day KE, Norris RH. 1995. Biological guidelines for freshwater sediment based on BEnthic Assessment of SedimenT (the BEAST) using a multivariate approach for predicting biological state. *Austral J Ecol* 20:198–219.

Reynoldson TB, Mudroch A, Edwards CJ. 1988. An overview of contaminated sediments in the Great Lakes with special reference to the international workshop held at Aberystwyth, Wales, UK. Report to the Great Lakes Science Advisory Board.

Reynoldson TB, Norris RH, Resh VH, Day KE, Rosenberg DM. 1997. The reference condition: A comparison of multimetric and multivariate approaches to assess water quality impairment using benthic macroinvertebrates. *J N Am Benthol Soc* 16:833–852.

Richards C, Haro RJ, Johnson LB, Host GE. 1997. Catchment and reach-scale properties as indicators of macroinvertebrate species traits. *Freshw Biol* 37:219–30.

Ringwood AH, Conners D, Dinovo A. 1998. The effects of copper exposures on cellular responses in oysters. *Mar Environ Res* 46:591–595.

Ringwood AH, Conners D, Keppler CJ, DiNovo AA. 1999. Biomarker studies with juvenile oysters (*Crassostrea virginica*) deployed in situ. *Biomarkers* 4:400–414.

Root RB. 1967. The niche exploitation pattern of the blue-gray gnatcatcher. *Ecol Monogr* 37:317–350.

Rudd JWM, Kelly CA, Schindler DW, Turner MA. 1988. Disruption of the nitrogen cycle in acidified lakes. *Science* (Washington DC) 240:1515–1517.

Russell RW, Hecnar SJ, Haffner GD. 1995. Organochlorine pesticide residues in Southern Ontario spring peepers. *Environ Toxicol Chem* 14:815–817.

Sallenave RM, Day KE, Kreutzweiser DP. 1994. The role of grazers and shredders in the retention and downstream transport of a PCB in lotic environments. *Environ Toxicol Chem* 13:1843–1847.

Samson FB, Perez-Trejo F, Salwasser H, Ruggiero LF, Shaffer ML. 1985. On determining and managing minimum population size. *Wildlife Soc Bull* 14:425–433.

Sanders BM, Martin LS, Nelson WG, Phelps DK, Welch W. 1991. Relationships between accumulation of a 60 kDa stress protein and scope-for-growth in *Mytilus edulis* exposed to a range of copper concentrations. *Mar Environ Res* 31:81–97.

Sanford E. 1999. Regulation of keystone predation by small changes in ocean temperature. *Science* 283:2095–2097.

Sarda N, Burton Jr GA. 1995. Ammonia variation in sediments: Spatial, temporal and method-related effects. *Environ Toxicol Chem* 14:1499–1506.

Sasson-Brickson G, Burton Jr GA. 1991. In situ and laboratory sediment toxicity testing with *Ceriodaphnia dubia*. *Environ Toxicol Chem* 10:201–207.

Saunier RE, Meganck RA. 1995. Conservation and biodiversity and the new regional planning. Organization of American States and IUCN. Washington DC, USA: The World Conservation Union.

Scheffer M. 1998. Ecology of shallow lakes. Population and community biology, Series 22. London UK: Chapman & Hall. 357 p.

Scherer E, McNicol RE, Evans RE. 1997. Impairment of lake trout foraging by chronic exposure to cadmium: A black box experiment. *Aquat Toxicol* 37:1–7.

Schindler DW. 1977. Evolution of phosphorus limitation in lakes: natural mechanisms compensate for deficiencies of nitrogen and carbon in eutrophied lakes. *Science* (Washington DC) 195:260–262.

Schindler DW. 1987. Detecting ecosystem responses in anthropogenic stress. *Can J Fish Aquat Sci* (suppl):6–25.

Schindler DW. 1988. Effects of acid rain on freshwater ecosystems. *Science* 239:149–157.

Schindler DW. 1998. Replication versus realism: The need for ecosystem-scale experiments. *Ecosystems* 1:323–334.

Schoellhamer DH. 1996. Factors affecting suspended-solids concentrations in South San Francisco Bay, California. *J Geophys Res* 101:12087–12095.

Schoener TW. 1974. Resource partitioning in ecological communities. *Science* 185:27–39.

Schumaker NH. 1998. A users guide to the PATCH model. Corvallis OR, USA: U.S. Environmental Protection Agency (USEPA). EPA/600/R-98/135.

Schwartz SE. 1989. Acid deposition: Unraveling a regional phenomenon. *Science* 243:753–763.

Seber GAF. 1982 The estimation of animal abundance. 2nd ed. London UK: Charles Griffin.

Shaffer M. 1987. Minimum viable populations: Coping with uncertainty. In: Soule ME, editor. Viable populations for conservation. Cambridge UK: Cambridge Univ Pr. p 69–86.

Shaffer ML, Samson FB. 1985 Population size and extinction: A note on determining critical population sizes. *Am Nat* 125:144–152.

Shaw J. 1999. The evolution of heavy metal tolerance in plants: Adaptations, limits, and costs. In: Forbes VE, editor. Genetics and ecotoxicology. Philadelphia PA, USA: Taylor & Francis. p 9–30.

Sheail J. 1987. Seventy-five years in ecology: The British Ecological Society. Oxford UK: Blackwell Scientific.

Sherratt TN, Jepson PC. 1993. A metapopulation approach to modelling the long-term impacts of pesticides on invertebrates. *J Appl Ecol* 30:696–705.

Sherratt TN, Roberts G, Williams P, Whitfield M, Biggs J, Shillabeer N, Maund SJ. 1999. A life-history approach to predicting the recovery of aquatic invertebrate populations after exposure to xenobiotic chemicals. *Environ Toxicol Chem* 18:2512–2518.

Shirazi MA, Lowrie LN. 1990. A probabilistic statement of the structure activity relationship for environmental risk analysis. *Arch Environ Contam Toxicol* 19:597–602.

Sibly RM. 1994. From organism to population: The role of life history theory. In: Sutcliffe DW, editor. Water quality and stress indicators in marine and freshwater systems: Linking levels of organisation. Ambleside UK: Freshwater Biological Assoc. p 63–74.

Sibly RM. 1996. Effects of populations on individual life histories and population growth rates. In: Newman MC, Jagoe CH, editors. Ecotoxicology. A hierarchical treatment. Boca Raton FL, USA: Lewis. p 197–223.

Sibly RM. 1999. Efficient experimental designs for studying stress and population density in animal populations. *Ecol Appl* 9:496–503.

Sih A, Crowley P, McPeek M, Petranka J, Strohmeier K. 1985. Predation, competition and prey communities: A review of field experiments. *Ann Rev Ecol System* 16:269–311.

Simkiss K, Mason AZ. 1984. Cellular responses of molluscan tissues to environmental metals. *Mar Environ Res* 14:103–118.

Sinclair ARE. 1996. Mammal populations: Fluctuation, regulation, life history theory and their implications for conservation. In: Floyd RB, Sheppard AW, De Barro PJ, editors. Frontiers of population ecology. Australia: Commonwealth Scientific and Industrial Research Organisation (CSIRO). p 127–154.

Sindermann CJ. 1996. Ocean pollution: Effects on living resources and humans. Boca Raton FL, USA: CRC Pr. 275 p.

Slobodkin LB, Richman S. 1956. The effect of removal of fixed percentages of the newborn on size and variability in populations of *Daphnia pulicaria* (Forbes). *Limnol Oceanog* 1:209–237.

Smith G, Weis JS. 1997. Predator/prey interactions in *Fundulus heteroclitus*: Effects of living in a polluted environment. *J Exp Mar Biol Ecol* 209:75–87.

Smith RJ, Hobson KA, Koopman HN, Lavigne DM. 1996. Distinguishing between populations of fresh and saltwater harbour seals (*Phoca vitulina*) using stable-isotope ratios and fatty acid profiles. *Can J Fish Aquat Sci* 53:272–279.

Sobral P, Widdows J. 1997. Effects of copper exposure on the scope for growth of the clam *Ruditapes decussatus* from Southern Portugal. *Mar Pollut Bull* 34:992–1000.

Soule ME. 1987. Introduction. In: Soule ML, editor. Viable populations for conservation. Cambridge UK: Cambridge Univ Pr. p 3–10.

Sousa WP. 1984. The role of disturbance in natural communities. *Ann Rev Ecol Syst* 15:353–392.

Southwood TR. 1977. Habitat, the template for ecological strategies. *J Anim Ecol* 46:337–365.

Stark JD, Tanigoshi L, Bounfour M, Antonelli A. 1997. Reproductive potential: Its influence on the susceptibility of a species to pesticides. *Ecotoxicol Environ Saf* 37:273–279.

Stay FS, Katko A, Rohm CM, Fix MA, Larsen DP. 1988. Effects of fluorene on microcosms developed from four natural communities. *Environ Toxicol Chem* 7:635–644.

Stearns SC. 1992. The evolution of life histories. Oxford UK: Oxford Univ Pr.

Stewart AR. 1999. Accumulation of Cd by a freshwater mussel (*Pyganodon grandis*) is reduced in the presence of Cu, Zn, Pb and Ni. *Can J Fish Aquat Sci* 56:467–478.

Sunda WG, Guillard RR. 1976. The relationship betweeen cupric ion activity and the toxicity of copper to phytoplankton. *J Mar Res* 34:411–422.

Susser M. 1986a. Rules of inference in epidemiology. *Regul Toxicol Pharmacol* 6:116–186.

Susser M. 1986b. The logic of Sir Carl Popper and the practice of epidemiology. *Am J Epidemiol* 124:711–718.

Suter II GW. 1993a. A critique of ecosystem health concepts and indexes. *Environ Toxicol Chem* 12:1533–1539.

Suter II GW. 1993b. Ecological risk assessment. Boca Raton FL, USA: Lewis.

Suter II GW. 1998. Retrospective assessment, ecoepidemiology, and ecological monitoring. In: Calow P, editor. Handbook of environmental risk assessment and management. Oxford UK: Blackwell Scientific. p 177–217.

Suter GW, Bartell S. 1993. Ecosystem-level effects. In: Suter GW, editor. Ecological risk assessment. Chelsea MI, USA: Lewis. p 275–310.

Suter II GW, Norton SB, Cormier S. A method for inferring the causes of observed impairments in aquatic ecosystems. *Environ Toxicol Chem* (in press).

Sutherland WJ, Reynolds JD. 1998. Sustainable and unsustainable exploitation. In: Sutherland WJ, editor. Conservation science and action. Oxford UK: Blackwell Science. p 90–115.

Swackhamer DL, Skoglund RS. 1993. Bioaccumulation of PCBs by algae: Kinetics versus equilibrium. *Environ Toxicol Chem* 12(5):831–838.

Swartz RC, Schults DW, Ozretich RJ, Lamberson JO, Cole FA, DeWitt TH, Redmond MS, Ferraro SP. 1995. SPAH: A model to predict the toxicity of polynuclear aromotic hydrocarbon mixtures in field-collected sediments. *Environ Toxicol Chem* 14:1977–1987.

Swartzman G, Rose KA. 1984. Simulating the biological effects of toxicants in aquatic microcosm systems. *Ecol Model* 22:123–134.

Taylor GJ, Baird DJ, Soares AMVM. 1998. Surface-binding of contaminants by algae: consequences for lethal toxicity and feeding to *Daphnia magna* Straus. *Environ Toxicol Chem* 17:412–419.

Terborgh J. 1974. Preservation of natural diversity: The problem of extinction prone species. *BioScience* 24:715–722.

Thacker JRM, Jepson PC. 1993. Pesticide risk assessment and non-target invertebrates: Integrating population depletion, population recovery and experimental design. *Bull Environ Contam Toxicol* 51:523–531.

Thom RM. 1995. Implications for restoration: Light, temperature and enriched CO_2 effects on eelgrass (*Zostera marina* L.). Proceedings, Puget Sound Research '95. Olympia WA, USA: Puget Sound Water Quality Authority. p 523–528.

Thom RM. 1997. System-development matrix for adaptive management of coastal ecosystem restoration projects. *Ecol Engineer* 8:219–232.

Thom RM. 2000. Adaptive management of coastal ecosystem restoration projects. *Ecol Engineer* 15:365–372.

Thom RM, Albright RG. 1990. Dynamics of benthic vegetation standing-stock, irradiance, and water properties in central Puget Sound. *Mar Biol* 104:129–141.

Thom RM, Antrim LD, Borde AB, Gardiner WW, Shreffler DK, Farley PG. 1998. Puget Sound's eelgrass meadows: Factors contributing to depth distribution and spatial patchiness. Proceedings, Puget Sound Research '98. Olympia WA, USA: Puget Sound Water Quality Action Team. p 363–370.

Thom RM, Shreffler DK, Simenstad CA, Olson AM, Wyllie Echeverria S, Cordell JR Schafer J. 1997. Mitigating impacts from ferry terminals on eelgrass (*Zostera marina* L.). Proceedings, Wetland and riparian restoration: Taking a broader view, of a conference. Seattle WA, USA: U.S. Environmental Protection Agency (USEPA), Region 10. EPA 910-R-97-007. p 95–107.

Thom RM, Simenstad CA, Cordell JR, Salo EO. 1989. Fish and their epibenthic prey in a marina and adjacent mudflats and eelgrass meadow in a small estuarine bay. Seattle WA, USA: Fisheries Research Institute, Univ Washington. FRI-UW-8901.

Thomann RV, Connolly JP 1984. Model of PCB in the Lake Michigan lake trout food chain. *Environ Sci Technol* 18(2):65–71.

Thomann RV, Connolly JP, Parkerton TF. 1992. An equilibrium model of organic chemical accumulation in aquatic foodwebs with sediment interaction. *Environ Toxicol Chem* 11:615–629.

Thomann RV, Mahoney JD, Mueller R. 1995. Steady-state model of biota-sediment accumulation factors for metals in two marine bivalves. *Environ Toxicol Chem* 14:1989–1998.

Thomas CFG, Hol EHA, Evert JW. 1990. Modelling the diffusion component of dispersal during the recovery of a population of linyphiid spiders from exosure to an insecticide. *Funct Ecol* 4:357–368.

Thompson B, Anderson B, Hunt J, Taberski K, Phillips B. 1999. Relationships between sediment contamination and toxicity in San Francisco Bay. *Mar Environ Res* 48:285–309.

Thomson EA, Luoma SN, Johansson CE, Cain DJ. 1984. Comparison of sediments and organisms in identifying sources of biologically available trace metal contamination. *Water Res* 18:755–765.

Thomson-Becker EA, Luoma SN. 1985. Temporal fluctuations in grain size, organic materials and iron concentrations in intertidal surface sediment. *Hydrobiologia* 129:91–109.

Tilman D, Downing JA. 1994. Biodiversity and stability in grasslands. *Nature* 367:363–365.

Tilman D, Kareiva P. 1997. Spatial ecology. Monographs in population biology 30. Princetown NJ, USA: Princeton Univ Pr.

Toppin SV, Heber M, Weis JS, Weis P. 1987. Changes in reproductive biology and life history in *Fundulus heteroclitus* in a polluted environment. In: Vernberg WB, Calabrese A, Thurberg F, Vernberg FJ, editors. Pollution physiology of estuarine organisms. Columbia SC, USA: Univ South Carolina Pr. p 171–184.

Traas TP, Aldenberg T, Janse JH. 1995. Hazardous concentrations for ecosystems: calculation and validation with CATS models. Bilthoven, Netherlands: RIVM. Publication Nr 719102037.

Traas TP, Janse JH, Aldenberg T, Brock TCM. 1998. A food web model for fate and direct and indirect effects of Dursban(r) 4E (active ingredient chlorpyrifos) in freshwater microcosms. *Aquat Ecol* 32:179–190.

Tucker KA, Burton Jr GA. 1999. Assessment of nonpoint source runoff in a stream using in situ and laboratory approaches. *Environ Toxicol Chem* 18:2797–2803.

Tuljapurkar SD, Orzack SH. 1980. Population dynamics in variable environments. I. Long-run growth rates and extinction. *Theor Popul Biol* 18:314–342.

Tuljapurkar SD. 1982. Population dynamics in variable environments. III Evolutionary dynamics of r-selection. *Theor Popul Biol* 21:141–165.

Tuljapurkar SD. 1996. Stochastic matrix models. In: Tuljapurkar S, Caswell H, editors. Structured-population models in marine, terrestrial and freshwater systems. Dordrecht, Netherlands: Kluwer Academic.

Turelli M. 1977. Random environments and stochastic calculus. *Theor Popul Biol* 12:140–178.

Turner L, Choplin F, Dugard P, Hermens J, Jaekh R, Marsmann M, Roberts D. 1997. Structure-activity relationships in toxicology and ecotoxicology: An assessment. *Toxic in vitro* 1(3):143–171.

Ulanowicz R. 1996. Trophic flow networks as indicators of ecosystem stress. In: Winemiller GA, Polis K, editors. Food webs: Integration of patterns and dynamics. New York NY, USA: Chapman & Hall. p 358–368.

[UNEP-IETC] United Nations Environmental Programme, International Environmental Technology Centre. 1996. Environmental risk assessment for sustainable cities. Osaka, Japan: UNEP-IETC. Technical Publication Series [3].

[USDOI] U.S. Department of the Interior. 1986. Natural resource damage assessments: Final rule. *Federal Register* 51:27674–27753.

[USDOI] U.S. Department of Interior. 1996. Natural resource damage assessment regulations, 43 CFR Part 11 (1995) as amended at 61 *Federal Register* 20609. May 7, 1996.

[USEPA] U.S. Environmental Protection Agency. 1989. Rapid bioassessment protocols for use in streams and rivers. Washington DC, USA: USEPA Office of Water. EPA/444/4-89-001.

[USEPA] U.S. Environmental Protection Agency. 1991. Methods for aquatic toxicity identification evaluations. Phase 1, Toxicity characterization procedures. 2nd ed. Washington DC, USA: Office of Research and Development. EPA/600/6-91/003.

[USEPA] U.S. Environmental Protection Agency. 1992. Framework for ecological risk assessment. Washington DC, USA: Risk Assessment Forum, USEPA. EPA/630/R-92/001.

[USEPA] U.S. Environmental Protection Agency. 1993. A review of ecological assessment case studies from a risk assessment perspective. Washington DC, USA: Risk Assessment Forum, USEPA. EPA/630/R-92/005.

[USEPA] U.S. Environmental Protection Agency. 1994a. A review of ecological assessment case studies from a risk assessment perspective, Volume II. Washington DC, USA: Risk Assessment Forum, USEPA. EPA/630/R-94/003.

[USEPA] U.S. Environmental Protection Agency. 1994b. Ecological risk assessment issue papers. Washington DC, USA: Risk Assessment Forum, USEPA. EPA/630/R-94/009.

[USEPA] U.S. Environmental Protection Agency. 1999a. Methods for measuring the toxicity and bioaccumulation of sediment-associated contaminants and freshwater invertebrates. Washington DC, USA: USEPA Office of Research and Development, Office of Water. EPA/600/R-99/064.

[USEPA] U.S. Environmental Protection Agency. 1999b. Rapid bioassessment protocols for use in streams and rivers. Washington DC, USA: USEPA Office of Water. EPA 841-B-99-002.

[USEPA] U.S. Environmental Protection Agency. 2000. Stressor identification guidance document. Washington DC, USA: USEPA. EPA/822/B-00/025.

Van den Brink PJ, Van Wijngaarden RPA, Lucassen WGH, Brock TCM, Leuuwangh P. 1996. Effects of the insecticide Dursban 4AE (active ingredient chlorpyrifos) in outdoor experimental ditches: II Invertebrate community responses and recovery. *Environ Toxicol Chem* 15:1143–1153.

Van der Zande MJ, Rasmussen JB. 1996. A trophic position model of pelagic food webs: Impact on contaminant bioaccumulation in lake trout. *Ecol Monogr* 66(4):451–477.

Van Donk E, Prins H, Voogd HM, Crum SJH, Brock TCM. 1995. Effects of nutrient loading and insecticide application on the ecology of Elodea-dominated freshwater microcosms. I. Responses of plankton and zooplanktivorous insects. *Arch Hydrobiol* 133:417–439.

van Geen A, Chase Z. 1998. Recent mine spill adds to contamination of Southern Spain. *EOS* 38:449–451.

van Geen A, Luoma SN. 1999. The impact of human activities on sediments of San Francisco Bay: An overview. *Mar Chem* 64:1–7.

Van Leeuwen CJ, Luttimer WJ, Griffioen PS. 1985. The use of cohorts and populations in chronic toxicity studies with *Daphnia magna*: A cadmium example. *Ecotoxicol Environ Saf* 9:26–39.

Vanni MJ. 1996.Nutrient transport and recycling by consumers in lake food webs: Implications for algal communities. In: Polis GA, Winemiller KO, editors. Food webs: Integration of patterns and dynamics. New York NY, USA: Chapman & Hall. p 81–95.

Vannote RL, Minshall GW, Cummins KW, Sedell JR, Cushing CE. 1980. The river continuum concept. *Can J Fish Aquat Sci* 37:130–137.

Van Straalen NM, Denneman CAJ. 1989. Ecotoxicological evaluation of soil quality criteria. *Ecotoxicol Environ Saf* 18:241–251.

Van Straalen NM, Kammenga JE. 1998. Assessment of ecotoxicity at the population level using demographic parameters. In: Schüürmann G, Markert B, editors. Ecotoxicology. Chichester UK: Wiley. p 621–644.

Van Straalen NM, Løkke H. 1997. Ecological risk assessment of contaminants in soil. London UK: Chapman & Hall.

Van Straalen NM, Van Rijn JP. 1998. Ecotoxicological risk assessment of soil fauna recovery from pesticide application. *Rev Environ Contam Toxicol* 154:83–141.

Van Wensem J. 1997. The use of models in ecological risk assessment. In: Straalen NM, van Løkke H, editors. Ecological risk assessment of contaminants in soil. London UK: Chapman & Hall.

Verhoef HA, Brussaard L. 1990. Decomposition and nitrogen mineralization in natural and agro-ecosystems: The contribution of animals. *Biogeochemistry* 11:175–211.

Viarengo A, Canesi L, Pertica M, Livingstone DR. 1991. Seasonal variations in the antioxidant defense systems and lipid peroxidation of the digestive gland of mussels. *Comp Biochem Physiol* 100C:187–190.

Viarengo A, Canesi L, Pertica M, Mancinelli G, Accomando R, Smaal AC, Orunesu M. 1995. Stress on stress response: A simple monitoring tool in the assessment of a general stress syndrome in mussels. *Mar Environ Res* 245–249.

Vickerman GP. 1992. The effects of different pesticide regimes on the invertebrate fauna of of winter wheat. In: Greig-Smith PW, Frampton GK, Hardy AR, editors. Pesticides, cereal farming and the environment. London UK: HMSO. p 82–108.

Vincent WF, Roy S. 1993. Solar ultraviolet-B radiation and aquatic primary production: Damage, protection, and recovery. *Environ Rev* 1:1–12.

Voyer RA, Cardin JA, Heltshe JF, Hoffman GL. 1982. Variability of embryos of the winter flounder Pseudopleuronectes americanus exposed to mixtures of cadmium and silver in combination with selected fixed salinities. *Aquat Toxicol* 2:223–233.

Wake DB. 1991. Declining amphibian populations. *Science* 253:860.

Walker CH, Hopkin SP, Sibly RM, Peakall D. 1996. Principles of ecotoxicology. London UK: Taylor & Francis.

Wallace JB, Grubaugh JW, Whiles MR. 1996. Biotic indices and stream ecosystem processes: Results from an experimental study. *Ecol Appl* 6:140–151.

Walthall WK, Stark JD. 1997a. Comparison of two population level ecotoxicological endpoints: The intrinsic and instantaneous rates of increase. *Environ Toxicol Chem* 16:1068–1073.

Walthall WK, Stark JD. 1997b. A comparison of acute mortality and population growth rate as endpoints of toxicological effects. *Ecotoxicol Environ Saf* 37:45–52.

Wang WX, Fisher NS, Luoma SN. 1996. Kinetic determinations of trace element bioaccumulation in the mussel *Mytilus edulis*. *Mar Ecol Prog Ser* 140:91–113.

Ward TJ, Young PC. 1984. Effects of metals and sediment particle size on the species composition of the epifauna of *Pinna bicolor* near a lead smelter, Spencer Gulf, South Australia. *Est Coast Shelf Sci* 18:79–95.

Wardle DA, Lavelle P. 1997. Linkages between soil biota, plant litter quality and decomposition. In: Cadisch G, Giller KE, editors. Driven by nature. Wallingford UK: CAB International.

Warren CE, Davis GE. 1967. Laboratory studies on the feeding, bioenergetics and growth of fish. In: Gerking SD, editor. The biological basis of freshwater fish production. Oxford UK: Blackwell Scientific. p 175–214.

Warwick RM. 1993. Environmental impact studies on marine communities: Pragmatical considerations. *Austral J Ecol* 18:63–80.

Wassenaar LI, Culp JM. 1996. The use of stable isotopic analyses to identify pulp mill effluent signatures in riverine food webs. In: Servos M, Carey J, Munkittrick K, Van Der Kraak G, editors. Environmental fate and effects of pulp mill effluents. Delray Beach FL, USA: St. Lucie Pr. p 413–424.

Weber DN. 1993. Exposure to sublethal levels of waterborne lead alters reproductive behavior patterns in fathead minnows (*Pimephales promelas*). *NeuroToxicol* 14:347–358.

Weins J. 1996. Coping with variability in environmental impact assessment. In: Baird DJ, Maltby L, Greig-Smith PW, Douben PET, editors. ECOtoxicology: Ecological dimensions. London UK: Chapman & Hall. p 55–70.

Weis JS, Mugue N, Weis P. 1999. Mercury tolerance, population effects and population genetics in the mummichog, *Fundulus heteroclitus*. In: Forbes VE, editor. Genetics and ecotoxicology. Philadelphia PA, USA: Taylor & Francis. p 31–54.

Weis JS, Weis P. 1989. Tolerance and stress in a polluted environment: The case of the mummichog. *BioScience* 39:89–95.

Weisberg SB, Frithsen JB, Holland AF, Paul JF, Scott KJ, Summers JK, Wilson HT, Valente R, Heimbuch DG, Gerritsen J, Schimmel SC, Latimer RW. 1993. EMAP-Estuaries Virginian Province 1990 Narragansett RI, USA: U.S. Environmental Protection Agency (USEPA) Environmental Research Laboratory. Demonstration project report. EPA/620/R-93/006.

Weisberg SB, Ranasinghe JA, Dauer DM, Schaffner LC, Diaz RJ, Frithsen JB. 1997. An estuarine bentic index of biotic integrity (B-IBI) for Chesapeake Bay. *Estuaries* 20:149–158.

Werner DD, Gilliam JF, Hall DJ, Mittelbach GG. 1983. An experimental test of the effects of predation risk on habitat use in fish. *Ecology* 64:1540–1548.

Weseloh DV, Ewins PJ, Struger J, Mineau P, Bishop CA, Postupalsky S, Ludwig JP. 1995. Double-crested cormorants of the Great Lakes: Changes in population size, breeding distribution and reproductive output between 1913 and 1991. *Colonial Waterbirds* 18 (Special Publication 1):48–59.

Weseloh DV, Hamr P, Bishop CA, Norstrom RJ. 1995. Organochlorine contaminant levels in waterbird species from Hamilton Harbour, Lake Ontario: An IJC area of concern. *J Great Lakes Res* 21(1):121–137.

Widdows J, Donkin P. 1989. The application of combined tissue residue chemistry and physiological measurement of mussel (*Mytilus edulis*) for the assessment of environmental pollution. *Hydrobiologia* 188/189:455–461.

Widdows J, Donkin P, Brinsley M, Evans SV, Page DS, Salkeld PN. 1995. Sublethal biological effects and contaminant monitoring of Sullom voe (Shetland) nusing mussels (*Mytilus edulis*). *Proc Roy Soc Edin* 103B:99–1112.

Widdows J, Donkin P, Brinsley MD, Evans SV, Salkeld PN, Franklin A, Law RJ, Waldock MJ. 1995. Scope for growth and contaminant levels in North Sea mussels, *Mytilus edulis*. *Mar Ecol Prog Ser* 127:131–148.

Widdows J, Nasci C, Fossato VU. 1996. Effects of pollution on the scope for growth of mussels (*Mytilus galloprovincialis*) from the Venice Lagoon, Italy. *Mar Environ Res* 43:69–81.

Wiegert RG, Wetzel RL. 1979. Simulation experiments with a 140-compartment salt marsh model. In: Dame RF, editor. Marsh-estuarine systems simulation. Columbia SC, USA: Univ South Carolina Pr. p 7–39.

Wiley MJ, Kohler SL, Seelbach PW. 1997. Reconciling landscape and local views of aquatic communities: Lessons from Michigan trout streams. *Freshw Biol* 37:133–48.

Wilson JB. 1988. The cost of heavy-metal tolerance: An example. *Evolution* 42:408–413.

Wood SN. 2000a. Partially specified ecological models. *Ecol Monogr* 71.

Wood SN. 2000b. Modelling and smoothing parameter estimation with multiple quadratic penalties. *J Roy Stat Soc(B)* 62:413–428.

Wood SN. 1994. Obtaining birth and death rate patterns from structured population trajectories. *Ecol Monogr* 64:23–44.

Wood SN, Nisbet RM. 1991. Estimation of mortality rates in stage- structured populations. Berlin, Germany: Springer-Verlag.

Wood SN, Thomas MB. 1999. Super-sensitivity to structure in biological models. *Proc Roy Soc (B)* 266:565–570.

Woodman JN, Cowling EB. 1987. Airborne chemicals and forest health. *Environ Sci Technol* 21:120–126.

Wootton J. 1994. Predicting direct and indirect effects: An integrated approach using experiments and path analysis. *Ecology* 75:151–165.

Wright JF, Furse MT, Armitage PD. 1993. RIVPACS: A technique for evaluating the biological quality of rivers in the UK. *Euro Wat Poll Contr* 3:15–25.

Wright JF, Furse MT, Clarke RT, Moss D, Gunn RJM, Blackburn JH, Symes KL, Winder JM, Grieve NJ, Bass JAB. 1995. Testing and further development of RIVPACS. Part 1, Main Report, Almondsbury UK: National Rivers Authority. p 1–72.

Wright JF, Moss D, Armitage PD, Furse MT. 1984. A preliminary classification of running-water sites in Great Britain based on macro-invertebrate species and the prediction of community type using environmental data. *Freshw Biol* 14:221–256.

Yerushalmy J, Palmer CE. 1959. On the methodology of investigations of etiologic factors in chronic disease. *J Chronic Disease* 10:27–40.

Yoccoz NG. 1991. Use, overuse, and misuse of significance tests in evolutionary biology and ecology. *Bull Ecol Soc Am* 72:106–111.

Yoder CO, Rankin ET. 1995. Biological criteria program development and implementation. In: Davis WS, Simon TP, editors. Biological assessment and criteria: Tools for water resource planning and decision making. Boca Raton FL, USA: Lewis. p 109–144.

Yodzis P. 1981. The stability of real ecosystems. *Nature* 289:674–676.

Yodzis P. 1988. The indeterminacy of ecological interactions, as perceived by perturbation experiments. *Ecology* 72:1964–1972.

Yodzis P. 1996. Food webs and perturbation experiments: Theory and practice. In: Polis GA, Winemiler K, editors. Food webs: Integration of patterns and dynamics. New York NY, USA: Chapman & Hall. p 192–200.

Yount JD, Niemi GJ. 1990. Recovery of lotic ecosystems from disturbance: A narrative review of case studies. *Environ Manage* 14:547–569.

Zaga A, Little EE, Rableni CF, Ellersieck MR. 1998. Photoenhanced toxicity of a carbamate insecticide to early life stage anuran amphibians. *Environ Toxicol Chem* 17:2543–2553.

Zwart KB, Burgers SLGE, Bloem J, Bouwman LA, Brussaard L, Lebbink G, Didden WAM, Marinissen JCY, De Vos JA, Vreeken-Buijs MJ, De Ruiter PC. 1994. Population dynamics in the belowground food webs in two different agricultural systems. *Agric Ecosyst Environ* 51:187–198.

Index

A

B

C

D

E

F

G

H

I

K

L

M

N

O

P

Q

R

S

T

U

V

Z

SETAC

A Professional Society for Environmental Scientists and Engineers and Related Disciplines Concerned with Environmental Quality

The Society of Environmental Toxicology and Chemistry (SETAC), with offices currently in North America and Europe, is a nonprofit, professional society established to provide a forum for individuals and institutions engaged in the study of environmental problems, management and regulation of natural resources, education, research and development, and manufacturing and distribution.

Specific goals of the society are:

- Promote research, education, and training in the environmental sciences.
- Promote the systematic application of all relevant scientific disciplines to the evaluation of chemical hazards.
- Participate in the scientific interpretation of issues concerned with hazard assessment and risk analysis.
- Support the development of ecologically acceptable practices and principles.
- Provide a forum (meetings and publications) for communication among professionals in government, business, academia, and other segments of society involved in the use, protection, and management of our environment.

These goals are pursued through the conduct of numerous activities which include:

- Hold annual meetings with study and workshop sessions, platform and poster papers, and achievement and merit awards.
- Sponsor a monthly scientific journal, a newsletter, and special technical publications.
- Provide funds for education and training through the SETAC Scholarship/Fellowship Program.
- Organize and sponsor chapters to provide a forum for the presentation of scientific data and for the interchange and study of information about local concerns.
- Provide advice and counsel to technical and nontechnical persons through a number of standing and ad hoc committees.

SETAC membership is currently composed of nearly 6,000 individuals from government, academia, business, and public-interest groups with technical backgrounds of chemistry, toxicology, biology, ecology, atmospheric sciences, health sciences, earth sciences, and engineering.

If you have training in these or related disciplines and are engaged in the study, use, or management of environmental resources, SETAC can fulfill your professional affiliation needs.

All members receive a newsletter highlighting environmental topics and SETAC activities, and reduced fees for the Annual Meeting and SETAC special publications.

All members except Students and Senior Active Members receive monthly issues of Environmental Toxicology and Chemistry (ET&C), a peer-reviewed journal of the Society. Student and Senior Active Members may subscribe to the journal. Members may hold office and, with the Emeritus Members, constitute the voting membership.

If you desire further information, contact the appropriate SETAC office.

1010 North 12th Avenue
Pensacola, Florida, USA 32501-3367
T 850 469 1500
F 850 469 9778
E setac@setac.org

Avenue de la Toison d'Or 67
B-1060 Brussels, Belgium
T 32 2 772 72 81
F 32 2 770 53 83
E setac@setaceu.org

http://www.setac.org

More titles from the Society of Environmental Toxicology and Chemistry (SETAC):

Impact of Low-Dose, High-Potency Herbicides on Nontarget and Unintended Plant Species
Ferenc, editor
2001

Risk Management: Ecological Risk-Based Decision Making
Stahl, Bachman, Barton, Clark, deFur, Ells, Pittinger, Slimak, Wentsel, editors
2001

Development of Methods for Effects-Driven Cumulative Effects Assessment Using Fish Populations: Moose River Project
Munkittrick, McMaster, Van der Kraak, Portt, Gibbons, Farwell, Gray, authors
2000

Environmental Contaminants and Terrestrial Vertebrates: Effects on Populations, Communities, and Ecosystems
Albers, Heinz, Ohlendorf, editors
2000

Evaluating and Communicating Subsistence Seafood Safety in a Cross-Cultural Context: Lessons Learned from the Exxon Valdez Oil Spill
Field, Fall, Nighswander, Peacock, Varanasi, editors
2000

Evaluation of Persistence and Long-Range Transport of Organic Chemicals in the Environment
Klecka, Boethling, Franklin, Grady, Graham, Howard, Kannan, Larson, Mackay, Muir, van de Meent, editors
2000

Natural Remediation of Environmental Contaminants: Its Role in Ecological Risk Assessment and Risk Management
Swindoll, Stahl, Ells, editors
2000

Ecotoxicology and Risk Assessment for Wetlands
Lewis, Mayer, Powell, Nelson, Klaine, Henry, Dickson, editors
1999

Endocrine Disruption in Invertebrates: Endocrinology, Testing and Assessment
DeFur, Crane, Ingersoll, Tattersfield, editors
1999

Reproductive and Developmental Effects of Contaminants in Oviparous Vertebrates
DiGiulio and Tillitt, editors
1999

Restoration of Lost Human Uses of the Environment
Grayson Cecil, editor
1999

Ecological Risk Assessment: A Meeting of Policy and Science
Peyster and Day, editors
1998

Ecotoxicology Risk Assessment of the Chlorinated Organic Chemicals
Carey, Cook, Giesy, Hodson, Muir, Owens, Solomon, editors
1998

Principles and Processes for Evaluating Endocrine Disruption in Wildlife
Kendall, Dickerson, Geisy, Suk, editors
1998

Radiotelemetry Applications for Wildlife Toxicology Field Studies
Brewer and Fagerstone, editors
1998

Sustainable Environmental Management
Barnthouse, Biddinger, Cooper, Fava, Gillett, Holland, Yosie, editors
1998

Uncertainty Analysis in Ecological Risk Assessment
Warren-Hicks and Moore, editors
1998

Atmospheric Deposition of Contaminants to the Great Lakes and Coastal Waters
Baker, editor
1997

Chemical Ranking and Scoring: Guidelines for Relative Assessments of Chemicals
Swanson and Socha, editors
1997

Chemically Induced Alterations in Functional Development and Reproduction of Fishes
Rolland, Gilbertson, Peterson, editors
1997

Life-Cycle Impact Assessment: The State-of-the-Art, 2nd edition
Barnthouse, Fava, Humphreys, Hunt, Laibson, Moesoen, Owens, Todd, Vigon, Weitz, Young, editors
1997

Public Policy Application of Life-Cycle Assessment
Allen and Consoli, editors
1997

Quantitative Structure-Activity Relationships (QSAR) in Environment Sciences VII
Chen and Schuurmann, editors
1997

Reassessment of Metals Criteria for Aquatic Life Protection: Priorities for Research and Implementation
Bergman and Dorward-King, editors
1997

Procedures for Assessing the Environmental Fate and Ecotoxicity of Pesticides
Mark Lynch, editor
1995

The Multi-Media Fate Model: A Vital Tool for Predicting the Fate of Chemicals
Cowan, D. Mackay, Feijtel, Meent, Di Guardo, Davies, N. Mackay, editors
1995

Aquatic Dialogue Group: Pesticide Risk Assessment and Mitigation
Baker, Barefoot, Beasley, Burns, Caulkins, Clark, Feulner, Giesy, Graney, Griggs, Jacoby, Laskowski, Maciorowski, Mihaich, Nelson, Parrish, Siefert, Solomon, van der Schalie, editors
1994

A Conceptual Framework for Life-Cycle Impact Assessment
Fava, Consoli, Denison, Dickson, Mohin, Vigon, editors
1993

Guidelines for Life-Cycle Assessment: A "Code of Practice"
Consoli, Allen, Boustead, Fava, Franklin, Jensen, Oude, Parrish, Perriman, Postlethewaite, Quay, Seguin, Vigon, editors
1993

A Technical Framework for Life-Cycle Assessment
Fava, Denison, Jones, Curran, Vigon, Selke, Barnum, editors
1991

Research Priorities in Environmental Risk Assessment
Fava, Adams, Larson, G. Dickson, K. Dickson, W. Bishop
1987

http://www.setac.org